ADVANCED
M⊙DERN
PHYSICS
Solutions to Problems

ADVANCED MODERN PHYSICS
Solutions to Problems

Paolo Amore
Universidad de Colima, Mexico

John Dirk Walecka
College of William and Mary, USA

 World Scientific

NEW JERSEY · LONDON · SINGAPORE · BEIJING · SHANGHAI · HONG KONG · TAIPEI · CHENNAI · TOKYO

Published by

World Scientific Publishing Co. Pte. Ltd.

5 Toh Tuck Link, Singapore 596224

USA office: 27 Warren Street, Suite 401-402, Hackensack, NJ 07601

UK office: 57 Shelton Street, Covent Garden, London WC2H 9HE

Library of Congress Cataloging-in-Publication Data
Amore, Paolo, 1968- author.
  Advanced modern physics : solutions to problems / Paolo Amore (Universidad de Colima, Mexico),
John Dirk Walecka (College of William and Mary, USA).
    pages cm
  Includes bibliographical references and index.
  ISBN 978-9814704519 (pbk. : alk. paper) -- ISBN 9814704512 (pbk. : alk. paper)
  1. Physics--Problems, exercises, etc. I. Walecka, John Dirk, 1932–   author. II. Title.
  QC32.A554 2015
  530.076--dc23

                                               2015028014

British Library Cataloguing-in-Publication Data
A catalogue record for this book is available from the British Library.

Printed in Singapore

Preface

Three books by the second author, aimed at the very best students, present the theoretical underpinnings of *modern physics* as developed in the twentieth century [Walecka (2008); Walecka (2010); Walecka (2013)]. The goal is to make the presentation clear, concise, and self-contained. While the initial volume assumes only a calculus-based freshman physics course, and a good one-year course in calculus, the three volumes eventually cover material up through graduate quantum mechanics. Although it is assumed that mathematical skills will continue to develop, necessary additional mathematics is provided along the way. When finished with these volumes, readers should be familiar with the essentials of quantum mechanics, special relativity, relativistic quantum mechanics, field theory, and the fundamentals of atomic, nuclear, particle, and condensed-matter physics.[1] Dedicated readers should obtain a working knowledge of twentieth-century physics, able to smoothly proceed to more detailed and advanced material, and be prepared to make their own significant contributions to the field.

Each book has an extensive set of problems, which provide an invaluable teaching and learning tool. Two other published volumes by the current authors provide solutions to the problems in *Introduction to Modern Physics: Solutions to Problems* [Amore and Walecka (2013)] and *Topics in Modern Physics: Solutions to Problems* [Amore and Walecka (2014)]. The present volume provides solutions to the over 180 problems in the third book *Advanced Modern Physics: Theoretical Foundations* [Walecka (2010)]. This is the most challenging material, ranging across advanced quantum mechanics, angular momentum, scattering theory, lagrangian field theory, symmetries, Feynman rules, quantum electrodynamics (QED), higher-order processes, path-integrals, and canonical transformations for quantum systems;

---

[1]And general relativity, if one includes [Walecka (2007)].

v

several appendices supply important details. The present book completes this *modern physics* series.

We were again delighted when World Scientific Publishing Company, which had done an exceptional job with several previous books, showed enthusiasm for publishing this new one. We would like to thank Dr. K. K. Phua, Executive Chairman of World Scientific Publishing Company, and our editor Ms. Lakshmi Narayanan, for their help and support on this project.

Twentieth-century physics provides the basis for our current understanding of the physical world. While obtaining familiarity with what has gone before would seem to be a daunting task, these volumes on *modern physics* should help the dedicated student to find that job less challenging, and even enjoyable. We certainly enjoyed preparing these solutions manuals. We hope students and instructors alike will find them valuable.

January 12, 2015   *Paolo Amore*
Facultad de Ciencias
Universidad de Colima
Colima, Mexico

John Dirk Walecka
College of William and Mary
Williamsburg, Virginia

Contents

Chapter 1

Introduction

Two previous books by the present authors, *Introduction to Modern Physics: Solutions to Problems* [Amore and Walecka (2013)] and *Topics in Modern Physics: Solutions to Problems* [Amore and Walecka (2014)], provide a complete set of solutions to the many problems in two of a three-volume series on twentieth-century physics. This *modern physics* series [Walecka (2008); Walecka (2010); Walecka (2013)], aimed at the very best students, begins at the undergraduate level, and continues through material generally accessed only at the advanced graduate level. While obtaining sufficient background to understand and contribute to today's physics might appear to be an overwhelming task, the goal of this modern physics series is to provide a clear, inclusive, self-contained roadmap, and elementary working knowledge of the field. Extension in depth and reach can then be obtained from more advanced courses, texts, and the literature.

The current book, *Advanced Modern Physics: Solutions to Problems*, completes this modern physics series. It provides solutions to the over 180 problems in the third volume, the most challenging of the three, in what we again believe to be a clear and concise manner.

Chapter 2, *Quantum Mechanics (Revisited)*, is concerned with formulating and interpreting quantum mechanics in abstract Hilbert space. There are several problems on basic operator methods. Ehrenfest's theorem, which provides the direct path to classical correspondence, is derived and applied. One problem shows that the usual many-body hamiltonian commutes with the number operator. Another problem examines the translation operator, and the overlap of the eigenstates of position and momentum is then derived. Matrix mechanics is derived in one problem. The abstract time-independent Schrödinger equation is projected onto eigenstates of momentum in another. Finally, the transition from wave functions to abstract

Hilbert space is examined in detail for one-dimensional rotations.

Chapter 3, on *Angular Momentum*, provides an extensive overview of this most basic and useful topic. The problems start with the verification of several commutation relations. Finite matrix representations of the angular momentum operators are then derived, including the Pauli matrices. Other problems deal with the calculation of the finite rotation matrices. Several problems focus on the properties of the Clebsch-Gordan (C-G) coefficients, used for the coupling of two angular momenta. One problem extends the derivation of these coefficients to $s = 1$. The relation to Wigner's 3-j coefficients is studied and used to deduce the symmetry properties of the C-G coefficients. Other problems deduce essential properties of the 6-j coefficients, employed in the coupling of three angular momenta. The definition of irreducible tensor operators (ITO) is examined in one problem, and, with the aid of the Wigner-Eckart theorem, valuable expressions for the reduced matrix elements of ITO's working on one part of a coupled scheme are derived in another. A final problem derives the addition law for angular momenta from the weight diagram.

Chapter 4 is on *Scattering Theory*. After problems demonstrating the symmetric form and group property of the time-dependent scattering operator in leading order, there are a variety of problems on the time-independent scattering analysis. One problem derives the form of the incoming-wave scattering state, and another develops the analysis of potential scattering in terms of these states. That the large semi-circle used to complete the contour for the scattering Green's function in potential scattering makes a vanishing contribution as $R \to \infty$ is demonstrated in a problem, while another works out the asymptotic form of the coordinate-space wave function in this same limit. A series of very useful problems then analyze the operator form of the general scattering Green's function, culminating in closed expressions for the scattering states. A subsequent problem obtains a general expression for the S-matrix following from these states.

Chapter 5, entitled *Lagrangian Field Theory*, provides the basis for the transition from classical continuum mechanics to quantum field theory, where a problem on the string provides the paradigm. The stress tensor is extended to several generalized coordinates in a problem. A series of problems take the reader, with the aid of Ehrenfest's theorem, to the quantum form of Hamilton's equations. There are two problems on constructing hermitian forms of radial momentum operators. Three problems deal with the transition from the Schrödinger picture to the interaction picture with time-dependent operators. One problem shows the canonical equal-time commutation relations are picture independent, Other problems deal with the massive scalar field, and the momentum in this field. An important series of problems deals with the complex scalar field. Here there is a conserved current, and the corresponding charge displays quanta of both signs, interpreted as particles and *antiparticles*. Several problems are concerned with Dirac field theory. One obtains the momentum in the Dirac field. Another shows how to symmetrize the stress tensor in this case. Still another problem demonstrates the Lorentz invariance of the Dirac lagrangian density. One problem shows that normal-ordering the Dirac current simply subtracts the number of negative-energy states. Additional problems expand on the discussion in the text of the interacting Dirac-scalar theory with a Yukawa coupling. One analyzes the processes described by the corresponding interaction-picture hamiltonian. Another demonstrates the coupling constant $g^2/4\pi\hbar c^3$ is dimensionless. Still another analyzes the field equations in the Schrödinger picture, and uses Ehrenfest's theorem to establish classical correspondence. Finally, a problem establishes the direct analogy between the Poisson bracket formulation of classical mechanics and quantum theory.

Chapter 6 is on *Symmetries*. There are problems concerned with the construction of the generators for various internal symmetries and the evaluation of their commutation relations. Other problems extend the analysis to Dirac fields obeying canonical *anticommutation* rules. The analysis of the massive scalar field is extended in a problem to the isovector case, where there are three equal-mass components to the field; another problem establishes the connection to the analysis of the complex scalar field. An isospin-invariant Yukawa coupling between isovector-scalar and isospinor-Dirac fields is constructed in a problem, and the consequences of this isospin invariance examined. An extended problem on the σ-model uses Noether's theorem to construct the conserved vector current and axial-vector currents in the model arising from isospin (CVC) and chiral symmetry. The partially conserved axial-vector current (PCAC), obtained when a small explicit chiral-symmetry-breaking term is included, is then derived. Finally, the lagrangian density obtained through an expansion around the new minimum in the chiral-symmetric potential of the σ-model, which leads to the spontaneous breaking of chiral symmetry and a nucleon mass, is found. Still another problem explicitly exhibits the $SU(2)_L \otimes SU(2)_R$ symmetry of the σ-model. The generators of the SU(3) internal symmetry of the Sakata model are constructed in still another problem, and manipulated to derive the Gell-Mann–Nishijima relation between charge, hypercharge, and third component of isospin. Finally, a problem explicitly verifies the fundamental SU(3) matrix relations in the case of infinitesimal transformations.

Chapter 7 is entitled *Feynman Rules*. Here the discussion is based on the use of Wick's theorem to derive S-matrix elements for various theories and processes from the scattering operator, and thereby to construct the Feynman diagrams and deduce the relevant Feynman rules. The Dirac-scalar theory serves as a prototype, and several problems deduce the S-matrix elements for various processes in this theory, extending the Feynman rules in the text to include antiparticles. One interesting problem displays all the relevant fourth-order Feynman diagrams for this theory. The decay rate for $\phi \to N + \bar{N}$, if kinematically allowed, is calculated as an informative exercise in another. The mass renormalization of the ϕ is studied in still another. For illustration, the theory is enlarged to include two independent scalars in several additional problems. The analysis is extended in the problems to a Dirac-scalar-vector theory, where a neutral, massive vector field is included along with the neutral, massive scalar. S-matrix elements for various processes, Feynman diagrams, and Feynman rules are again obtained. One problem relaxes the restriction to a normal-ordered scalar density, and demonstrates that the result is an additional fermion mass renormalization. Finally, there is a useful problem concerned with calculating various matrix elements of the creation and destruction operators.

Chapter 8 is on *Quantum Electrodynamics (QED)*. First, there are problems on current conservation and commutation relations, the form of the Coulomb interaction, and basic relations satisfied by the Dirac matrices. The density of states required in the calculation of the Compton cross section is evaluated in one problem, while the contribution of the square of the direct term to the Klein-Nishina formula for this process is evaluated in another. While the details may become more complicated, these exercises provide the reader with the tools required to calculate *any* cross section. Limiting forms of various cross sections are examined. The amplitudes for pair production and bremsstrahlung are derived in one problem. The covariant photon polarization sum is derived in another. The amplitude and Feynman diagrams for Bhabha scattering, $e^- + e^+ \to e^- + e^+$, are derived in a problem. As an exercise, the Dirac equation for the external legs is used to remove the q_μ terms in the photon propagator for some specific processes. A pair of problems derive Furry's theorem, which states that loops with an odd number of photon interactions are absent in QED. Finally, a pair of problems demonstrate that neither one possible modification of the current, nor the addition of a Schwinger term to the current commutator, alter the lowest-order S-matrix in QED.

Chapter 9, entitled *Higher-Order Processes*, concerns the higher-order contributions in QED. One problem examines the role of the mass counter-term. A second derives the important Feynman parametrization relations, which allow one to complete the square and do momentum integrals. Still others examine the $\epsilon \to 0$ limit in the dimensional regulation of these momentum integrals in $n = 4 - \epsilon$ dimensions. Another important problem shows how, by letting a photon end up everywhere else on a charged electron line running through a diagram, one can explicitly eliminate those terms in the photon propagator proportional to q_μ. The use of current conservation and equal-time commutation relations to eliminate such terms in a Dirac-vector theory is demonstrated in another. The amplitude and Feynman diagrams for the scattering of light-by-light through an electron loop are investigated in a problem. The properties of the adiabatic damping term are studied in another problem. The properties of the cut-off charge renormalization constant are examined in still another. One problem verifies Ward's identity in fourth order. An important problem performs a counting of the powers of momentum in an arbitrary Feynman diagram, which, in a second problem, leads to an enumeration of the four classes of *primitively-divergent* diagrams. Additional problems demonstrate that one obtains the proper lowest-order results upon iteration of Dyson's equation for the vertex and Ward's equation for his vertex-construct. It is demonstrated in a problem that one obtains the correct result when the renormalized charge is used in skeleton diagrams into which the finite, renormalized photon and electron propagators have been inserted. It is shown in an additional problem that the Schwinger term makes no contribution to the third-order S-matrix in an external field. Finally, the origin of the concept of a running coupling constant is demonstrated in a QED theory which contains both muons and electrons.

Chapter 10 is on *Path Integrals*, which provide an essential alternative for quantizing classical systems. Two problems deal with indices and left-variational derivatives for Dirac fields. An important exercise details the calculation of the generating functional for a scalar field with a $\lambda\phi^4$ interaction, from which the propagator and Feynman rules follow. One problem shows how the sum over the diagonal matrix elements with a complete set of states, the "Trace", of the thermal operator $e^{-\beta \hat{H}}$ is independent of basis. Then, with an analysis that parallels that in the text for the propagator, three additional problems consider a particle in a potential and demonstrate how that Trace can be expressed as a path integral over an exponential of the action evaluated for imaginary time. An additional problem shows how the thermal average of a quantity is obtained from that path integral. The relation expressing a determinant as an integral over Grassmann variables is verified in fourth order in another problem. A final problem studies an effective field theory involving nucleon, pion, and scalar fields, and shows how one limit of this effective theory reproduces the σ-model.

Chapter 11 is entitled *Canonical Transformations for Quantum Systems*. In this solutions manual, the methods developed in the text to describe quantum fluids are used to solve two basic problems in quantum field theory. The first problem involves a real, massive scalar field interacting with a time-independent, localized c-number source. It is demonstrated that the exact interaction of two point sources is given by the Yukawa potential. The overlap between the new and old vacuua is then evaluated analytically. The second Bloch-Nordsieck problem concerns the quantized radiation field interacting with a time-independent, c-number, external current. It is shown that the overlap of the new and old vacuua is given by e^{-N} where N is the number of photons, and there are an infinite number of long-wavelength photons present due to the infrared divergence of QED. One implication is that the long-wavelength limit cannot be properly treated with low-order perturbation theory. There are three problems on boson interactions, one on the symmetry of the scattering amplitude and two on the low-energy form of that amplitude. The integral for the depletion of the boson ground state is verified in a problem. General properties of the two-body matrix element are found in another. One problem provides the details of the transformation of the fermion hamiltonian, and another then examines the properties of a specific excited state.

Several appendices supply essential details:

- Appendix A is on *Multipole Analysis of the Radiation Field*. Here, in addition to verifying some aspects of the vector spherical harmonics and the transformation to circular polarization, the formula for the transition rate is reproduced starting from the general decay amplitude with the photon going off in an arbitrary direction;

- In appendix B, on *Functions of a Complex Variable*, some implications of Cauchy's theorem are explored in a problem, and there are two additional exercises on residue theory. The symbolic replacement of $(x - x_0 + i\eta)^{-1}$ is derived in one problem, and another provides an example of analytic continuation. A problem then finds the analytic properties of the magnetic form factor of the electron in QED;

- Appendix C is on the *Electromagnetic Field*. One problem shows the QED action is gauge invariant. Other problems concern the Poynting vector and momentum in the field. A problem examines the processes described by the interaction hamiltonian with an external current. The general decomposition of a vector field into longitudinal and transverse parts is established in a problem, and the consequences explored. Two problems then develop the quantum theory of a massive, neutral vector field;

- Appendix D is on *Irreducible Representations of SU(n)*. After problems on the singlet state and the representations available to an octet of mesons and the triplet of fermions in the Sakata model with SU(3) symmetry, there are two interesting applications to physical systems. The first finds the representations available to two nucleons outside closed shell in Wigner's SU(4) theory, and then applies that to the spectra of A=6 nuclei. The second problem finds the states available in the Fermi-Yang model of bound states in the Sakata model, and then applies that to the observed octets of pseudoscalar and vector mesons;

- In appendix E on *Lorentz Transformations in Quantum Field Theory*, one problem derives the commutation relations for the generators of the inhomogeneous Lorentz group and then demonstrates how the addition of a Dirac matrix term leaves those relations unaltered. Other important problems examine in detail the Lorentz transformation properties of the Dirac field. Another problem relates the field representation of a Lie group to the Lie algebra of its generators;

- Appendix F is on *Green's Functions and Other Singular Functions*. A valuable problem derives the positive- and negative-energy Dirac projection matrices. There are additional problems on the Dirac Green's function and transverse photon propagator. Two more challenging problems derive the propagator for a massive, neutral vector meson;

- Appendix G is entitled *Dimensional Regularization*. Two problems focus on the regulated form of the vacuum-polarization insertion in QED and the implied mass of the photon. A third problem derives a property of the Dirac matrices in n-dimensions;

- In Appendix H, on *Path Integrals and the Electromagnetic Field*, the first problem derives the generating functional for the field.

A second problem then demonstrates the required modification in going from H-L to S-I units. Another problem verifies the form of the inverse of the kernel in the generating functional and hence derives the photon propagator in an arbitrary gauge. As a final exercise, the Faddeev-Popov determinant is derived for the ghost fields.

Our understanding of the physical world was revolutionized in the twentieth century. The aim of this *modern physics* series is to provide a path through those developments in sufficient detail so that the very best students can obtain an understanding and an elementary working knowledge of the field. We certainly hope that these solutions manuals aid in that effort. Further depth of understanding can then be achieved through more detailed and advanced courses and texts.

Some suggested texts for an exposure to the interesting and challenging areas of *today's physics* are as follows: for quantum field theory [Weinberg (2005); Banks (2008)]; for the status of the standard model of the strong and electroweak interactions in particle physics [Schwartz (2013); Donoghue, Golowich, and Holstein (2014)]; for the frontiers of nuclear physics [Walecka (2004)]; for neutrino physics, where more surprises certainly lie, [Giunti and Kim (2007)]; for application of quantum field theory to condensed-matter systems [Fetter and Walecka (2003a); Fradkin (2013)]; and for general relativity and cosmology [Walecka (2007); Roos (2015)].

Dedicated readers of the *modern physics* series should now be able to handle most of these books. Good luck! Enjoy! And may you make your own significant contribution to the field.

Chapter 2

Quantum Mechanics (Revisited)

Problem 2.1 Repeat ProbsI. 4.17–4.18 in abstract Hilbert space.

Solution to Problem 2.1 This problem is solved in section 3.8.3 of [Walecka (2013)].

Problem 2.2 Take a matrix element of the completeness relation in Eq. (2.65) between the states $|\xi'\rangle$ and $|\xi''\rangle$, and show that one obtains the correct expression for $\langle\xi'|\xi''\rangle$.

Solution to Problem 2.2

The completeness relation for the eigenstates of position in Eq. (2.65) states that

$$\int d\xi\, |\xi\rangle\langle\xi| = 1_{\mathrm{op}}$$

The continuum orthonormality statement for these states is given in Eqs. (2.51) as

$$\langle\xi'|\xi\rangle = \delta(\xi' - \xi)$$

The matrix element of the completeness relation between the states $|\xi'\rangle$ and $|\xi''\rangle$ then gives

$$\langle\xi''|\xi'\rangle = \int d\xi\, \langle\xi''|\xi\rangle\langle\xi|\xi'\rangle$$
$$= \int d\xi\, \delta(\xi'' - \xi)\delta(\xi - \xi') = \delta(\xi'' - \xi')$$

which is the proper result.

Problem 2.3 (a) Let $\hat{O}$ be some time-independent linear hermitian operator in abstract Hilbert space. Prove Ehrenfest's theorem[1]

$$\frac{d}{dt}\langle\Psi(t)|\hat{O}|\Psi(t)\rangle = \frac{i}{\hbar}\langle\Psi(t)|[\hat{H},\hat{O}]|\Psi(t)\rangle$$

(b) Hence conclude that if $[\hat{H},\hat{O}] = 0$, then the observable O represented by $\hat{O}$ will be a constant the motion.

Solution to Problem 2.3

(a) The Schrödinger equation in abstract Hilbert space is given in Eq. (2.70)

$$i\hbar\frac{\partial}{\partial t}|\Psi(t)\rangle = \hat{H}|\Psi(t)\rangle$$

The adjoint of this relation, which has a well-defined meaning in matrix elements, can be written formally as

$$-i\hbar\frac{\partial}{\partial t}\langle\Psi(t)| = \langle\Psi(t)|\hat{H}^{\dagger} = \langle\Psi(t)|\hat{H}$$

where we have used the fact that $\hat{H}^{\dagger} = \hat{H}$ is hermitian.[2]

Now take the time derivative of the expectation value of the operator $\hat{O}$, assumed here to have no explicit time dependence,

$$\frac{d}{dt}\langle\Psi(t)|\hat{O}|\Psi(t)\rangle = \left\{\frac{\partial}{\partial t}\langle\Psi(t)|\right\}\hat{O}|\Psi(t)\rangle + \langle\Psi(t)|\hat{O}\frac{\partial}{\partial t}|\Psi(t)\rangle$$

$$= -\frac{1}{i\hbar}\langle\Psi(t)|\hat{H}\hat{O} - \hat{O}\hat{H}|\Psi(t)\rangle$$

$$= \frac{i}{\hbar}\langle\Psi(t)|[\hat{H},\hat{O}]|\Psi(t)\rangle$$

This is Ehrenfest's theorem.[3]

---

[1]Compare Probl. 4.9.

[2]If we are working in the coordinate representation, for example, then the quantity $\langle\Psi(t)|x\rangle = \Psi^{\star}(x,t)$ appears on the left in the matrix element, and

$$-i\hbar\frac{\partial}{\partial t}\Psi^{\star}(x,t) = [H(p,x)\Psi^{\star}(x,t)]^{\star} \rightarrow \Psi^{\star}(x,t)H(p,x)$$

where the last relation follows from the hermiticity of $H(p,x)$.

[3]If $\hat{O}$ has an explicit time dependence, this relation reads

$$\frac{d}{dt}\langle\Psi(t)|\hat{O}|\Psi(t)\rangle = \langle\Psi(t)|\frac{\partial\hat{O}}{\partial t}|\Psi(t)\rangle + \frac{i}{\hbar}\langle\Psi(t)|[\hat{H},\hat{O}]|\Psi(t)\rangle$$

(b) Suppose the operator $\hat{O}$ commutes with the hamiltonian so that $[\hat{H}, \hat{O}] = 0$. It then follows from Ehrenfest's theorem that the expectation value of the operator $\hat{O}$ will not change with time

$$\frac{d}{dt}\langle \Psi(t)|\hat{O}|\Psi(t)\rangle = 0 \qquad ; [\hat{H}, \hat{O}] = 0$$

Problem 2.4 The number operator that is companion to the hamiltonian $\hat{H}$ in Eq. (2.113) is

$$\hat{N} = \int d^3x\, \hat{\psi}^\dagger(\mathbf{x})\hat{\psi}(\mathbf{x}) = \sum_{i=1}^{\infty} \hat{N}_i \qquad ; \text{number operator}$$

(a) Show $[\hat{N}, \hat{H}] = 0$ in this case;

(b) Use the results in Prob. 2.3 to conclude that if the relation in part (a) is satisfied, then $N = \sum_{i=1}^{\infty} n_i$ is a constant of the motion.

Solution to Problem 2.4

(a) The hamiltonian is

$$\hat{H} = \int d^3x\, \hat{\psi}^\dagger(\mathbf{x})T\hat{\psi}(\mathbf{x}) + \frac{1}{2}\int d^3x \int d^3y\, \hat{\psi}^\dagger(\mathbf{x})\hat{\psi}^\dagger(\mathbf{y})V(\mathbf{x},\mathbf{y})\hat{\psi}(\mathbf{y})\hat{\psi}(\mathbf{x})$$

where the quantum field is

$$\hat{\psi}(\mathbf{x}) = \sum_k \hat{a}_k \phi_k(\mathbf{x})$$

Here the $\phi_k(\mathbf{x})$ form a complete set of solutions to a one-body Schrödinger equation. Moreover, the creation and destruction operators $\hat{a}_k^\dagger$ and $\hat{a}_j$ satisfy the commutation relations[4]

$$[\hat{a}_j, \hat{a}_k^\dagger] = \delta_{jk}$$

We will first obtain the expression for the number operator directly in terms of the creation and destruction operators, using the definitions given

---

[4] We leave it to the reader to extend the proof to the case of fermions; for clarity, we here explicitly include the hats on the creation and destruction operators.

above

$$\hat{N} = \int d^3x \, \hat{\psi}^\dagger(\mathbf{x})\hat{\psi}(\mathbf{x})$$

$$= \sum_j \sum_k \hat{a}_j^\dagger \hat{a}_k \int d^3x \, \phi_j^\star(\mathbf{x})\phi_k(\mathbf{x})$$

$$= \sum_j \hat{a}_j^\dagger \hat{a}_j \equiv \sum_{j=1}^\infty \hat{n}_j$$

Let us now come to the hamiltonian, which we write as

$$\hat{H} \equiv \hat{T} + \hat{V}$$

We will work separately with the kinetic-energy and potential-energy operators. The kinetic-energy operator reads

$$\hat{T} = \int d^3x \, \hat{\psi}^\dagger(\mathbf{x}) T \hat{\psi}(\mathbf{x})$$

$$= \sum_j \sum_k \hat{a}_j^\dagger \hat{a}_k \int d^3x \, \phi_j^\star(\mathbf{x}) \left(-\frac{\hbar^2 \nabla^2}{2m} \right) \phi_k(\mathbf{x})$$

$$= \sum_j \sum_k \hat{a}_j^\dagger \hat{a}_k \langle j|T|k \rangle$$

where we have used the Dirac notation to express the matrix element of the one-body kinetic energy.

Our first goal is to prove that the number operator commutes with the kinetic-energy operator, and therefore we write

$$[\hat{T}, \hat{N}] = \sum_j \sum_k \sum_l \langle j|T|k \rangle [\hat{a}_j^\dagger \hat{a}_k, \hat{a}_l^\dagger \hat{a}_l]$$

It is convenient to reduce the commutator appearing in this expression, which involves composite operators, to a simpler form in terms of commutators of the creation and destruction operators alone. To this purpose we consider the generic operators $\hat{A}$, $\hat{B}$, $\hat{C}$ and $\hat{D}$ and the commutators

$$[\hat{A}, \hat{B}\hat{C}] = [\hat{A}, \hat{B}]\hat{C} + \hat{B}[\hat{A}, \hat{C}]$$
$$[\hat{A}\hat{B}, \hat{C}\hat{D}] = \hat{A}[\hat{B}, \hat{C}]\hat{D} + [\hat{A}, \hat{C}]\hat{B}\hat{D} + \hat{C}\hat{A}[\hat{B}, \hat{D}] + \hat{C}[\hat{A}, \hat{D}]\hat{B}$$

The proof of these expressions is straightforward and follows from explicitly writing the commutators on the left- and right-hand sides of the equations.

In particular, the second relation is just what we need for evaluating the commutator

$$[\hat{a}_j^\dagger \hat{a}_k, \, \hat{a}_l^\dagger \hat{a}_l] = \hat{a}_j^\dagger [\hat{a}_k, \hat{a}_l^\dagger] \hat{a}_l + [\hat{a}_j^\dagger, \hat{a}_l^\dagger] \hat{a}_k \hat{a}_l + \hat{a}_l^\dagger \hat{a}_j^\dagger [\hat{a}_k, \hat{a}_l] + \hat{a}_l^\dagger [\hat{a}_j^\dagger, \hat{a}_l] \hat{a}_k$$

$$= \delta_{kl} \, \hat{a}_j^\dagger \hat{a}_l - \delta_{jl} \, \hat{a}_l^\dagger \hat{a}_k$$

Therefore

$$[\hat{T}, \hat{N}] = \sum_j \sum_k \langle j|T|k \rangle \left(\hat{a}_j^\dagger \hat{a}_k - \hat{a}_j^\dagger \hat{a}_k \right) = 0$$

We now consider the potential-energy operator and express it directly in terms of the creation and destruction operators

$$\hat{V} = \frac{1}{2} \sum_j \sum_k \sum_l \sum_m \hat{a}_j^\dagger \hat{a}_k^\dagger \hat{a}_l \hat{a}_m \int d^3x \int d^3y \, \phi_j^\star(\mathbf{x}) \phi_k^\star(\mathbf{y}) V(\mathbf{x}, \mathbf{y}) \phi_l^\star(\mathbf{y}) \, \phi_m^\star(\mathbf{x})$$

$$= \frac{1}{2} \sum_j \sum_k \sum_l \sum_m \hat{a}_j^\dagger \hat{a}_k^\dagger \hat{a}_l \hat{a}_m \langle jk|V|ml \rangle$$

We then use this expression to calculate the commutator of the potential-energy operator with the number operator

$$[\hat{V}, \hat{N}] = \frac{1}{2} \sum_j \sum_k \sum_l \sum_m \sum_i \langle jk|V|ml \rangle [\hat{a}_j^\dagger \hat{a}_k^\dagger \hat{a}_l \hat{a}_m, \, \hat{a}_i^\dagger \hat{a}_i]$$

To evaluate the commutator on the r.h.s., we may write

$$\hat{a}_j^\dagger \hat{a}_k^\dagger \hat{a}_l \hat{a}_m = \hat{a}_j^\dagger \left([\hat{a}_k^\dagger, \hat{a}_l] + \hat{a}_l \hat{a}_k^\dagger \right) \hat{a}_m$$

$$= -\delta_{kl} \, \hat{a}_j^\dagger \hat{a}_m + \hat{a}_j^\dagger \hat{a}_l \hat{a}_k^\dagger \hat{a}_m$$

and therefore

$$[\hat{a}_j^\dagger \hat{a}_k^\dagger \hat{a}_l \hat{a}_m, \, \hat{a}_i^\dagger \hat{a}_i] = -\delta_{kl} [\hat{a}_j^\dagger \hat{a}_m, \, \hat{a}_i^\dagger \hat{a}_i] + [\hat{a}_j^\dagger \hat{a}_l \hat{a}_k^\dagger \hat{a}_m, \, \hat{a}_i^\dagger \hat{a}_i]$$

The last commutator may still appear intimidating, but it can put in a nicer form using the expression for the commutator $[\hat{A}, \hat{B}\hat{C}]$ given earlier

$$[\hat{a}_j^\dagger \hat{a}_l \hat{a}_k^\dagger \hat{a}_m, \, \hat{a}_i^\dagger \hat{a}_i] = \hat{a}_j^\dagger \hat{a}_l [\hat{a}_k^\dagger \hat{a}_m, \, \hat{a}_i^\dagger \hat{a}_i] + [\hat{a}_j^\dagger \hat{a}_l, \, \hat{a}_i^\dagger \hat{a}_i] \hat{a}_k^\dagger \hat{a}_m$$

Thus

$$[\hat{a}_j^\dagger \hat{a}_k^\dagger \hat{a}_l \hat{a}_m, \, \hat{a}_i^\dagger \hat{a}_i] = -\delta_{kl} [\hat{a}_j^\dagger \hat{a}_m, \, \hat{a}_i^\dagger \hat{a}_i] + \hat{a}_j^\dagger \hat{a}_l [\hat{a}_k^\dagger \hat{a}_m, \, \hat{a}_i^\dagger \hat{a}_i] + [\hat{a}_j^\dagger \hat{a}_l, \, \hat{a}_i^\dagger \hat{a}_i] \hat{a}_k^\dagger \hat{a}_m$$

Now the r.h.s. contains the same kind of commutators that we have already evaluated in the calculation concerning the kinetic-energy operator

$$[\hat{a}_j^\dagger \hat{a}_k^\dagger \hat{a}_l \hat{a}_m, \hat{a}_i^\dagger \hat{a}_i] = -\delta_{kl}\left(\delta_{mi}\,\hat{a}_j^\dagger \hat{a}_i - \delta_{ji}\,\hat{a}_i^\dagger \hat{a}_m\right) +$$
$$\hat{a}_j^\dagger \hat{a}_l \left(\delta_{mi}\,\hat{a}_k^\dagger \hat{a}_i - \delta_{ki}\,\hat{a}_i^\dagger \hat{a}_m\right) + \left(\delta_{li}\,\hat{a}_j^\dagger \hat{a}_i - \delta_{ji}\,\hat{a}_i^\dagger \hat{a}_l\right)\hat{a}_k^\dagger \hat{a}_m$$

If this relation is substituted inside the expression for $[\hat{V}, \hat{N}]$, one demonstrates that this commutator also vanishes

$$[\hat{V}, \hat{N}] = \frac{1}{2}\sum_{j,k,m} \langle jk|V|mk\rangle \left(-\hat{a}_j^\dagger \hat{a}_m + \hat{a}_j^\dagger \hat{a}_m\right) +$$
$$\frac{1}{2}\sum_{j,k,l,m} \langle jk|V|ml\rangle \left(\hat{a}_j^\dagger \hat{a}_l \hat{a}_k^\dagger \hat{a}_m - \hat{a}_j^\dagger \hat{a}_l \hat{a}_k^\dagger \hat{a}_m + \hat{a}_j^\dagger \hat{a}_l \hat{a}_k^\dagger \hat{a}_m - \hat{a}_j^\dagger \hat{a}_l \hat{a}_k^\dagger \hat{a}_m\right) = 0$$

We have thus shown that the hamiltonian commutes with the number operator

$$[\hat{H}, \hat{N}] = 0$$

(b) From Ehrenfest's theorem, proven in Prob. 2.3, it follows that

$$\frac{dN}{dt} = \frac{d}{dt}\langle \Psi(t)|\hat{N}|\Psi(t)\rangle$$
$$= \frac{i}{\hbar}\langle \Psi(t)|[\hat{H}, \hat{N}]|\Psi(t)\rangle = 0$$

Therefore $N = \sum_{i=1}^{\infty} n_i$ is a constant of the motion. The kinetic-energy and potential-energy operators both create as many particles as they destroy, thus leaving the total number of particles unchanged.

Problem 2.5 One of the most useful operator relations in quantum mechanics is the following

$$e^{i\hat{A}}\hat{B}e^{-i\hat{A}} = \hat{B} + i[\hat{A}, \hat{B}] + \frac{i^2}{2!}[\hat{A}, [\hat{A}, \hat{B}]] + \frac{i^3}{3!}[\hat{A}, [\hat{A}, [\hat{A}, \hat{B}]]] + \cdots$$

This expresses the l.h.s. as a series of repeated commutators that can be evaluated using the canonical commutation relations.

Expand the exponentials on the l.h.s., and prove this relation through the indicated order.

Solution to Problem 2.5

An expansion of the exponentials through $O(\hat{A}^3)$ gives

$$e^{i\hat{A}}\hat{B}e^{-i\hat{A}} \approx \left(1 + i\hat{A} + \frac{i^2}{2!}\hat{A}^2 + \frac{i^3}{3!}\hat{A}^3\right)\hat{B}\left(1 - i\hat{A} + \frac{i^2}{2!}\hat{A}^2 - \frac{i^3}{3!}\hat{A}^3\right)$$

Now collect terms on the r.h.s.:

- The terms of $O(\hat{A})$ give

$$i(\hat{A}\hat{B} - \hat{B}\hat{A}) = i[\hat{A}, \hat{B}]$$

- The terms of $O(\hat{A}^2)$ give

$$\frac{i^2}{2!}(\hat{A}^2\hat{B} - 2\hat{A}\hat{B}\hat{A} + \hat{B}\hat{A}^2) = \frac{i^2}{2!}(\hat{A}[\hat{A}, \hat{B}] - [\hat{A}, \hat{B}]\hat{A}) = \frac{i^2}{2!}[\hat{A}, [\hat{A}, \hat{B}]]$$

- The terms of $O(\hat{A}^3)$ give

$$\frac{i^3}{3!}(\hat{A}^3\hat{B} - 3\hat{A}^2\hat{B}\hat{A} + 3\hat{A}\hat{B}\hat{A}^2 - \hat{B}\hat{A}^3)$$

$$= \frac{i^3}{3!}\{\hat{A}(\hat{A}[\hat{A}, \hat{B}] - [\hat{A}, \hat{B}]\hat{A}) - (\hat{A}[\hat{A}, \hat{B}] - [\hat{A}, \hat{B}]\hat{A})\hat{A}\}$$

$$= \frac{i^3}{3!}[\hat{A}, [\hat{A}, [\hat{A}, \hat{B}]]]$$

This proves the relation through $O(\hat{A}^3)$.

Problem 2.6 The translation operator is defined as $\hat{U}(a) \equiv \exp\{-i\hat{p}a/\hbar\}$ where $\hat{p}$ is the momentum operator with $[\hat{p}, \hat{x}] = \hbar/i$, and a is a real number.

(a) Show that $\hat{U}(a)$ is a unitary operator $\hat{U}(a)^\dagger = \hat{U}(a)^{-1}$;

(b) Use the result in Prob. (2.5) to show $\hat{U}(a)\hat{x}\hat{U}(a)^{-1} = \hat{x} - a$ and $\hat{U}(a)^{-1}\hat{x}\hat{U}(a) = \hat{x} + a$;

(c) Apply the last relation in part (b) to an eigenstate of position $|\xi\rangle$ and show $\hat{U}(a)|\xi\rangle = |\xi + a\rangle$.

Solution to Problem 2.6

(a) Take the adjoint of $\hat{U}(a)$, and use the fact that $\hat{p}$ is hermitian

$$\hat{U}(a)^\dagger = \exp\{i\hat{p}^\dagger a/\hbar\} = \exp\{i\hat{p}a/\hbar\}$$

Therefore

$$\hat{U}(a)^\dagger\hat{U}(a) = 1$$

Hence the operator $\hat{U}(a) \equiv \exp\{-i\hat{p}a/\hbar\}$ is unitary

$$\hat{U}(a)^\dagger = \hat{U}(a)^{-1} \qquad ; \text{ unitary}$$

(b) Use the result in Prob. 2.4 to compute

$$e^{i\hat{p}a/\hbar}\hat{x}\,e^{-i\hat{p}a/\hbar} = \hat{x} + \frac{ia}{\hbar}[\hat{p},\hat{x}] + \frac{1}{2!}\left(\frac{ia}{\hbar}\right)^2[\hat{p},[\hat{p},\hat{x}]] + \cdots$$

Now use the canonical commutation relation

$$[\hat{p},\hat{x}] = \frac{\hbar}{i}$$

Since this is a c-number, all the further commutators vanish, and only the first two terms in the above relation are non-zero. Thus

$$\hat{U}(a)^{-1}\,\hat{x}\,\hat{U}(a) = \hat{x} + a$$

(c) Now apply the position operator to the state $\hat{U}(a)|\xi\rangle$, and use the previous result in the form $\hat{x}\,\hat{U}(a) = \hat{U}(a)(\hat{x}+a)$

$$\hat{x}\,\hat{U}(a)|\xi\rangle = \hat{U}(a)(\hat{x}+a)|\xi\rangle$$
$$= (\xi+a)\hat{U}(a)|\xi\rangle$$

Hence $\hat{U}(a)|\xi\rangle$ is an eigenstate of the position operator with eigenvalue $\xi+a$

$$\hat{U}(a)|\xi\rangle = |\xi+a\rangle$$

This state remains normalized since $\hat{U}(a)$ is unitary

$$\langle \xi+a|\xi+a\rangle = \langle \xi|\hat{U}(a)^\dagger\hat{U}(a)|\xi\rangle = \langle \xi|\xi\rangle$$

Problem 2.7 Consider the matrix element $\langle k|\hat{U}(a)|\xi\rangle$.

(a) Use the results from Prob. 2.6(c) and the analysis in the text to show that

$$\langle k|\hat{U}(a)|\xi\rangle = \langle k|\xi+a\rangle = e^{-ika}\langle k|\xi\rangle$$

(b) Now let $\xi \to 0$, and then re-label $a \to \xi$. Hence prove the last of Eqs. (2.51) entirely with operator methods.

Solution to Problem 2.7

(a) Consider the matrix element $\langle k|\hat{U}(a)|\xi\rangle$, with $\hat{U}(a) = \exp\{-i\hat{p}a/\hbar\}$ the translation operator in Prob. 2.6. This can be evaluated in two ways:

- First, use the result in Prob. 2.6(c) to replace $\hat{U}(a)|\xi\rangle$ by the state with the translated position eigenvalue $|\xi + a\rangle$;
- Second, let the hermitian momentum operator $\hat{p}$ act on the eigenstate on the left, where it gets replaced by the eigenvalue $\hbar k$.

This leads to the relations

$$\langle k|\hat{U}(a)|\xi\rangle = \langle k|\xi + a\rangle = e^{-ika}\langle k|\xi\rangle$$

(b) Now let $\xi \to 0$, and then re-label $a \to \xi$. The last relation then gives

$$\langle k|\xi\rangle = e^{-ik\xi}\langle k|0\rangle$$

Choose the normalization $\langle k|0\rangle$ to correspond to a big box of size L, with periodic boundary conditions,

$$\langle k|0\rangle = \frac{1}{\sqrt{L}}$$

This provides an operator derivation of the last of Eqs. (2.51)

$$\langle \xi|k\rangle = \langle k|\xi\rangle^\star = \frac{1}{\sqrt{L}}e^{ik\xi}$$

Problem 2.8 Suppose one looks for an approximate solution to the time-independent Schrödinger equation $Hu(x) = Eu(x)$ as a finite sum of a complete orthonormal set of wave functions $\phi_n(x)$ which satisfy boundary conditions appropriate to the problem at hand

$$u(x) = \sum_{n=1}^{N} c_n \phi_n(x)$$

The completeness statement in Eq. (2.14) indicates that in the limit $N \to \infty$, the true solution can, in fact, be reproduced this way.[5]

(a) Substitute this expression in the Schrödinger equation, use the orthonormality of the eigenfunctions, and derive the following set of matrix equations

$$\sum_{n=1}^{N}[\langle m|H|n\rangle - E\delta_{mn}]c_n = 0 \qquad ; m = 1, 2, \cdots, N$$

(b) Linear algebra teaches one that this set of linear homogeneous algebraic equations for the set of amplitudes $\{c_n\}$ will only have a non-trivial

---

[5]Here n labels an ordered set of states.

solution if the determinant of their coefficients vanishes. Show this leads to a polynomial expression for the eigenvalues E with precisely N roots $E^{(s)}$, where $s = 1, 2, \cdots, N$;

(c) Show that if H is hermitian with $H_{mn}^{\star} = H_{nm}$, then the eigenvalues $E^{(s)}$ will be real;

(d) Prove that the solutions $\{c_n^{(s)}\}$ for the amplitudes corresponding to different values of $E^{(s)}$ are orthogonal. Since degenerate solutions can always be orthogonalized, and one is free to choose the normalization of the $\{c_n\}$, show that

$$\sum_{n=1}^{N} c_n^{(s)\star} c_n^{(t)} = \delta_{st}$$

This analysis provides a powerful and widely-used approximation method for any problem in quantum mechanics.[6]

Solution to Problem 2.8

(a) Consider the time-independent Schrödinger equation

$$Hu(x) = Eu(x)$$

and express $u(x)$ in terms of the wave functions $\phi_n(x)$ which satisfy the appropriate boundary conditions

$$u(x) = \sum_{n=1}^{N} c_n \phi_n(x)$$

The Schrödinger equation becomes

$$\sum_{n=1}^{N} c_n H \phi_n(x) = E \sum_{n=1}^{N} c_n \phi_n(x)$$

If we multiply both sides by $\phi_m^{\star}(x)$ with $m = 1, \cdots, N$, and integrate over x, we have

$$\sum_{n=1}^{N} \langle m|H|n \rangle c_n = E \sum_{n=1}^{N} c_m \delta_{nm} \qquad ; m = 1, 2, \ldots N$$

---

[6]An early discussion of this method, which follows from the Rayleigh-Ritz variational principle (see the solution to Prob. 4.14 in [Amore and Walecka (2014)]) can be found in [MacDonald (1933)]. See also Sec. 3.4 in [Walecka (2013)].

Notice that the Kronecker delta on the r.h.s. follows from the orthonormality of the $\phi_m(x)$. These equations can be written as

$$\sum_{n=1}^{N} [\langle m|H|n \rangle - E\delta_{nm}] \, c_n = 0 \qquad ; \; m = 1, 2, \cdots, N$$

(b) In part (a) we obtained a set of N linear, homogeneous, algebraic equations for the amplitudes c_n. In matrix notation, we may write

$$\begin{bmatrix} \langle 1|H|1 \rangle - E & \langle 1|H|2 \rangle & \cdots & \langle 1|H|N \rangle \\ \langle 2|H|1 \rangle & \langle 2|H|2 \rangle - E & \cdots & \langle 2|H|N \rangle \\ \vdots & \vdots & \vdots & \vdots \\ \langle N|H|1 \rangle & \langle N|H|2 \rangle & \cdots & \langle N|H|N \rangle - E \end{bmatrix} \begin{bmatrix} c_1 \\ c_2 \\ \vdots \\ c_N \end{bmatrix} = \begin{bmatrix} 0 \\ 0 \\ \vdots \\ 0 \end{bmatrix}$$

The condition to have a non-trivial solution to these equations is that the determinant of the matrix on the l.h.s. of this equation be zero. It is easy to convince oneself that this determinant will be a polynomial of order N in E, since it contains the product of the diagonal elements. This polynomial will therefore have N roots, $E^{(s)}$, with $s = 1, \ldots, N$.

(c) Assume now that H is hermitian, and therefore $H_{mn}^\star = H_{nm}$. Consider the determinant

$$f(E) \equiv \begin{vmatrix} \langle 1|H|1 \rangle - E & \langle 1|H|2 \rangle & \cdots & \langle 1|H|N \rangle \\ \langle 2|H|1 \rangle & \langle 2|H|2 \rangle - E & \cdots & \langle 2|H|N \rangle \\ \vdots & \vdots & \vdots & \vdots \\ \langle N|H|1 \rangle & \langle N|H|2 \rangle & \cdots & \langle N|H|N \rangle - E \end{vmatrix} = \sum_{n=0}^{N} c_n E^n$$

where the coefficients c_n are real numbers, because of the hermiticity of the hamiltonian.

We will assume that the roots E are complex, and therefore

$$f(E) = \operatorname{Re} f(E) + i \operatorname{Im} f(E)$$

We then take the complex conjugate of $f(E)$

$$f^\star(E) = \sum_{n=0}^{N} c_n (E^n)^\star = f(E^\star)$$

As a result, if $E^{(s)}$ are the roots of $f(E)$, then the complex conjugates $E^{(s)\star}$ must be roots as well. When E is a root, both the real and imaginary parts of $f(E)$ must vanish; in particular

$$0 = \text{Im}\, f(E) = \frac{f(E) - f^\star(E)}{2i} = \frac{f(E) - f(E^\star)}{2i}$$

$$= \text{Im}\, E \sum_{n=1}^{N} c_n \frac{[E^n - (E^\star)^n]}{E - E^\star}$$

$$= \text{Im}\, E \sum_{n=1}^{N} c_n \sum_{j=0}^{n-1} E^j (E^\star)^{n-1-j}$$

Thus we see that $\text{Im}\, E = 0$ allows one to obtain $\text{Im}\, f(E) = 0$, and therefore the roots $E^{(s)}$ are real.[7]

(d) Write the matrix equation as

$$\begin{bmatrix} \langle 1|H|1\rangle & \langle 1|H|2\rangle & \cdots & \langle 1|H|N\rangle \\ \langle 2|H|1\rangle & \langle 2|H|2\rangle & \cdots & \langle 2|H|N\rangle \\ \vdots & \vdots & \vdots & \vdots \\ \langle N|H|1\rangle & \langle N|H|2\rangle & \cdots & \langle N|H|N\rangle \end{bmatrix} \begin{bmatrix} c_1^{(s)} \\ c_2^{(s)} \\ \vdots \\ c_N^{(s)} \end{bmatrix} = E^{(s)} \begin{bmatrix} c_1^{(s)} \\ c_2^{(s)} \\ \vdots \\ c_N^{(s)} \end{bmatrix}$$

Upon multiplying both sides of the equation on the left with the adjoint vector $\underline{C}^{(r)\dagger} = [c_1^{(r)\star}, c_2^{(r)\star}, \cdots, c_N^{(r)\star}]$, with $r \neq s$, we have

$$\underline{C}^{(r)\dagger}\, \underline{H}\, \underline{C}^{(s)} = E^{(s)} \underline{C}^{(r)\dagger} \underline{C}^{(s)}$$

Here $\underline{H}$ is the $N \times N$ matrix containing the matrix elements of the hamiltonian, and $\underline{C}^{(s)}$ is the column vector with elements the coefficients $c_n^{(s)}$.

In a similar way we have

$$\underline{C}^{(s)\dagger}\, \underline{H}\, \underline{C}^{(r)} = E^{(r)} \underline{C}^{(s)\dagger} \underline{C}^{(r)}$$

If we take the adjoint of this last equation, using the hermiticity of $\underline{H}$, we obtain

$$\underline{C}^{(r)\dagger}\, \underline{H}\, \underline{C}^{(s)} = E^{(r)} \underline{C}^{(r)\dagger} \underline{C}^{(s)}$$

After subtracting this equation from the first equation, we obtain

$$[E^{(s)} - E^{(r)}]\underline{C}^{(r)\dagger} \underline{C}^{(s)} = 0$$

which, for $E^{(r)} \neq E^{(s)}$, implies[8]

$$\underline{C}^{(r)\dagger} \underline{C}^{(s)} = 0 \qquad ;\ r \neq s$$

---

[7]For an alternate proof, see the solution to Prob. 2.4 in [Amore and Walecka (2014)].
[8]One can always diagonalize in the degenerate subspaces.

Since one is free to choose the normalization of the $\underline{C}$, we finally have

$$\underline{C}^{(r)\dagger}\underline{C}^{(s)} = \sum_{n=1}^{N} c_n^{(r)\star} c_n^{(s)} = \delta_{rs}$$

Problem 2.9 This problem concerns the projection of the abstract time-independent Schrödinger Eq. (2.66) onto eigenstates of momentum; it makes liberal use of the completeness of both the eigenstates of momentum [Eq. (2.63)] and of position [Eq. (2.65)].

(a) Show that the projection of the state vector $|\psi\rangle$ on the state $|k\rangle$ is[9]

$$\langle k|\psi\rangle = \frac{1}{\sqrt{L}} \int d\xi e^{-ik\xi} \psi(\xi) \equiv A(k)$$

(b) Define $\tilde{V}(q) \equiv (2\pi)^{-1/2} \int d\xi\, e^{-iq\xi} V(\xi)$ and show

$$\langle k|\hat{H}|k'\rangle = \frac{\hbar^2 k^2}{2m} \delta_{kk'} + \frac{\sqrt{2\pi}}{L} \tilde{V}(k - k')$$

(c) Hence show that $\langle k|H|\psi\rangle = E\langle k|\psi\rangle$ implies

$$\left(E - \frac{\hbar^2 k^2}{2m}\right) A(k) = \frac{\sqrt{2\pi}}{L} \sum_{k'} \tilde{V}(k - k') A(k')$$

(d) Define $\tilde{\psi}(k) \equiv (L/2\pi)^{1/2} A(k)$, and take the limit $L \to \infty$. Show the Schrödinger equation in momentum space is an *integral equation*

$$\left(E - \frac{\hbar^2 k^2}{2m}\right) \tilde{\psi}(k) = \frac{1}{\sqrt{2\pi}} \int dk'\, \tilde{V}(k - k') \tilde{\psi}(k')$$

where $\tilde{\psi}(k)$ is the Fourier transform of the coordinate-space wave function, and $\tilde{V}(q)$ is the Fourier transform of the potential;

(e) Re-derive this result by taking the Fourier transform of the coordinate-space Schrödinger equation.

Solution to Problem 2.9

(a) The time-independent Schrödinger Eq. (2.66) is

$$\hat{H}|\psi\rangle = E|\psi\rangle$$

---

[9]We first use the periodic boundary conditions to convert the coordinate interval to $[-L/2, L/2]$.

The coordinate-space wave function is defined in Eq. (2.61)

$$\langle \xi | \psi \rangle = \psi(\xi)$$

The completeness relation for the eigenstates of position in Eq. (2.65) is

$$\int d\xi \, |\xi\rangle\langle\xi| = \hat{1}$$

Now compute the projection of the Schrödinger state vector on an eigenstate of momentum as

$$\langle k | \psi \rangle = \int d\xi \, \langle k | \xi \rangle \langle \xi | \psi \rangle$$
$$= \frac{1}{\sqrt{L}} \int d\xi \, e^{-ik\xi} \psi(\xi) \equiv A(k)$$

which defines the quantity $A(k)$. Here use has been made of Eqs. (2.51), and we impose periodic boundary conditions on the coordinate interval $[-L/2, L/2]$.

(b) Define the Fourier transform of the potential as

$$\tilde{V}(q) \equiv \frac{1}{\sqrt{2\pi}} \int d\xi \, e^{-iq\xi} V(\xi)$$
$$V(\xi) = \frac{1}{\sqrt{2\pi}} \int dq \, e^{iq\xi} \, \tilde{V}(q)$$

The matrix element of the potential in momentum space then follows as

$$\langle k | \hat{H} | k' \rangle = \langle k | \frac{\hat{p}^2}{2m} + V(\hat{x}) | k' \rangle$$
$$= \frac{\hbar^2 k^2}{2m} \delta_{kk'} + \int d\xi \, \langle k | V(\hat{x}) | \xi \rangle \langle \xi | k' \rangle$$
$$= \frac{\hbar^2 k^2}{2m} \delta_{kk'} + \frac{1}{L} \int d\xi \, V(\xi) \, e^{-i(k-k')\xi}$$
$$= \frac{\hbar^2 k^2}{2m} \delta_{kk'} + \frac{\sqrt{2\pi}}{L} \tilde{V}(k - k')$$

(c) The completeness relation for the eigenstates of momentum in Eq. (2.63) states

$$\sum_k |k\rangle\langle k| = \hat{1}$$

The projection of the time-independent Schrödinger equation on an eigenstate of momentum thus takes the form

$$\langle k|\hat{H}|\psi\rangle = \sum_{k'} \langle k|\hat{H}|k'\rangle \langle k'|\psi\rangle$$

$$= \frac{\hbar^2 k^2}{2m} A(k) + \sum_{k'} \frac{\sqrt{2\pi}}{L} \tilde{V}(k - k') A(k')$$

$$= E A(k)$$

(d) Define the Fourier transform of the wave function by

$$\tilde{\psi}(k) \equiv \left(\frac{L}{2\pi}\right)^{1/2} A(k) = \frac{1}{\sqrt{2\pi}} \int d\xi \, e^{-ik\xi} \psi(\xi)$$

$$\psi(\xi) = \frac{1}{\sqrt{2\pi}} \int dk \, e^{ik\xi} \, \tilde{\psi}(k)$$

The result in part (c) is then written as

$$\frac{\hbar^2 k^2}{2m} \tilde{\psi}(k) + \sum_{k'} \frac{\sqrt{2\pi}}{L} \tilde{V}(k - k') \tilde{\psi}(k') = E \tilde{\psi}(k)$$

In the large L limit, the sum over wave numbers can be replaced by an integral according to

$$\sum_k \rightarrow \frac{L}{2\pi} \int dk \qquad ; L \rightarrow \infty$$

Hence the Schrödinger equation in momentum space becomes an integral equation

$$\left(E - \frac{\hbar^2 k^2}{2m}\right) \tilde{\psi}(k) = \frac{1}{\sqrt{2\pi}} \int dk' \, \tilde{V}(k - k') \tilde{\psi}(k')$$

where $\tilde{\psi}(k)$ is the Fourier transform of the coordinate-space wave function, and $\tilde{V}(q)$ is the Fourier transform of the potential.

(e) The coordinate-space Schrödinger Eq. (2.69) is

$$\left[-\frac{\hbar^2}{2m} \frac{\partial^2}{\partial \xi^2} + V(\xi)\right] \psi(\xi) = E \, \psi(\xi)$$

Now perform the operation $(2\pi)^{-1/2} \int d\xi \, e^{-ik\xi}$ on this equation. This yields

$$\int \frac{d\xi}{\sqrt{2\pi}} e^{-ik\xi} \left[-\frac{\hbar^2}{2m} \frac{\partial^2}{\partial \xi^2}\right] \psi(\xi) + \int \frac{dk'}{2\pi} \int d\xi \, e^{-i(k-k')\xi} V(\xi) \tilde{\psi}(k') = E \tilde{\psi}(k)$$

Two partial integrations on the first term,[10] and identification of the Fourier transform of the potential, reproduce the result in part (d)

$$\left(E - \frac{\hbar^2 k^2}{2m}\right)\tilde{\psi}(k) = \frac{1}{\sqrt{2\pi}}\int dk'\, \tilde{V}(k-k')\tilde{\psi}(k')$$

Problem 2.10 Consider the rotations of a particle in the x-y plane. The abstract eigenstates of the z-component of angular momentum satisfy

$$\hat{L}_z|m\rangle = m|m\rangle \qquad ; \langle m|m'\rangle = \delta_{mm'} \qquad ; \sum_m |m\rangle\langle m| = \hat{1}$$

The eigenfunctions satisfying p.b.c. in this case were found in Vol. I

$$\langle\phi|m\rangle = \frac{1}{\sqrt{2\pi}}e^{im\phi} \qquad ; m = 0, \pm 1, \pm 2, \cdots$$

where ϕ is the polar angle. As in the text, use these wave functions as a basis for arriving at each of the above abstractions.

Solution to Problem 2.10

Let $|\phi\rangle$ represent an abstract eigenstate of azimuthal position. The coordinate-space wave functions satisfying p.b.c. are as given

$$\langle\phi|m\rangle = \frac{1}{\sqrt{2\pi}}e^{im\phi} \qquad ; m = 0, \pm 1, \pm 2, \cdots$$

We know from our discussion of angular momentum that the explicit representation of the third component of the angular momentum operator in coordinate space is

$$\langle\phi|\hat{L}_z|\phi'\rangle = \frac{1}{i}\frac{\partial}{\partial\phi}\delta(\phi-\phi')$$

This produces the familiar coordinate-space relation

$$\int_0^{2\pi} d\phi'\langle\phi|\hat{L}_z|\phi'\rangle\langle\phi'|m\rangle = \frac{1}{i}\frac{\partial}{\partial\phi}\langle\phi|m\rangle = m\langle\phi|m\rangle$$

Now the inner product in the abstract Hilbert space is defined in terms of the coordinate-space wave functions, which implies that the abstract eigenstates of azimuthal position satisfy the completeness relation

$$\int_0^{2\pi} d\phi\, |\phi\rangle\langle\phi| = \hat{1}$$

---

[10]Note that $\hat{p}^2$ is hermitian.

If we multiply the previous equation on the left by $|\phi\rangle$, integrate over ϕ, and use the completeness of these states, we obtain

$$\int_0^{2\pi} d\phi \, |\phi\rangle\langle\phi|\hat{L}_z|m\rangle = m \int_0^{2\pi} d\phi \, |\phi\rangle\langle\phi|m\rangle$$

Another use of completeness produces the equation

$$\hat{L}_z|m\rangle = m|m\rangle$$

Hence we deduce that the abstract states $|m\rangle$ are indeed eigenstates of the linear hermitian angular momentum operator $\hat{L}_z$ with eigenvalue m.

We observe that the wave functions $\langle\phi|m\rangle$ are orthogonal since

$$\int_0^{2\pi} d\phi \, \langle m'|\phi\rangle\langle\phi|m\rangle = \frac{1}{2\pi} \int_0^{2\pi} d\phi \, e^{i(m-m')\phi} = \delta_{mm'}$$

It follows from the completeness of the eigenstates $|\phi\rangle$, that we have

$$\int_0^{2\pi} d\phi \, \langle m'|\phi\rangle\langle\phi|m\rangle = \langle m'|m\rangle$$

Hence the abstract eigenstates of angular momentum are orthonormal

$$\langle m'|m\rangle = \delta_{mm'}$$

Consider now the completeness of the wave functions $\langle\phi|m\rangle$

$$\sum_m \langle\phi|m\rangle\langle m|\phi'\rangle = \frac{1}{2\pi} \sum_{m=-\infty}^{\infty} e^{im(\phi-\phi')} = \delta(\phi - \phi')$$

The eigenstates of position $|\phi\rangle$ are orthonormal

$$\langle\phi|\phi'\rangle = \delta(\phi - \phi')$$

and thus

$$\sum_m \langle\phi|m\rangle\langle m|\phi'\rangle = \langle\phi|\phi'\rangle$$

It follows that the abstract eigenstates of angular momentum are also complete

$$\sum_m |m\rangle\langle m| = \hat{1}$$

The argument can now be reversed to reproduce the previous representation of the third component of the angular momentum operator in coordinate space

$$\langle\phi|\hat{L}_z|\phi'\rangle = \sum_m \sum_{m'} \langle\phi|m\rangle\langle m|\hat{L}_z|m'\rangle\langle m'|\phi'\rangle$$

$$= \sum_m m \langle\phi|m\rangle \langle m|\phi'\rangle$$

$$= \frac{1}{i}\frac{\partial}{\partial\phi} \sum_m \frac{1}{2\pi} e^{im(\phi-\phi')} = \frac{1}{i}\frac{\partial}{\partial\phi}\delta(\phi - \phi')$$

Chapter 3

Angular Momentum

Problem 3.1 Operator commutation relations play a central role here:
(a) Verify Eqs. (3.19) for the vectors $\hat{\mathbf{v}} = (\hat{\mathbf{x}}, \hat{\mathbf{p}})$;
(b) Verify the re-writing in Eqs. (3.25);
(c) Verify Eqs. (3.30) for the raising and lowering operators.

Solution to Problem 3.1

(a) We are asked to verify Eqs. (3.19)

$$i[\hat{J}_3, \hat{v}_1] = -\hat{v}_2 \qquad ; \; i[\hat{J}_3, \hat{v}_2] = \hat{v}_1$$

where from Eq. (3.16)

$$\hbar \hat{J}_3 = \hat{x}_1 \hat{p}_2 - \hat{x}_2 \hat{p}_1$$

We make use of the C.C.R. in Eq. (3.17)

$$[\hat{p}_i, \hat{x}_j] = \frac{\hbar}{i}\delta_{ij}$$

Consider

$$i[\hat{J}_3, \hat{x}_1] = -\frac{i}{\hbar}\hat{x}_2[\hat{p}_1, \hat{x}_1] = -\hat{x}_2$$

$$i[\hat{J}_3, \hat{p}_1] = \frac{i}{\hbar}[\hat{x}_1, \hat{p}_1]\hat{p}_2 = -\hat{p}_2$$

In a similar fashion

$$i[\hat{J}_3, \hat{x}_2] = \frac{i}{\hbar}\hat{x}_1[\hat{p}_2, \hat{x}_2] = \hat{x}_1$$

$$i[\hat{J}_3, \hat{p}_2] = -\frac{i}{\hbar}[\hat{x}_2, \hat{p}_2]\hat{p}_1 = \hat{p}_1$$

This verifies Eqs. (3.19).

(b) We must verify that Eqs. (3.24)

$$[\hat{J}_1, \hat{J}_2] = i\hat{J}_3 \qquad ; \text{ and cyclic permutations of } (1,2,3)$$
$$[\hat{\mathbf{J}}^2, \hat{J}_i] = 0 \qquad ; i = (1,2,3)$$

can be re-written as[1]

$$[\hat{J}_i, \hat{J}_j] = i\epsilon_{ijk}\hat{J}_k \qquad ; (i,j,k) = (1,2,3)$$
$$[\hat{\mathbf{J}}^2, \hat{J}_i] = 0$$

where ϵ_{ijk} denotes the completely antisymmetric Levi-Civita tensor in three-dimensions

$$\epsilon_{ijk} = +1 \qquad ; (i,j,k) \text{ an even permutation of } (1,2,3)$$
$$= -1 \qquad ; (i,j,k) \text{ an odd permutation of } (1,2,3)$$
$$= 0 \qquad ; \text{ otherwise}$$

consider

$$[\hat{J}_1, \hat{J}_2] = -[\hat{J}_2, \hat{J}_1] = i\epsilon_{123}\hat{J}_3 = i\hat{J}_3$$
$$[\hat{J}_2, \hat{J}_3] = -[\hat{J}_3, \hat{J}_2] = i\epsilon_{231}\hat{J}_1 = i\hat{J}_1$$
$$[\hat{J}_3, \hat{J}_1] = -[\hat{J}_1, \hat{J}_3] = i\epsilon_{312}\hat{J}_2 = i\hat{J}_2$$

Hence we reproduce Eqs. (3.24).

(c) We are asked to verify Eqs. (3.30)

$$[\hat{J}_\pm, \hat{J}_3] = \mp\hat{J}_\pm$$
$$[\hat{J}_+, \hat{J}_-] = 2\hat{J}_3$$

where

$$\hat{J}_\pm \equiv \hat{J}_1 \pm i\hat{J}_2$$

Consider

$$[\hat{J}_+, \hat{J}_3] = [\hat{J}_1 + i\hat{J}_2, \hat{J}_3] = -\hat{J}_1 - i\hat{J}_2 = -\hat{J}_+$$
$$[\hat{J}_-, \hat{J}_3] = [\hat{J}_1 - i\hat{J}_2, \hat{J}_3] = \hat{J}_1 - i\hat{J}_2 = \hat{J}_-$$

Also

$$[\hat{J}_+, \hat{J}_-] = [\hat{J}_1 + i\hat{J}_2, \hat{J}_1 - i\hat{J}_2] = -2i[\hat{J}_1, \hat{J}_2] = 2\hat{J}_3$$

This verifies Eqs. (3.30).

---

[1]Recall that repeated Latin indices are summed from 1 to 3.

Problem 3.2 Finite transformations follow from the commutation relations:

(a) Verify the 3-D translation operator in Eq. (3.10);
(b) Verify the finite form of the rotation in Eq. (3.20);
(c) Verify the finite rotation in Eqs. (3.22).

Solution to Problem 3.2

(a) The operator in Eq. (3.10) is

$$\hat{U}(\mathbf{a}) = \exp\left\{-\frac{i}{\hbar}\hat{\mathbf{p}}\cdot\mathbf{a}\right\} = \exp\left\{-\frac{i}{\hbar}\hat{p}_j a_j\right\}$$

From Eq. (3.9)

$$e^{i\hat{p}_j a_j/\hbar}\,\hat{x}_k\,e^{-i\hat{p}_j a_j/\hbar} = \hat{x}_k + \frac{i}{\hbar}a_j[\hat{p}_j,\hat{x}_k] + \left(\frac{i}{\hbar}\right)^2\frac{1}{2!}a_j a_l[\hat{p}_j,[\hat{p}_l,\hat{x}_k]] + \cdots$$

$$= \hat{x}_k + a_k$$

Therefore $\hat{U}(\mathbf{a})$ is indeed the 3-D translation operator[2]

$$\exp\left\{\frac{i}{\hbar}\hat{\mathbf{p}}\cdot\mathbf{a}\right\}\hat{\mathbf{x}}\exp\left\{-\frac{i}{\hbar}\hat{\mathbf{p}}\cdot\mathbf{a}\right\} = \hat{\mathbf{x}} + \mathbf{a}$$

(b) As stated in the text, the generator of rotations about the third axis is $\hat{K} = \hbar\hat{J}_3$. Therefore the infinitesimal form of the rotation operator in Eq. (3.13) is

$$\hat{\mathcal{R}} = 1 - i\varepsilon\hat{J}_3$$

The transformation through a finite angle ω is obtained by repeating the infinitesimal transformation N times, so that $\omega = N\varepsilon$, and then using the binomial theorem to write

$$\left(1 - i\varepsilon\hat{J}_3\right)^N = 1 + N\left(-i\varepsilon\hat{J}_3\right) + \frac{N(N-1)}{2!}\left(-i\varepsilon\hat{J}_3\right)^2 + \cdots$$

$$= 1 + N\left(-i\varepsilon\hat{J}_3\right) + \frac{N^2}{2!}\left(-i\varepsilon\hat{J}_3\right)^2 + \cdots + O(\varepsilon\omega)$$

Up to $O(\varepsilon\omega)$ this is identical with the exponential series, and hence in the limit as $N \to \infty$ at fixed $\omega = N\varepsilon$ (which implies $\varepsilon\omega \to 0$), we have the important relation

$$\mathrm{Lim}_{N\to\infty}\left(1 - i\varepsilon\hat{J}_3\right)^N = \exp\left\{-i\omega\hat{J}_3\right\} \qquad ; \omega = N\varepsilon \text{ fixed}$$

---

[2]Compare Prob. 2.6(b).

Thus the operator producing finite rotations is

$$\hat{R}(\omega) = \exp\left\{-i\omega\hat{J}_3\right\} \qquad ; \text{ finite rotation}$$

(c) Consider

$$\hat{R}(\omega)\,\hat{v}_1\,\hat{R}(\omega)^{-1} = \hat{v}_1 - i\omega[\hat{J}_3,\,\hat{v}_1] + \frac{(-i\omega)^2}{2!}[\hat{J}_3,\,[\hat{J}_3,\,\hat{v}_1]] +$$
$$\frac{(-i\omega)^3}{3!}[\hat{J}_3,\,[\hat{J}_3,\,[\hat{J}_3,\,\hat{v}_1]]] + \cdots$$

Now use the commutation relations for a vector operator in Eqs. (3.19)

$$\hat{R}(\omega)\,\hat{v}_1\,\hat{R}(\omega)^{-1} = \hat{v}_1\left(1 - \frac{\omega^2}{2!} + \cdots\right) + \hat{v}_2\left(\omega - \frac{\omega^3}{3!} + \cdots\right)$$
$$= \hat{v}_1\cos\omega + \hat{v}_2\sin\omega$$

In a similar manner

$$\hat{R}(\omega)\,\hat{v}_2\,\hat{R}(\omega)^{-1} = \hat{v}_2 - i\omega[\hat{J}_3,\,\hat{v}_2] + \frac{(-i\omega)^2}{2!}[\hat{J}_3,\,[\hat{J}_3,\,\hat{v}_2]] +$$
$$\frac{(-i\omega)^3}{3!}[\hat{J}_3,\,[\hat{J}_3,\,[\hat{J}_3,\,\hat{v}_2]]] + \cdots$$

This gives

$$\hat{R}(\omega)\,\hat{v}_2\,\hat{R}(\omega)^{-1} = \hat{v}_2\left(1 - \frac{\omega^2}{2!} + \cdots\right) - \hat{v}_1\left(\omega - \frac{\omega^3}{3!} + \cdots\right)$$
$$= \hat{v}_2\cos\omega - \hat{v}_1\sin\omega$$

Since $\hat{R}(\omega)\,\hat{v}_3\,\hat{R}(\omega)^{-1} = \hat{v}_3$, these are Eqs. (3.22).

Problem 3.3 Use $\hat{J}_1 = (\hat{J}_+ + \hat{J}_-)/2$, $\hat{J}_2 = (\hat{J}_+ - \hat{J}_-)/2i$, and Eqs. (3.59):

(a) Compute the matrix $\underline{\mathbf{s}} \equiv \boldsymbol{\sigma}/2$ where

$$[\underline{\mathbf{s}}]_{m'm} \equiv \left\langle \frac{1}{2}m'\left|\hat{\mathbf{J}}\right|\frac{1}{2}m\right\rangle \qquad ; \underline{\mathbf{s}} \equiv \frac{1}{2}\boldsymbol{\sigma}$$

Show that $(\sigma_x, \sigma_y, \sigma_z)$ are just the Pauli matrices[3]

$$\sigma_x \equiv \begin{pmatrix} 0 & 1 \\ 1 & 0 \end{pmatrix} \qquad \sigma_y \equiv \begin{pmatrix} 0 & -i \\ i & 0 \end{pmatrix} \qquad \sigma_z \equiv \begin{pmatrix} 1 & 0 \\ 0 & -1 \end{pmatrix} \qquad ; \text{ Pauli matrices}$$

---

[3] Here $(x, y, z) \equiv (1, 2, 3)$; as with the Dirac matrices, the underlining of the Pauli matrices is suppressed.

(b) Show that as *matrices*, the $\underline{s}_i$ obey the commutation relations of angular momentum

$$[\underline{s}_i, \underline{s}_j] = i\epsilon_{ijk}\underline{s}_k \qquad ; \text{ spin}$$

(c) Hence conclude that for the quantum mechanics of a single non-relativistic particle, one can append a direct product two-component spin wave function and extend the set of vectors to $(\mathbf{p}, \mathbf{x}, \mathbf{s})$ (see Vol. I).

Solution to Problem 3.3

(a) The effects of the raising and lowering operators on the eigenstates of angular momentum are given in Eqs. (3.59) as

$$\hat{J}_+|jm\rangle = A(j, -m)|j, m+1\rangle$$
$$\hat{J}_-|jm\rangle = A(j, m)|j, m-1\rangle$$

where from Eqs. (3.57)

$$A(j, m) = \sqrt{(j+m)(j-m+1)}$$

It follows that for angular momentum $j = 1/2$

$$\hat{J}_-|1/2, 1/2\rangle = |1/2, -1/2\rangle \qquad ; \hat{J}_-|1/2, -1/2\rangle = 0$$
$$\hat{J}_+|1/2, 1/2\rangle = 0 \qquad ; \hat{J}_+|1/2, -1/2\rangle = |1/2, 1/2\rangle$$

Now write

$$\hat{J}_x = (\hat{J}_+ + \hat{J}_-)/2 \qquad ; \hat{J}_y = (\hat{J}_+ - \hat{J}_-)/2i$$

and compute the matrix elements of the angular momentum operator in the basis $|1/2, m\rangle$

$$\langle 1/2, m'|\hat{\mathbf{J}}|1/2, m\rangle \equiv \frac{1}{2}[\boldsymbol{\sigma}]_{m'm}$$

In a matrix representation where the rows are labeled with m' and the columns are labeled with m, this gives the *Pauli matrices*[4]

$$\sigma_x \equiv \begin{pmatrix} 0 & 1 \\ 1 & 0 \end{pmatrix} \qquad \sigma_y \equiv \begin{pmatrix} 0 & -i \\ i & 0 \end{pmatrix} \qquad \sigma_z \equiv \begin{pmatrix} 1 & 0 \\ 0 & -1 \end{pmatrix}$$

---

[4] Recall $\hat{J}_z|1/2, m\rangle = m|1/2, m\rangle$.

(b) By direct matrix multiplication

$$\frac{1}{4}\left[\begin{pmatrix} 0 & 1 \\ 1 & 0 \end{pmatrix}\begin{pmatrix} 0 & -i \\ i & 0 \end{pmatrix} - \begin{pmatrix} 0 & -i \\ i & 0 \end{pmatrix}\begin{pmatrix} 0 & 1 \\ 1 & 0 \end{pmatrix}\right] = \frac{i}{2}\begin{pmatrix} 1 & 0 \\ 0 & -1 \end{pmatrix}$$

Hence

$$[\frac{1}{2}\sigma_x, \frac{1}{2}\sigma_y] = i\frac{1}{2}\sigma_z$$

This result is readily extended to

$$[\frac{1}{2}\sigma_i, \frac{1}{2}\sigma_j] = i\epsilon_{ijk}\frac{1}{2}\sigma_k$$

where ϵ_{ijk} is the completely antisymmetric Levi-Civita tensor in three dimensions.[5] Hence

$$\underline{\mathbf{J}} = \frac{1}{2}\boldsymbol{\sigma}$$

provides a 2×2 matrix representation of the angular momentum commutation relations in Eqs. (3.25).[6]

(c) To describe spin-1/2 in non-relativistic quantum mechanics, one can then simply append a two-dimensional space with basis vectors that are two-component column matrices

$$\eta_\uparrow = \begin{pmatrix} 1 \\ 0 \end{pmatrix} \qquad ; \eta_\downarrow = \begin{pmatrix} 0 \\ 1 \end{pmatrix}$$

and operators that are $(1, \boldsymbol{\sigma})$.

Problem 3.4 Repeat Prob. 3.3 for spin $s = 1$, and derive the 3×3 matrix representation of the commutation relations.

Solution to Problem 3.4

It follows from the solution to Prob. 3.3(a) that for angular momentum $j = 1$

$$\hat{J}_-|1,1\rangle = \sqrt{2}|1,0\rangle \qquad ; \hat{J}_+|1,1\rangle = 0$$
$$\hat{J}_-|1,0\rangle = \sqrt{2}|1,-1\rangle \qquad ; \hat{J}_+|1,0\rangle = \sqrt{2}|1,1\rangle$$
$$\hat{J}_-|1,-1\rangle = 0 \qquad ; \hat{J}_+|1,-1\rangle = \sqrt{2}|1,0\rangle$$

---

[5]Here repeated Latin indices are summed from 1 to 3.
[6]Note $\underline{\mathbf{J}}^2 = \boldsymbol{\sigma}^2/4 = 3/4$ is a unit matrix.

Now write

$$\hat{J}_x = (\hat{J}_+ + \hat{J}_-)/2 \qquad ; \ \hat{J}_y = (\hat{J}_+ - \hat{J}_-)/2i$$

and compute the matrix elements of the angular momentum operator in the basis $|1m\rangle$

$$\langle 1m'|\hat{\mathbf{J}}|1m\rangle \equiv [\boldsymbol{S}]_{m'm}$$

In a matrix representation where the rows are labeled with m' and the columns are labeled with m, this gives[7]

$$\boldsymbol{S}_x \equiv \frac{1}{\sqrt{2}} \begin{pmatrix} 0 & 1 & 0 \\ 1 & 0 & 1 \\ 0 & 1 & 0 \end{pmatrix} \qquad \boldsymbol{S}_y \equiv \frac{1}{\sqrt{2}} \begin{pmatrix} 0 & -i & 0 \\ i & 0 & -i \\ 0 & i & 0 \end{pmatrix} \qquad \boldsymbol{S}_z \equiv \begin{pmatrix} 1 & 0 & 0 \\ 0 & 0 & 0 \\ 0 & 0 & -1 \end{pmatrix}$$

(b) By direct matrix multiplication

$$\frac{1}{2} \left[\begin{pmatrix} 0 & 1 & 0 \\ 1 & 0 & 1 \\ 0 & 1 & 0 \end{pmatrix} \begin{pmatrix} 0 & -i & 0 \\ i & 0 & -i \\ 0 & i & 0 \end{pmatrix} - \begin{pmatrix} 0 & -i & 0 \\ i & 0 & -i \\ 0 & i & 0 \end{pmatrix} \begin{pmatrix} 0 & 1 & 0 \\ 1 & 0 & 1 \\ 0 & 1 & 0 \end{pmatrix} \right] = i \begin{pmatrix} 1 & 0 & 0 \\ 0 & 0 & 0 \\ 0 & 0 & -1 \end{pmatrix}$$

Hence

$$[\boldsymbol{S}_x, \boldsymbol{S}_y] = i\boldsymbol{S}_z$$

This result is readily extended to

$$[\boldsymbol{S}_i, \boldsymbol{S}_j] = i\epsilon_{ijk}\boldsymbol{S}_k$$

where ϵ_{ijk} is the completely antisymmetric Levi-Civita tensor in three dimensions.[8] Hence

$$\underline{\mathbf{J}} = \boldsymbol{S}$$

provides a 3×3 matrix representation of the angular momentum commutation relations in Eqs. (3.25).[9]

Problem 3.5 Derive the orthonormality and completeness relations for the C-G coefficients in Eqs. (3.75) on the basis of the general assumptions concerning operators and state vectors.

---

[7] Recall $\hat{J}_z|1, m\rangle = m|1, m\rangle$. We suppress the underlining of the matrices $\boldsymbol{S}$.
[8] Here repeated Latin indices are summed from 1 to 3.
[9] Note $\underline{\mathbf{J}}^2 = \boldsymbol{S}^2 = 2$ is a unit matrix.

Solution to Problem 3.5

The Clebsch-Gordan coupling coefficients are defined in Eq. (3.74)

$$|lsjm_j\rangle = \sum_{m_l,m_s} \langle lm_lsm_s|lsjm_j\rangle \, |lm_lsm_s\rangle$$

Both sets of basis states are orthonormal

$$\langle lsjm_j|lsj'm'_j\rangle = \delta_{jj'}\delta_{m_jm'_j}$$
$$\langle lm_lsm_s|lm'_lsm'_s\rangle = \delta_{m_lm'_l}\delta_{m_sm'_s}$$

This immediately leads to the first relation for the C-G coefficients

$$\sum_{m_l,m_s} \langle lm_lsm_s|lsjm_j\rangle\langle lm_lsm_s|lsj'm'_j\rangle^\star = \delta_{jj'}\delta_{m_jm'_j}$$

The basis states are also complete in the subspace of given (l,s)

$$\sum_{j,m_j} |lsjm_j\rangle\langle lsjm_j| = \sum_{m_l,m_s} |lm_lsm_s\rangle\langle lm_lsm_s|$$

This implies

$$\sum_{\underline{m}_l,\underline{m}_s} \sum_{\underline{m}'_l,\underline{m}'_s} \left[\sum_{j,m_j}\langle l\underline{m}_ls\underline{m}_s|lsjm_j\rangle\langle l\underline{m}'_ls\underline{m}'_s|lsjm_j\rangle^\star\right] |l\underline{m}_ls\underline{m}_s\rangle\langle l\underline{m}'_ls\underline{m}'_s|$$
$$= \sum_{\underline{m}_l,\underline{m}_s} |l\underline{m}_ls\underline{m}_s\rangle\langle l\underline{m}_ls\underline{m}_s|$$

The matrix element $\langle lm_lsm_s|\cdots|lm'_lsm'_s\rangle$ of this expression then leads to the second relation for the C-G coefficients

$$\sum_{j,m_j}\langle lm_lsm_s|lsjm_j\rangle\langle lm'_lsm'_s|lsjm_j\rangle^\star = \delta_{m_lm'_l}\delta_{m_sm'_s}$$

Problem 3.6 (a) Verify the orthonormality and completeness relations in Eqs. (3.97);

(b) Verify that Eqs. (3.107) and (3.110) define ITO's of rank one;

(c) Verify the finite rotation in Eq. (3.123).

Solution to Problem 3.6

(a) The first set of coupled states with (j_{12}, j, m) constructed from three angular momenta (j_1, j_1, j_3) is defined in Eq. (3.96)

$$|(j_1 j_2) j_{12} j_3 j m\rangle = \sum_{m_1, m_2, m_3, m_{12}} \langle j_1 m_1 j_2 m_2 | j_1 j_2 j_{12} m_{12} \rangle \times$$
$$\langle j_{12} m_{12} j_3 m_3 | j_{12} j_3 j m \rangle \, |j_1 m_1 j_2 m_2 j_3 m_3\rangle$$

The properties of these states follow immediately from those of the (real) C-G coefficients established in the previous problem. The basis states are orthonormal

$$\langle j_1 m_1 j_2 m_2 j_3 m_3 | j_1 m_1' j_2 m_2' j_3 m_3' \rangle = \delta_{m_1 m_1'} \delta_{m_2 m_2'} \delta_{m_3 m_3'}$$

It follows that

$$\langle (j_1 j_2) j_{12} j_3 j m | (j_1 j_2) j_{12}' j_3 j' m' \rangle = \sum_{m_1, m_2, m_3, m_{12}, m_{12}'} \langle j_1 m_1 j_2 m_2 | j_1 j_2 j_{12} m_{12} \rangle \times$$
$$\langle j_{12} m_{12} j_3 m_3 | j_{12} j_3 j m \rangle \langle j_1 m_1 j_2 m_2 | j_1 j_2 j_{12}' m_{12}' \rangle \langle j_{12}' m_{12}' j_3 m_3 | j_{12}' j_3 j' m' \rangle$$

Now perform the sum over (m_1, m_2)[10]

$$\sum_{m_1, m_2} \langle j_1 m_1 j_2 m_2 | j_1 j_2 j_{12} m_{12} \rangle \langle j_1 m_1 j_2 m_2 | j_1 j_2 j_{12}' m_{12}' \rangle = \delta_{j_{12} j_{12}'} \delta_{m_{12} m_{12}'}$$

With the equality of j_{12} and j_{12}' established, perform the sum over m_{12}', and then over (m_3, m_{12})

$$\sum_{m_3, m_{12}} \langle j_{12} m_{12} j_3 m_3 | j_{12} j_3 j m \rangle \langle j_{12} m_{12} j_3 m_3 | j_{12} j_3 j' m' \rangle = \delta_{jj'} \delta_{mm'}$$

Hence the orthogonality statement

$$\langle (j_1 j_2) j_{12} j_3 j m | (j_1 j_2) j_{12}' j_3 j' m' \rangle = \delta_{j_{12} j_{12}'} \delta_{j j'} \delta_{m m'}$$

The completeness is exhibited in a similar fashion

$$\sum_{j, m, j_{12}} |(j_1 j_2) j_{12} j_3 j m\rangle \langle (j_1 j_2) j_{12} j_3 j m| = \sum_{j, m, j_{12}} \sum_{m_1, m_2, m_3, m_{12}} \sum_{m_1', m_2', m_3', m_{12}'} \times$$
$$\langle j_1 m_1 j_2 m_2 | j_1 j_2 j_{12} m_{12} \rangle \langle j_{12} m_{12} j_3 m_3 | j_{12} j_3 j m \rangle \langle j_1 m_1' j_2 m_2' | j_1 j_2 j_{12} m_{12}' \rangle \times$$
$$\langle j_{12} m_{12}' j_3 m_3' | j_{12} j_3 j m \rangle \, |j_1 m_1 j_2 m_2 j_3 m_3\rangle \langle j_1 m_1' j_2 m_2' j_3 m_3'|$$

---

[10]One has to take some care in these problems to establish just where the quantum numbers appear in the sums.

Perform the sum over (j, m)

$$\sum_{j,m} \langle j_{12} m_{12} j_3 m_3 | j_{12} j_3 j m \rangle \langle j_{12} m'_{12} j_3 m'_3 | j_{12} j_3 j m \rangle = \delta_{m_{12} m'_{12}} \delta_{m_3 m'_3}$$

Then, after performing the sum over m'_{12}, sum over (j_{12}, m_{12})

$$\sum_{j_{12}, m_{12}} \langle j_1 m_1 j_2 m_2 | j_1 j_2 j_{12} m_{12} \rangle \langle j_1 m'_1 j_2 m'_2 | j_1 j_2 j_{12} m_{12} \rangle = \delta_{m_1 m'_1} \delta_{m_2 m'_2}$$

Hence

$$\sum_{j,m,j_{12}} |(j_1 j_2) j_{12} j_3 j m \rangle \langle (j_1 j_2) j_{12} j_3 j m | =$$

$$\sum_{m_1, m_2, m_3} |j_1 m_1 j_2 m_2 j_3 m_3 \rangle \langle j_1 m_1 j_2 m_2 j_3 m_3 |$$

This reproduces the completeness statement in the (j_1, j_2, j_3) subspace.

(b) Equation (3.107) defines the spherical components of the vector $\mathbf{x}$

$$x_{1, \pm 1} = \mp \frac{1}{\sqrt{2}} (x \pm iy) \qquad ; x_{1, 0} = z$$

The angular momentum operator in the coordinate representation, in units of $\hbar$, is given in Eq. (3.16)

$$L_z = \frac{1}{i} \left(x \frac{\partial}{\partial y} - y \frac{\partial}{\partial x} \right) \quad ; L_y = \frac{1}{i} \left(z \frac{\partial}{\partial x} - x \frac{\partial}{\partial z} \right) \quad ; L_x = \frac{1}{i} \left(y \frac{\partial}{\partial z} - z \frac{\partial}{\partial y} \right)$$

In order to establish that $x_{1, q}$ is an ITO of rank 1, we must verify Eqs. (3.106)

$$[L_\pm, x_{1, q}] = A(1, \mp q) x_{1, q \pm 1} \qquad ; A(1, \mp q) = \sqrt{(1 \mp q)(1 \pm q + 1)}$$
$$[L_z, x_{1, q}] = q x_{1, q}$$

The proof lies in simply working out the required commutators. For example,

$$[L_z, x_{1, 0}] = 0$$
$$[L_z, x_{1, 1}] = -\frac{1}{\sqrt{2} i} [x \frac{\partial}{\partial y} - y \frac{\partial}{\partial x}, x + iy] = -\frac{1}{\sqrt{2}} (x + iy) = x_{1, 1}$$

Furthermore

$$L_+ = L_x + iL_y = -(x + iy) \frac{\partial}{\partial z} + iz \frac{\partial}{\partial y} + z \frac{\partial}{\partial x}$$

Then

$$[L_+, x_{1,1}] = -\frac{1}{\sqrt{2}}(-z + z) = 0$$

$$[L_+, x_{1,0}] = -(x + iy) = \sqrt{2}\,x_{1,1}$$

$$[L_+, x_{1,-1}] = \sqrt{2}\,z = \sqrt{2}\,x_{1,0}$$

The remaining required commutators are obtained from the adjoints of the above.

The previous results are obtained from the basic commutation relation that defines a vector operator in quantum mechanics

$$[L_i,\ x_j] = i\epsilon_{ijk}\,x_k$$

where ϵ_{ijk} is the completely antisymmetric Levi-Civita tensor, and repeated Latin indices are summed from 1 to 3. The Pauli matrices satisfy an identical commutation relation

$$[\frac{1}{2}\sigma_i,\ \sigma_j] = i\epsilon_{ijk}\,\sigma_k$$

Hence the proof that $\sigma_{1,q}$ is an ITO of rank 1 in Eqs. (3.111) is completely analogous to the above.

(c) Use the result from Prob. 2.5 to write

$$e^{-i\beta\hat{J}_y}\hat{J}_z\,e^{i\beta\hat{J}_y} = \hat{J}_z - i\beta[\hat{J}_y, \hat{J}_z] + \frac{(-i\beta)^2}{2!}[\hat{J}_y, [\hat{J}_y, \hat{J}_z]] +$$

$$\frac{(-i\beta)^3}{3!}[\hat{J}_y, [\hat{J}_y, [\hat{J}_y, \hat{J}_z]]] + \cdots$$

Now use the angular momentum commutation relations

$$[\hat{J}_y, \hat{J}_z] = i\hat{J}_x \qquad\qquad ;\ [\hat{J}_y, \hat{J}_x] = -i\hat{J}_z$$

Thus

$$e^{-i\beta\hat{J}_y}\hat{J}_z\,e^{i\beta\hat{J}_y} = \hat{J}_z\left(1 - \frac{\beta^2}{2!} + \cdots\right) + \hat{J}_x\left(\beta - \frac{\beta^3}{3!} + \cdots\right)$$

$$= \hat{J}_z\cos\beta + \hat{J}_x\sin\beta$$

This is Eq. (3.123). Note that this result for the finite rotation follows entirely from the commutation relations for the angular momenta.

Problem 3.7 (a) Use the relation between the C-G coefficients and 3-j symbols given in Eq. (3.94), and the symmetry properties of the 3-j symbols stated there, to verify the following properties of the C-G coefficients

$$\langle j_1 m_1 j_2 m_2 | j_1 j_2 j m \rangle = (-1)^{j_1+j_2-j} \langle j_2 m_2 j_1 m_1 | j_2 j_1 j m \rangle$$

$$\langle j_1 m_1 j_2 m_2 | j_1 j_2 j m \rangle = (-1)^{j_1+j_2-j} \langle j_1, -m_1, j_2, -m_2 | j_1 j_2 j, -m \rangle$$

(b) Extend the results in part (a) to include the relations

$$\langle j_1 m_1 j_2 m_2 | j_1 j_2 j_3 m_3 \rangle = (-1)^{j_2+m_2} \left(\frac{2j_3+1}{2j_1+1} \right)^{1/2} \langle j_2, -m_2, j_3 m_3 | j_2 j_3 j_1 m_1 \rangle$$

$$\langle j_1 m_1 j_2 m_2 | j_1 j_2 j_3 m_3 \rangle = (-1)^{j_1-m_1} \left(\frac{2j_3+1}{2j_2+1} \right)^{1/2} \langle j_3 m_3 j_1, -m_1 | j_3 j_1 j_2 m_2 \rangle$$

Solution to Problem 3.7

There are various *symmetry properties* of the C-G coefficients that follow from the defining Eqs. (3.87) and the phase conventions. These symmetry properties are most usefully summarized in terms of Wigner's 3-j symbol defined by

$$\begin{pmatrix} j_1 & j_2 & j_3 \\ m_1 & m_2 & m_3 \end{pmatrix} \equiv (-1)^{j_1-j_2-m_3} \frac{1}{\sqrt{2j_3+1}} \langle j_1 m_1 j_2 m_2 | j_1 j_2 j_3, -m_3 \rangle$$

The properties of the 3-j symbols are [Edmonds (1974)]

- They are invariant under any even permutation of the columns;
- Any odd permutation of the columns is equivalent to multiplication of the 3-j symbol by $(-1)^{j_1+j_2+j_3}$;
- They vanish unless $m_1 + m_2 + m_3 = 0$;
- The reflection $m_i \to -m_i$ is equivalent to multiplication of the 3-j symbol by $(-1)^{j_1+j_2+j_3}$;[11]
- They vanish unless $j_1 + j_2 \geq j_3 \geq |j_1 - j_2|$, which is the *addition law* for angular momenta.[12]

(a) It follows that the C-G coefficient is expressed as a 3-j symbol by

$$\langle j_1 m_1 j_2 m_2 | j_1 j_2 j_3, -m_3 \rangle = (-1)^{j_1-j_2-m_3} \sqrt{2j_3+1} \begin{pmatrix} j_1 & j_2 & j_3 \\ m_1 & m_2 & m_3 \end{pmatrix}$$

---

[11] We will also need this one [Edmonds (1974)].

[12] Problem 3.15 deals with this addition law; note that $j_1 + j_2 + j_3$ must be an integer.

The interchange $(j_1 m_1) \rightleftharpoons (j_2 m_2)$ then gives

$$\langle j_2 m_2 j_1 m_1 | j_2 j_1 j_3, -m_3 \rangle = (-1)^{j_2 - j_1 - m_3} \sqrt{2j_3 + 1} \begin{pmatrix} j_2 & j_1 & j_3 \\ m_2 & m_1 & m_3 \end{pmatrix}$$

$$= (-1)^{j_1 + j_2 + j_3 + j_2 - j_1 - m_3} \sqrt{2j_3 + 1} \begin{pmatrix} j_1 & j_2 & j_3 \\ m_1 & m_2 & m_3 \end{pmatrix}$$

$$= (-1)^{3j_2 - j_1 + j_3} \langle j_1 m_1 j_2 m_2 | j_1 j_2 j_3, -m_3 \rangle$$

Since $4j_2$ is an even integer, and $j_3 - j_1 - j_2$ is an integer, this gives

$$\langle j_2 m_2 j_1 m_1 | j_2 j_1 j_3, -m_3 \rangle = (-1)^{j_1 + j_2 - j_3} \langle j_1 m_1 j_2 m_2 | j_1 j_2 j_3, -m_3 \rangle$$

which is the first result.

The interchange $m_i \rightleftharpoons -m_i$ gives

$$\langle j_1, -m_1 j_2, -m_2 | j_1 j_2 j_3, m_3 \rangle = (-1)^{j_1 - j_2 + m_3} \sqrt{2j_3 + 1} \begin{pmatrix} j_1 & j_2 & j_3 \\ -m_1 & -m_2 & -m_3 \end{pmatrix}$$

$$= (-1)^{j_1 + j_2 + j_3 + j_1 - j_2 + m_3} \sqrt{2j_3 + 1} \begin{pmatrix} j_1 & j_2 & j_3 \\ m_1 & m_2 & m_3 \end{pmatrix}$$

$$= (-1)^{j_1 + j_2 - j_3} \langle j_1 m_1 j_2 m_2 | j_1 j_2 j_3, -m_3 \rangle$$

where the last relation follows since $2(j_3 + m_3)$ is an even integer. This is the second result.

(b) A re-ordering of the coupling gives

$$\langle j_2, -m_2 j_3 m_3 | j_2 j_3 j_1 m_1 \rangle = (-1)^{j_2 - j_3 + m_1} \sqrt{2j_1 + 1} \begin{pmatrix} j_2 & j_3 & j_1 \\ -m_2 & m_3 & -m_1 \end{pmatrix}$$

$$= (-1)^{j_1 + j_2 + j_3 + j_2 - j_3 + m_3 - m_2} \sqrt{2j_1 + 1} \begin{pmatrix} j_1 & j_2 & j_3 \\ m_1 & m_2 & -m_3 \end{pmatrix}$$

$$= (-1)^{j_2 + m_2} \left(\frac{2j_1 + 1}{2j_3 + 1} \right)^{1/2} \langle j_1 m_1 j_2 m_2 | j_1 j_2 j_3 m_3 \rangle$$

The last result follows since $4j_2$ is an even integer, and $(j_2 + m_2)$ is an integer. This is the third relation.

A similar re-ordering gives

$$\langle j_3 m_3 j_1, -m_1 | j_3 j_1 j_2 m_2 \rangle = (-1)^{j_3 - j_1 + m_2} \sqrt{2j_2 + 1} \begin{pmatrix} j_3 & j_1 & j_2 \\ m_3 & -m_1 & -m_2 \end{pmatrix}$$

$$= (-1)^{-j_1 - j_2 - j_3 + j_3 - j_1 - m_1 + m_3} \sqrt{2j_2 + 1} \begin{pmatrix} j_1 & j_2 & j_3 \\ m_1 & m_2 & -m_3 \end{pmatrix}$$

$$= (-1)^{j_1 - m_1} \left(\frac{2j_2 + 1}{2j_3 + 1} \right)^{1/2} \langle j_1 m_1 j_2 m_2 | j_1 j_2 j_3 m_3 \rangle$$

The last result follows since $(-1)^{-3j_1 - m_1} = (-1)^{-4j_1 + j_1 - m_1} = (-1)^{j_1 - m_1}$. This is the fourth relation.

Problem 3.8 Use the definition of the 3-j symbol in Eq. (3.94) to express the 6-j symbol in Eqs. (3.100) as a sum over four 3-j symbols. Verify with [Edmonds (1974)].

Solution to Problem 3.8

Equation (3.100) expresses the 6-j coefficient as[13]

$$(-1)^{j_1 + j_2 + l_1 + l_2} \sqrt{(2j_3 + 1)(2l_3 + 1)} \begin{Bmatrix} j_1 & j_2 & j_3 \\ l_1 & l_2 & l_3 \end{Bmatrix} =$$

$$\sum_{m_1, m_2, m_3, \mu_1, \mu_3} \langle j_1 m_1 j_2 m_2 | j_1 j_2 j_3 m_3 \rangle \langle j_3 m_3 l_1 \mu_1 | j_3 l_1 l_2 \mu_2 \rangle \times$$
$$\langle j_1 m_1 l_3 \mu_3 | j_1 l_3 l_2 \mu_2 \rangle \langle j_2 m_2 l_1 \mu_1 | j_2 l_1 l_3 \mu_3 \rangle$$

Since the r.h.s. is independent of μ_2, we can simply sum this over μ_2

$$(-1)^{j_1 + j_2 + l_1 + l_2} (2l_2 + 1) \sqrt{(2j_3 + 1)(2l_3 + 1)} \begin{Bmatrix} j_1 & j_2 & j_3 \\ l_1 & l_2 & l_3 \end{Bmatrix} =$$

$$\sum_{m_1, m_2, m_3, \mu_1, \mu_2, \mu_3} \langle j_1 m_1 j_2 m_2 | j_1 j_2 j_3 m_3 \rangle \langle j_3 m_3 l_1 \mu_1 | j_3 l_1 l_2 \mu_2 \rangle \times$$
$$\langle j_1 m_1 l_3 \mu_3 | j_1 l_3 l_2 \mu_2 \rangle \langle j_2 m_2 l_1 \mu_1 | j_2 l_1 l_3 \mu_3 \rangle$$

Now use the relation between the C-G coefficients and the 3-j symbols in Eq. (3.94)

$$\begin{pmatrix} j_1 & j_2 & j_3 \\ m_1 & m_2 & m_3 \end{pmatrix} \equiv (-1)^{j_1 - j_2 - m_3} \frac{1}{\sqrt{2j_3 + 1}} \langle j_1 m_1 j_2 m_2 | j_1 j_2 j_3, -m_3 \rangle$$

---

[13]Here we use (j_i, l_i) for the angular momenta to obtain a more symmetric expression; note that l_i need not be an integer.

This gives

$$(-1)^{j_1+j_2+l_1+l_2} \begin{Bmatrix} j_1 & j_2 & j_3 \\ l_1 & l_2 & l_3 \end{Bmatrix} = \sum_{m_1,m_2,m_3,\mu_1,\mu_2,\mu_3} \times$$

$$(-1)^{j_1-j_2+m_3}(-1)^{j_3-l_1+\mu_2}(-1)^{j_1-l_3+\mu_2}(-1)^{j_2-l_1+\mu_3} \times$$

$$\begin{pmatrix} j_1 & j_2 & j_3 \\ m_1 & m_2 & -m_3 \end{pmatrix} \begin{pmatrix} j_3 & l_1 & l_2 \\ m_3 & \mu_1 & -\mu_2 \end{pmatrix} \begin{pmatrix} j_1 & l_3 & l_2 \\ m_1 & \mu_3 & -\mu_2 \end{pmatrix} \begin{pmatrix} j_2 & l_1 & l_3 \\ m_2 & \mu_1 & -\mu_3 \end{pmatrix}$$

Start the analysis of this expression by focusing on the phase:

- The combination $j_1-l_3+\mu_2$ is an integer, so its sign can be reversed;
- The second 3-j symbol allows us to replace $m_3 \to \mu_2 - \mu_1$;
- Then $-(\mu_1 + l_1)$ is integer, so the sign can also be reversed;
- The combination $j_1 + j_2 + l_1 + l_2$ is an integer, so the phase on the l.h.s. can be simply moved to the r.h.s.

With a change in dummy summation variables $(m_1, m_2) \to (-m_1, -m_2)$, the result of these steps is

$$\begin{Bmatrix} j_1 & j_2 & j_3 \\ l_1 & l_2 & l_3 \end{Bmatrix} = \sum_{m_1,m_2,m_3,\mu_1,\mu_2,\mu_3} (-1)^{j_1+j_2+j_3+l_1+l_2+l_3+\mu_1+\mu_2+\mu_3} \times$$

$$\begin{pmatrix} j_1 & j_2 & j_3 \\ -m_1 & -m_2 & -m_3 \end{pmatrix} \begin{pmatrix} j_3 & l_1 & l_2 \\ m_3 & \mu_1 & -\mu_2 \end{pmatrix} \begin{pmatrix} j_1 & l_3 & l_2 \\ -m_1 & \mu_3 & -\mu_2 \end{pmatrix} \begin{pmatrix} j_2 & l_1 & l_3 \\ -m_2 & \mu_1 & -\mu_3 \end{pmatrix}$$

Now use the symmetry properties of the 3-j symbols given in the text and the solution to Prob. 3.7 to obtain

$$\begin{Bmatrix} j_1 & j_2 & j_3 \\ l_1 & l_2 & l_3 \end{Bmatrix} = \sum_{m_1,m_2,m_3,\mu_1,\mu_2,\mu_3} (-1)^{l_1+l_2+l_3+\mu_1+\mu_2+\mu_3} \times$$

$$\begin{pmatrix} j_1 & j_2 & j_3 \\ m_1 & m_2 & m_3 \end{pmatrix} \begin{pmatrix} j_1 & l_2 & l_3 \\ m_1 & \mu_2 & -\mu_3 \end{pmatrix} \begin{pmatrix} l_1 & j_2 & l_3 \\ -\mu_1 & m_2 & \mu_3 \end{pmatrix} \begin{pmatrix} l_1 & l_2 & j_3 \\ \mu_1 & -\mu_2 & m_3 \end{pmatrix}$$

This result follows directly from Eq. 6.2.8 in [Edmonds (1974)].[14]

Problem 3.9 Use the construction of the states in Eqs. (3.96) and (3.98), and the definition of the 6-j symbol in Eq. (3.99), to prove the

---

[14] After correcting an obvious misprint in the phase there.

following recoupling relation

$$\sum_{m_{12}} \langle j_1 m_1 j_2 m_2 | j_1 j_2 j_{12} m_{12} \rangle \langle j_{12} m_{12} j_3 m_3 | j_{12} j_3 j m \rangle =$$

$$\sum_{j_{23}, m_{23}} (-1)^{j_1 + j_2 + j_3 + j} \sqrt{(2j_{12} + 1)(2j_{23} + 1)} \begin{Bmatrix} j_1 & j_2 & j_{12} \\ j_3 & j & j_{23} \end{Bmatrix} \times$$

$$\langle j_2 m_2 j_3 m_3 | j_2 j_3 j_{23} m_{23} \rangle \langle j_1 m_1 j_{23} m_{23} | j_1 j_{23} j m \rangle$$

Solution to Problem 3.9

One coupled basis is expressed in terms of the other through Eq. (3.102)

$$| (j_1 j_2) j_{12} j_3 j m \rangle = \sum_{j_{23}} \langle j_1 (j_2 j_3) j_{23} j | (j_1 j_2) j_{12} j_3 j \rangle \times$$

$$| j_1 (j_2 j_3) j_{23} j m \rangle$$

As proven in the text, the transformation coefficients are independent of m. The 6-j symbols are defined in terms of these coefficients in Eq. (3.99)

$$\langle ((j_1 j_2) j_{12} j_3 j | j_1 (j_2 j_3) j_{23} j \rangle \equiv$$

$$(-1)^{j_1 + j_2 + j_3 + j} \sqrt{(2j_{12} + 1)(2j_{23} + 1)} \begin{Bmatrix} j_1 & j_2 & j_{12} \\ j_3 & j & j_{23} \end{Bmatrix}$$

The states are obtained from the direct-product basis $| j_1 m_1 j_2 m_2 j_3 m_3 \rangle$ through Eqs. (3.96) and (3.98)

$$| (j_1 j_2) j_{12} j_3 j m \rangle = \sum_{m_1, m_2, m_3, m_{12}} \langle j_1 m_1 j_2 m_2 | j_1 j_2 j_{12} m_{12} \rangle \times$$

$$\langle j_{12} m_{12} j_3 m_3 | j_{12} j_3 j m \rangle | j_1 m_1 j_2 m_2 j_3 m_3 \rangle$$

$$| j_1 (j_2 j_3) j_{23} j m \rangle = \sum_{m_1, m_2, m_3, m_{23}} \langle j_1 m_1 j_{23} m_{23} | j_1 j_{23} j m \rangle \times$$

$$\langle j_2 m_2 j_3 m_3 | j_2 j_3 j_{23} m_{23} \rangle | j_1 m_1 j_2 m_2 j_3 m_3 \rangle$$

Now take the matrix element of the initial relation with the state $| j_1 m_1 j_2 m_2 j_3 m_3 \rangle$ for a given m_1, m_2, m_3, and use the orthonormality of the basis states. This yields

$$\sum_{m_{12}} \langle j_1 m_1 j_2 m_2 | j_1 j_2 j_{12} m_{12} \rangle \langle j_{12} m_{12} j_3 m_3 | j_{12} j_3 j m \rangle =$$

$$\sum_{j_{23}, m_{23}} (-1)^{j_1 + j_2 + j_3 + j} \sqrt{(2j_{12} + 1)(2j_{23} + 1)} \left\{ \begin{matrix} j_1 & j_2 & j_{12} \\ j_3 & j & j_{23} \end{matrix} \right\} \times$$

$$\langle j_2 m_2 j_3 m_3 | j_2 j_3 j_{23} m_{23} \rangle \langle j_1 m_1 j_{23} m_{23} | j_1 j_{23} j m \rangle$$

We have added a redundant sum over m_{23}; it is in fact determined through the C-G coefficients. This is the desired recoupling relation for two C-G coefficients.

Problem 3.10 Start from the result in Prob. 3.9, and prove the additional recoupling relation[15]

$$\sum_{m_2, m_3, m_{12}} \langle j_1 m_1 j_2 m_2 | j_1 j_2 j_{12} m_{12} \rangle \langle j_{12} m_{12} j_3 m_3 | j_{12} j_3 j m \rangle \langle j_2 m_2 j_3 m_3 | j_2 j_3 j_{23} m_{23} \rangle$$

$$= (-1)^{j_1 + j_2 + j_3 + j} \sqrt{(2j_{12} + 1)(2j_{23} + 1)} \left\{ \begin{matrix} j_1 & j_2 & j_{12} \\ j_3 & j & j_{23} \end{matrix} \right\} \langle j_1 m_1 j_{23} m_{23} | j_1 j_{23} j m \rangle$$

Solution to Problem 3.10

Make use of the orthonormality of the C-G coefficients in Eqs. (3.75)

$$\sum_{m_2, m_3} \langle j_2 m_2 j_3 m_3 | j_2 j_3 j_{23} m_{23} \rangle \langle j_2 m_2 j_3 m_3 | j_2 j_3 j'_{23} m'_{23} \rangle = \delta_{j_{23} j'_{23}} \delta_{m_{23} m'_{23}}$$

Multiply the result in Prob. 3.9 by $\langle j_2 m_2 j_3 m_3 | j_2 j_3 j_{23} m_{23} \rangle$ with a given $j_{23} m_{23}$ and perform $\sum_{m_2} \sum_{m_3}$. The result is

$$\sum_{m_2, m_3, m_{12}} \langle j_1 m_1 j_2 m_2 | j_1 j_2 j_{12} m_{12} \rangle \langle j_{12} m_{12} j_3 m_3 | j_{12} j_3 j m \rangle \langle j_2 m_2 j_3 m_3 | j_2 j_3 j_{23} m_{23} \rangle$$

$$= (-1)^{j_1 + j_2 + j_3 + j} \sqrt{(2j_{12} + 1)(2j_{23} + 1)} \left\{ \begin{matrix} j_1 & j_2 & j_{12} \\ j_3 & j & j_{23} \end{matrix} \right\} \langle j_1 m_1 j_{23} m_{23} | j_1 j_{23} j m \rangle$$

This is the desired recoupling relation for three C-G coefficients.

Problem 3.11 Write out the linear algebraic Eqs. (3.87) that determine the C-G coefficients for $\mathbf{j} = \mathbf{l} + \mathbf{s}$ in the case $s = 1$.

---

[15] Note that there is no easy way to get these results just starting from Eqs. (3.100).

(a) Show the analogue of the eigenvalue Eq. (3.92) for α_j is now

$$(\alpha_j - 2l)(\alpha_j + 2)(\alpha_j + 2l + 2) = 0$$

This equation has the roots $\alpha_j = (2l, -2, -2l - 2)$. Show this implies $j = (l + 1, l, l - 1)$;

(b) The solutions to the linear equations in each case determine the C-G coefficients in Table 3.1.[16] Pick any row in this table, and verify the result.

Table 3.1   Clebsch-Gordan coefficients $\langle l, m - m_s, 1, m_s | l, 1, j, m \rangle$ for $s = 1$.

	$m_s = 1$	$m_s = 0$	$m_s = -1$
$j = l + 1$	$\left[\frac{(l+m)(l+m+1)}{(2l+1)(2l+2)}\right]^{1/2}$	$\left[\frac{(l+m+1)(l-m+1)}{(2l+1)(l+1)}\right]^{1/2}$	$\left[\frac{(l-m)(l-m+1)}{(2l+1)(2l+2)}\right]^{1/2}$
$j = l$	$-\left[\frac{(l+m)(l-m+1)}{2l(l+1)}\right]^{1/2}$	$\frac{m}{\sqrt{l(l+1)}}$	$\left[\frac{(l-m)(l+m+1)}{2l(l+1)}\right]^{1/2}$
$j = l - 1$	$\left[\frac{(l-m)(l-m+1)}{2l(2l+1)}\right]^{1/2}$	$-\left[\frac{(l-m)(l+m)}{l(2l+1)}\right]^{1/2}$	$\left[\frac{(l+m)(l+m+1)}{2l(2l+1)}\right]^{1/2}$

Solution to Problem 3.11

(a) In this case we have $s = 1$, $m_s = 0, \pm 1$, and $m_l = m_j - m_s$. With the use of Eq. (3.88), we see that

$$A(1, 2) = 0 \qquad ; A(1, 1) = \sqrt{2} \qquad ; A(1, 0) = \sqrt{2}$$

Also

$$A(l, m_j) = A(l, 1 - m_j) = \sqrt{(l + m_j)(1 + l - m_j)}$$

$$A(l, m_j + 1) = A(l, -m_j) = \sqrt{(l - m_j)(1 + l + m_j)}$$

$$A(l, m_j + 2) = \sqrt{(l - m_j - 1)(2 + l + m_j)}$$

$$A(l, -m_j + 2) = \sqrt{(l + m_j - 1)(2 + l - m_j)}$$

From Eqs. (3.87), we obtain three relations for the C-G coefficients

$$[2(m_j-1)-\alpha_j]\,\langle m_j-1,1|jm_j\rangle+\sqrt{2(l+m_j)(1+l-m_j)}\langle m_j,0|jm_j\rangle=0$$

$$\sqrt{2(l+m_j)(1+l-m_j)}\langle m_j-1,1|jm_j\rangle-\alpha_j\langle m_j,0|jm_j\rangle+$$
$$\sqrt{2(l-m_j)(1+l+m_j)}\langle m_j+1,-1|jm_j\rangle=0$$

$$\sqrt{2(l-m_j)(1+l+m_j)}\langle m_j,0|jm_j\rangle-[2(m_j+1)+\alpha_j]\,\langle m_j+1,-1|jm_j\rangle=0$$

---

[16]See [Edmonds (1974)].

These linear, algebraic equations can be written in matrix form

$$\underline{M}\,\underline{V} = 0$$

where the matrix $\underline{M}$ is

$$\underline{M} \equiv \begin{bmatrix} 2(m_j - 1) - \alpha_j & \sqrt{2(l - m_j + 1)(l + m_j)} & 0 \\ \sqrt{2(l - m_j + 1)(l + m_j)} & -\alpha_j & \sqrt{2(l - m_j)(l + m_j + 1)} \\ 0 & \sqrt{2(l - m_j)(l + m_j + 1)} & -2(m_j + 1) - \alpha_j \end{bmatrix}$$

and the column vector of C-G coefficients is

$$\underline{V} \equiv \begin{bmatrix} \langle m_j - 1, 1 | j m_j \rangle \\ \langle m_j, 0 | j m_j \rangle \\ \langle m_j + 1, -1 | j m_j \rangle \end{bmatrix}$$

A non-trivial solution to the equations is possible only if the determinant of $\underline{M}$ vanishes. An expansion in minors along the first row gives

$$\det \underline{M} = \alpha_j[2(m_j - 1) - \alpha_j][2(m_j + 1) + \alpha_j] -$$
$$2\,[2(m_j - 1) - \alpha_j](l - m_j)(l + m_j + 1) +$$
$$2\,[2(m_j + 1) + \alpha_j](l + m_j)(l - m_j + 1)$$

This is evaluated as

$$\det \underline{M} = \alpha_j[4m_j^2 - (\alpha_j + 2)^2] + (\alpha_j + 2)[4l(l + 1) - 4m_j^2] + 8m_j^2$$
$$= (\alpha_j + 2)[4l(l + 1) - \alpha_j(\alpha_j + 2)]$$
$$= -(\alpha_j + 2)(\alpha_j - 2l)(\alpha_j + 2l + 2)$$

Hence the condition for a non-trivial solution, $\det \underline{M} = 0$, is

$$(\alpha_j - 2l)(\alpha_j + 2)(\alpha_j + 2l + 2) = 0$$

With the use of Eq. (3.89)

$$\alpha_j = j(j + 1) - l(l + 1) - s(s + 1)$$

and substituting the obvious roots of the characteristic polynomial $\alpha_j = (2l, -2, -2l - 2)$, we see that with $s = 1$, this implies $j = (l + 1, l, l - 1)$.

(b) We are now in position to determine the C-G coefficients corresponding to each of the three different solutions of the characteristic equation. Since the equations are now linearly dependent, we can only use two of them and thus express two of the C-G coefficients in terms of the remaining one; the remaining coefficient is then determined using the normalization in Eq. (3.75).

Consider, for instance, the case $\alpha_j = 2l$ which implies $j = l + 1$. The first and last equations become

$$[2(m_j-1)-2l]\langle m_j-1,1|jm_j\rangle + \sqrt{2(l+m_j)(1+l-m_j)}\langle m_j,0|jm_j\rangle = 0$$

$$\sqrt{2(l-m_j)(1+l+m_j)}\langle m_j,0|jm_j\rangle - [2(m_j+1)+2l]\langle m_j+1,-1|jm_j\rangle = 0$$

The solutions to these equations provide

$$\langle m_j - 1,1|jm_j\rangle = \frac{\sqrt{(l+m_j)}}{\sqrt{2(l+1-m_j)}}\langle m_j,0|jm_j\rangle$$

$$\langle m_j - 1,-1|jm_j\rangle = \frac{\sqrt{(l-m_j)}}{\sqrt{2(l+1+m_j)}}\langle m_j,0|jm_j\rangle$$

If these ratios are substituted into the normalization condition in Eq. (3.75), one obtains

$$\left[\frac{l+m_j}{2(l+1-m_j)} + \frac{l-m_j}{2(l+1+m_j)} + 1\right]|\langle m_j,0|jm_j\rangle|^2 = 1$$

This simplifies to

$$\frac{(l+1)(2l+1)}{(l+1)^2 - m_j^2}|\langle m_j,0|jm_j\rangle|^2 = 1$$

Hence, with an appropriate choice of phase, one finally obtains

$$\langle m_j - 1,1|jm_j\rangle = \sqrt{\frac{(l+m_j)(l+m_j+1)}{(2l+1)(2l+2)}}$$

$$\langle m_j,0|jm_j\rangle = \sqrt{\frac{(l+1-m_j)(l+1+m_j)}{(2l+1)(l+1)}}$$

$$\langle m_j + 1,-1|jm_j\rangle = \sqrt{\frac{(l+1-m_j)(l-m_j)}{(2l+1)(2l+2)}}$$

These are, indeed, the C-G coefficients reported in the first row of the table. The cases corresponding to $j = l$ and $j = l - 1$ can be obtained in an analogous way.

Problem 3.12 The rotation matrix $[\underline{d}^j(\beta)]$ is defined through its matrix elements $d^j_{m'm}(\beta) = \langle jm'|e^{i\beta\hat{J}_2}|jm\rangle$. Write $\hat{J}_2 = (\hat{J}_+ - \hat{J}_-)/2i$, use

Eqs. (3.59), and show that for $j = 1/2$ the rotation matrix takes the form

$$\underline{d}^{1/2}(\beta) = \begin{bmatrix} \cos\beta/2 & \sin\beta/2 \\ -\sin\beta/2 & \cos\beta/2 \end{bmatrix}$$

What is this expression if $\beta = 2\pi$? If $\beta = 4\pi$? Discuss.

Solution to Problem 3.12

Expand the exponential operator

$$e^{i\beta \hat{J}_2} = \sum_{n=0}^{\infty} \frac{1}{n!} \left(i\beta \hat{J}_2 \right)^n$$

Now separate the sum into the contributions from even and odd powers of the operator $\hat{J}_2 = (\hat{J}_+ - \hat{J}_-)/2i$

$$e^{i\beta \hat{J}_2} = \hat{S}_E + \hat{S}_O$$

$$\hat{S}_E \equiv \sum_{n=0}^{\infty} \frac{1}{(2n)!} \left[\frac{\beta}{2} \left(\hat{J}_+ - \hat{J}_- \right) \right]^{2n}$$

$$\hat{S}_O \equiv \sum_{n=0}^{\infty} \frac{1}{(2n+1)!} \left[\frac{\beta}{2} \left(\hat{J}_+ - \hat{J}_- \right) \right]^{2n+1}$$

The effect of the operator

$$(\hat{J}_+ - \hat{J}_-)^2 = \hat{J}_+^2 + \hat{J}_-^2 - \hat{J}_+\hat{J}_- - \hat{J}_-\hat{J}_+$$

on the states $|1/2, m\rangle$ follows from the results in Prob. 3.3(a). The operators $\hat{J}_+^2$, $\hat{J}_-^2$ give zero, and one of the operators $\hat{J}_-\hat{J}_+$, $\hat{J}_+\hat{J}_-$ reproduces the state, while the other vanishes. Hence

$$(\hat{J}_+ - \hat{J}_-)^2 |1/2, m\rangle = -|1/2, m\rangle$$

It follows that

$$\hat{S}_E |1/2, m\rangle = \sum_{n=0}^{\infty} \frac{(-1)^n}{(2n)!} \left(\frac{\beta}{2} \right)^{2n} |1/2, m\rangle$$

$$= \cos\left(\frac{\beta}{2} \right) |1/2, m\rangle$$

In the case of $\hat{S}_O$, there is one additional power of $(\hat{J}_+ - \hat{J}_-)$ remaining.

Therefore

$$\hat{S}_O|1/2, -1/2\rangle = \sum_{n=0}^{\infty} \frac{(-1)^n}{(2n+1)!} \left(\frac{\beta}{2}\right)^{2n+1} |1/2, 1/2\rangle$$

$$= \sin\left(\frac{\beta}{2}\right) |1/2, 1/2\rangle$$

and similarly

$$\hat{S}_O|1/2, 1/2\rangle = -\sin\left(\frac{\beta}{2}\right) |1/2, -1/2\rangle$$

Thus in this case, the rotation matrix takes the form

$$\underline{d}^{1/2}(\beta) = \begin{bmatrix} \cos\beta/2 & \sin\beta/2 \\ -\sin\beta/2 & \cos\beta/2 \end{bmatrix}$$

Under a rotation $\beta = 2\pi$, the rotation matrix becomes

$$\underline{d}^{1/2}(2\pi) = -\begin{bmatrix} 1 & 0 \\ 0 & 1 \end{bmatrix}$$

This is simply a change in sign, and hence the spin-1/2 state vectors are *double-valued under rotations*. This is acceptable for physics, since observables are bilinear in the state vectors.

With $\beta = 4\pi$, the rotation matrix is

$$\underline{d}^{1/2}(4\pi) = \begin{bmatrix} 1 & 0 \\ 0 & 1 \end{bmatrix}$$

and the state vector returns to its initial value.

Problem 3.13 Repeat Prob. 3.12 for $j = 1$. Show through the first two contributing powers in β in each term that

$$\underline{d}^1(\beta) = \begin{bmatrix} (1+\cos\beta)/2 & (\sin\beta)/\sqrt{2} & (1-\cos\beta)/2 \\ -(\sin\beta)/\sqrt{2} & \cos\beta & (\sin\beta)/\sqrt{2} \\ (1-\cos\beta)/2 & -(\sin\beta)/\sqrt{2} & (1+\cos\beta)/2 \end{bmatrix}$$

Solution to Problem 3.13

From Eqs. (3.59), the effect of the operators $\hat{J}_+$, $\hat{J}_-$ on the states $|1, m\rangle$ are

$$\hat{J}_-|1,1\rangle = \sqrt{2}\,|1,0\rangle \quad ; \quad \hat{J}_-|1,0\rangle = \sqrt{2}\,|1,-1\rangle \quad ; \quad \hat{J}_-|1,-1\rangle = 0$$
$$\hat{J}_+|1,1\rangle = 0 \quad ; \quad \hat{J}_+|1,0\rangle = \sqrt{2}\,|1,1\rangle \quad ; \quad \hat{J}_+|1,-1\rangle = \sqrt{2}\,|1,0\rangle$$

As in Prob. 3.12, the expansion of the exponential operator through $O(\beta^3)$ gives

$$e^{i\beta\hat{J}_2} \approx 1 + \frac{\beta}{2}(\hat{J}_+ - \hat{J}_-) + \frac{1}{2!}\left(\frac{\beta}{2}\right)^2 (\hat{J}_+ - \hat{J}_-)^2 + \frac{1}{3!}\left(\frac{\beta}{2}\right)^3 (\hat{J}_+ - \hat{J}_-)^3$$

Use

$$(\hat{J}_+ - \hat{J}_-)^2 = \hat{J}_+^2 + \hat{J}_-^2 - \hat{J}_+\hat{J}_- - \hat{J}_-\hat{J}_+$$
$$(\hat{J}_+ - \hat{J}_-)^3 = \hat{J}_+^3 - \hat{J}_-^3 - \hat{J}_+\hat{J}_+\hat{J}_- - \hat{J}_+\hat{J}_-\hat{J}_+ - \hat{J}_-\hat{J}_+\hat{J}_+ +$$
$$\hat{J}_+\hat{J}_-\hat{J}_- + \hat{J}_-\hat{J}_+\hat{J}_- + \hat{J}_-\hat{J}_-\hat{J}_+$$

When applied to the state $|1,1\rangle$, these give

$$(\hat{J}_+ - \hat{J}_-)|1,1\rangle = -\sqrt{2}\,|1,0\rangle$$
$$(\hat{J}_+ - \hat{J}_-)^2|1,1\rangle = 2\,|1,-1\rangle - 2\,|1,1\rangle$$
$$(\hat{J}_+ - \hat{J}_-)^3|1,1\rangle = 4\sqrt{2}\,|1,0\rangle$$

Hence when the exponential $e^{i\beta\hat{J}_2}$ is applied to the state $|1,1\rangle$ one obtains to this order

$$e^{i\beta\hat{J}_2}|1,1\rangle \approx \left(1 - \frac{\beta^2}{4}\right)|1,1\rangle - \frac{\beta}{\sqrt{2}}\left(1 - \frac{\beta^2}{6}\right)|1,0\rangle + \frac{\beta^2}{4}|1,-1\rangle$$

A similar calculation gives

$$(\hat{J}_+ - \hat{J}_-)|1,0\rangle = \sqrt{2}\,|1,1\rangle - \sqrt{2}\,|1,-1\rangle$$
$$(\hat{J}_+ - \hat{J}_-)^2|1,0\rangle = -4\,|1,0\rangle$$
$$(\hat{J}_+ - \hat{J}_-)^3|1,0\rangle = -4\sqrt{2}\,|1,1\rangle + 4\sqrt{2}\,|1,-1\rangle$$

Therefore

$$e^{i\beta\hat{J}_2}|1,0\rangle \approx \frac{\beta}{\sqrt{2}}\left(1 - \frac{\beta^2}{6}\right)|1,1\rangle + \left(1 - \frac{\beta^2}{2}\right)|1,0\rangle - \frac{\beta}{\sqrt{2}}\left(1 - \frac{\beta^2}{6}\right)|1,-1\rangle$$

For the state $|1, -1\rangle$ we have

$$(\hat{J}_+ - \hat{J}_-)|1, -1\rangle = \sqrt{2}|1, 0\rangle$$
$$(\hat{J}_+ - \hat{J}_-)^2|1, -1\rangle = -2|1, -1\rangle + 2|1, 1\rangle$$
$$(\hat{J}_+ - \hat{J}_-)^3|1, -1\rangle = -4\sqrt{2}|1, 0\rangle$$

Hence

$$e^{i\beta\hat{J}_2}|1, -1\rangle \approx \frac{\beta^2}{4}|1, 1\rangle + \frac{\beta}{\sqrt{2}}\left(1 - \frac{\beta^2}{6}\right)|1, 0\rangle + \left(1 - \frac{\beta^2}{4}\right)|1, -1\rangle$$

These results are identical with those obtained from the first two contributing powers in β in each term of the rotation matrix

$$\underline{d}^1(\beta) = \begin{bmatrix} (1 + \cos\beta)/2 & (\sin\beta)/\sqrt{2} & (1 - \cos\beta)/2 \\ -(\sin\beta)/\sqrt{2} & \cos\beta & (\sin\beta)/\sqrt{2} \\ (1 - \cos\beta)/2 & -(\sin\beta)/\sqrt{2} & (1 + \cos\beta)/2 \end{bmatrix}$$

where $d^j_{m'm}(\beta) = \langle jm'|e^{i\beta\hat{J}_2}|jm\rangle$.

Problem 3.14 If $\hat{T}(\kappa, q)$ is an ITO working on the first part of a coupled scheme $|\gamma j_1 j_2 jm\rangle$, and $\hat{U}(\kappa, q)$ is an ITO working on the second part, then [Edmonds (1974)]

$$\langle\gamma' j_1' j_2 j'||T(\kappa)||\gamma j_1 j_2 j\rangle = (-1)^{j_1 + j_2 + j + \kappa}\sqrt{(2j + 1)(2j' + 1)}\begin{Bmatrix} j_1' & j' & j_2 \\ j & j_1 & \kappa \end{Bmatrix} \times$$
$$\langle\gamma' j_1'||T(\kappa)||\gamma j_1\rangle$$

$$\langle\gamma' j_1 j_2' j'||U(\kappa)||\gamma j_1 j_2 j\rangle = (-1)^{j_1 + j_2 + j' + \kappa}\sqrt{(2j + 1)(2j' + 1)}\begin{Bmatrix} j_2' & j' & j_1 \\ j & j_2 & \kappa \end{Bmatrix} \times$$
$$\langle\gamma' j_2'||U(\kappa)||\gamma j_2\rangle$$

These relations are proven by writing out the coupling of the direct-product states using C-G coefficients, using the W-E theorem on the one-body matrix element, then using a recoupling relation from Probs. 3.7–3.10 on the sum over the three C-G coefficients, and then identifying the reduced matrix element. Prove either one of these relations.[17]

Solution to Problem 3.14

We shall prove the second relation. Write out the required matrix element, explicitly exhibiting the coupling of the states and using the fact

---

[17] This problem requires somewhat more algebra, but the results are extremely useful.

that the first set of states is orthonormal[18]

$$\langle\gamma'j_1j_2'j'm'|U(\kappa,q)|\gamma j_1j_2jm\rangle = \sum_{m_2',m_1,m_2} \langle j_1m_1j_2'm_2'|j_1j_2'j'm'\rangle \times$$
$$\langle\gamma'j_2'm_2'|U(\kappa,q)|\gamma j_2m_2\rangle\langle j_1m_1j_2m_2|j_1j_2jm\rangle$$

Use the Wigner-Eckart theorem on the matrix element [Edmonds (1974)]

$$\langle\gamma'j_2'm_2'|U(\kappa,q)|\gamma j_2m_2\rangle = (-1)^{j_2-m_2}\langle j_2'm_2'j_2,-m_2|j_2'j_2\kappa q\rangle\frac{\langle\gamma'j_2'||U(\kappa)||\gamma j_2\rangle}{(2\kappa+1)^{1/2}}$$

Now re-arrange the C-G coefficients, using the relations in Prob. 3.7, to get them into the order in Prob. 3.10

$$\Sigma \equiv \sum_{m_2',m_1,m_2} \langle j_1m_1j_2'm_2'|j_1j_2'j'm'\rangle\langle j_1m_1j_2m_2|j_1j_2jm\rangle \times$$
$$(-1)^{j_2-m_2}\langle j_2'm_2'j_2,-m_2|j_2'j_2\kappa q\rangle$$
$$= \sum_{m_2',m_1,m_2} (-1)^{j_1-m_1}\left(\frac{2j+1}{2j_2+1}\right)^{1/2}\langle jmj_1,-m_1|jj_1j_2m_2\rangle(-1)^{j_2-m_2} \times$$
$$\langle j_2m_2j_2',-m_2'|j_2j_2'\kappa,-q\rangle(-1)^{j_1+j_2'-j'}\langle j_1,-m_1j_2',-m_2'|j_1j_2'j',-m'\rangle$$
$$= (-1)^{j_1+j_2'-j'+j_1+j_2-m}\left(\frac{2j+1}{2j_2+1}\right)^{1/2}\sum_{m_2'm_1,m_2}\langle jmj_1,m_1|jj_1j_2m_2\rangle \times$$
$$\langle j_2m_2j_2'm_2'|j_2j_2'\kappa,-q\rangle\langle j_1m_1j_2'm_2'|j_1j_2'j',-m'\rangle$$

Use the result in Prob. 3.10

$$\Sigma = (-1)^{j_1+j_2'-j'+j_1+j_2-m}\left(\frac{2j+1}{2j_2+1}\right)^{1/2} \times$$
$$(-1)^{j+j_1+j_2'+\kappa}\sqrt{(2j_2+1)(2j'+1)}\begin{Bmatrix} j & j_1 & j_2 \\ j_2' & \kappa & j' \end{Bmatrix}\langle jmj',-m'|jj'\kappa,-q\rangle$$

The Wigner-Eckart theorem applied to the full matrix element gives

$$\langle\gamma'j_1j_2'j'm'|U(\kappa,q)|\gamma j_1j_2jm\rangle = (-1)^{j-m}\langle j'm'j,-m|j'j\kappa q\rangle \times$$
$$\frac{\langle\gamma'j_1j_2'j'||U(\kappa)||\gamma j_1j_2j\rangle}{(2\kappa+1)^{1/2}}$$

---

[18] Use $\langle j_1m_1'|j_1m_1\rangle = \delta_{m_1'm_1}$. Here γ refers to the additional quantum numbers in the second set of states.

Hence we can can identify the reduced matrix element as

$$\langle \gamma' j_1 j_2' j' || U(\kappa) || \gamma j_1 j_2 j \rangle = (-1)^{j_1 + j_2 + j' + \kappa} \sqrt{(2j+1)(2j'+1)} \begin{Bmatrix} j_2' & j' & j_1 \\ j & j_2 & \kappa \end{Bmatrix} \times$$
$$\langle \gamma' j_2' || U(\kappa) || \gamma j_2 \rangle$$

where we have used the fact that $j_1 + j_2' - j'$ is integer, and the symmetry properties of the 6-j symbol. This is the result stated in the problem.

Problem 3.15 The general result for the addition of two angular momenta can be obtained from the *weight diagram*. Suppose one wants to add the angular momenta of the states $|j_1 m_1\rangle$ and $|j_2 m_2\rangle$, where, for concreteness, we assume that both are half-integral and $j_1 \geq j_2$. Make a plot where the ordinate is j and the abscissa is m, and place a cross on it for each state. From the general theory of angular momentum one knows that $m = m_1 + m_2$ for the state $|j_1 j_2 j m\rangle$. One also knows that the allowed values of m for a given j are $-j \leq m \leq j$ in integer steps, with each value occuring once.

(a) The maximum value of m is $j_1 + j_2$. Place a cross at this point, and then at all of the other requisite values of m at the appropriate value of j on the weight diagram.[19] What is the next highest possible value of m? Show that it occurs twice in the direct-product basis, and observe that it has already been used once in the above. Place a cross on this m, and all the other requisite values of m at the appropriate value of j;

(b) Repeat the process in (a);

(c) Show that the value $m = 0$ occurs $2j_2 + 1$ times in the direct-product basis. Since one of these is used once in each step, show that the process in (b) must terminate after $2j_2 + 1$ steps. Hence conclude that the correct addition law for angular momenta in this case is[20]

$$j_1 + j_2 \geq j \geq |j_1 - j_2| \qquad ; \text{ integer steps}$$
$$\text{each value occurs once}$$

Solution to Problem 3.15

(a) Consider the states $|j_1 m_1\rangle$ and $|j_2 m_2\rangle$, where j_1 and j_2 are both half-integral and $j_1 \geq j_2$. The allowed values for m_1 and m_2 are $|m_1| \leq j_1$

---

[19]Note that these states can actually be constructed by applying $\hat{J}_- = (\hat{j}_1 + \hat{j}_2)_-$ repeatedly to the state with $m = m_{\text{max}}$.

[20]We correct an obvious misprint.

and $|m_2| \leq j_2$, with $\Delta m = 1$, and the direct-product basis $|j_1 m_1\rangle |j_2 m_2\rangle$ contains $(2j_1 + 1) \times (2j_2 + 1)$ states.

The maximum value of m is $m_{max} = j_1 + j_2$, and the corresponding state in the direct-product basis is $|j_1 j_1\rangle |j_2 j_2\rangle$. This single state will be an eigenstate of $\hat{\mathbf{J}}^2$ with eigenvalue $j_{max}(j_{max} + 1)$ where

$$j_{max} = j_1 + j_2$$

If we apply the operator $\hat{J}_- = (\hat{j}_1 + \hat{j}_2)_-$ to this state, we obtain a state with $j = j_{max}$ and $m = m_{max} - 1$. This is the next highest possible value of m. In the direct-product basis there are *two* states which allow for this value of m, $|j_1, j_1 - 1\rangle |j_2 j_2\rangle$ and $|j_1 j_1\rangle |j_2, j_2 - 1\rangle$. The lowering process produces one linear combination of these states.

(b) The lowering process can be repeated as many times as needed, thus generating all the $2j + 1$ possible values of m. If these states are now removed from the direct-product basis, one is left with a single state with $m_{max} = j_1 + j_2 - 1$, and the process can be repeated.

(c) The value $m = 0$ requires that $m_1 = -m_2$. Since $j_1 \geq j_2$, there are only $2j_2 + 1$ possible values of m_1 for which $m_1 + m_2 = 0$. We conclude that the value $m = 0$ occurs $2j_2 + 1$ times in the direct-product basis. Thus the process in (b) must terminate after producing $2j_2 + 1$ values of j, and hence

$$j_{min} = j_1 - j_2$$

The number of states sorted according to (j, m) is

$$\sum_{j_{min}}^{j_{max}} (2j + 1) = (2j_1 + 1)(2j_2 + 1)$$

which exhausts the states in the direct-product basis.

Thus the addition law for angular momenta, obtained from the weight diagram is

$$j_1 + j_2 \geq j \geq |j_1 - j_2| \qquad ; \text{integer steps}$$

each value occurs once

This analysis is illustrated for $j_1 = 5/2$ and $j_2 = 3/2$ in Fig. 3.1. Here $m = m_1 + m_2$, and the states are sorted according to the resulting values of j. In this case $m_{max} = 4$, and therefore $j_{max} = 4$. If all the necessary accompanying m-states are removed, one is left with $m_{max} = 3$, and thus $j = 3$. This process can be repeated down to $m_{max} = 1$ and $j = 1$. Thus, in this case, $j = (4, 3, 2, 1)$, each value occuring once, which utilizes all 24 direct-product states.

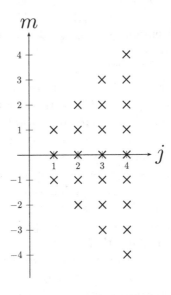

Fig. 3.1   Weight diagram of the available m-states in the direct-product basis $|j_1 m_1\rangle |j_2 m_2\rangle$ in the case of $j_1 = 5/2$ and $j_2 = 3/2$, where $m = m_1 + m_2$, and the states are sorted according to the resulting values of j. In this case $m_{max} = 4$, and therefore $j_{max} = 4$. If all the necessary accompanying m-states are removed, one is left with $m_{max} = 3$, and thus $j = 3$. This process can be repeated down to $m_{max} = 1$ and $j = 1$. Thus, in this case, $j = (4, 3, 2, 1)$, each value occuring once, which utilizes all 24 direct-product states.

Problem 3.16 (a) Show that the definitions of an ITO in Eqs. (3.106) and (3.112) are equivalent;

(b) Show that the tensor product in Eq. (3.147) satisfies the defining relation for an ITO in Eq. (3.148).

Solution to Problem 3.16

(a) The definition of an ITO in Eq. (3.112) is

$$[\hat{J}_i, \hat{T}(\kappa, q)] = \sum_{q'} \langle \kappa q' | \hat{J}_i | \kappa q \rangle \hat{T}(\kappa, q') \qquad ; i = 1, 2, 3$$

From Eqs. (3.29) and (3.59)

$$\hat{J}_\pm \equiv \hat{J}_1 \pm i\hat{J}_2$$
$$\hat{J}_\pm|\kappa q\rangle = A(k, \mp q)|\kappa, q \pm 1\rangle$$
$$\hat{J}_3|\kappa q\rangle = q|\kappa q\rangle$$

It then follows from linear combinations of the first equation in (a) that

$$[\hat{J}_+, \hat{T}(\kappa, q)] = \sum_{q'} \langle \kappa q'|\hat{J}_+|\kappa q\rangle \hat{T}(\kappa, q') = A(\kappa, -q)\hat{T}(\kappa, q+1)$$

$$[\hat{J}_-, \hat{T}(\kappa, q)] = \sum_{q'} \langle \kappa q'|\hat{J}_-|\kappa q\rangle \hat{T}(\kappa, q') = A(\kappa, +q)\hat{T}(\kappa, q-1)$$

$$[\hat{J}_3, \hat{T}(\kappa, q)] = \sum_{q'} \langle \kappa q'|\hat{J}_3|\kappa q\rangle \hat{T}(\kappa, q') = q\hat{T}(\kappa, q)$$

These are Eqs. (3.106).

(b) The tensor product is defined in Eq. (3.147)

$$\hat{X}(K, Q) \equiv \sum_{q_1,q_2} \langle \kappa_1 q_1 \kappa_2 q_2|\kappa_1 \kappa_2 K Q\rangle \hat{T}(\kappa_1, q_1)\hat{U}(\kappa_2, q_2)$$

Consider its behavior under rotation

$$\hat{\mathcal{R}}_{\alpha\beta\gamma}\hat{X}(K, Q)\hat{\mathcal{R}}_{\alpha\beta\gamma}^{-1} = \sum_{q_1,q_2} \langle \kappa_1 q_1 \kappa_2 q_2|\kappa_1 \kappa_2 K Q\rangle \times$$
$$\hat{\mathcal{R}}_{\alpha\beta\gamma}\hat{T}(\kappa_1, q_1)\hat{\mathcal{R}}_{\alpha\beta\gamma}^{-1}\hat{\mathcal{R}}_{\alpha\beta\gamma}\hat{U}(\kappa_2, q_2)\hat{\mathcal{R}}_{\alpha\beta\gamma}^{-1}$$

Each operator on the r.h.s. is an ITO satisfying Eq. (3.136). Hence

$$\hat{\mathcal{R}}_{\alpha\beta\gamma}\hat{X}(K, Q)\hat{\mathcal{R}}_{\alpha\beta\gamma}^{-1} = \sum_{q_1,q_2} \sum_{q_1',q_2'} \langle \kappa_1 q_1 \kappa_2 q_2|\kappa_1 \kappa_2 K Q\rangle \times$$
$$\mathcal{D}_{q_1' q_1}^{\kappa}(\alpha\beta\gamma)\mathcal{D}_{q_2' q_2}^{\kappa}(\alpha\beta\gamma)\hat{T}(\kappa_1, q_1')\hat{U}(\kappa_2, q_2')$$

Now use the composition law for the rotation matrices in Eq. (3.135)

$$\sum_{q_1,q_2} \langle \kappa_1 q_1 \kappa_2 q_2|\kappa_1 \kappa_2 K Q\rangle \mathcal{D}_{q_1' q_1}^{\kappa_1}(\omega)\mathcal{D}_{q_2' q_2}^{\kappa_2}(\omega) =$$
$$\sum_{Q'} \langle \kappa_1 q_1' \kappa_2 q_2'|\kappa_1 \kappa_2 K Q'\rangle \mathcal{D}_{Q' Q}^{K}(\omega)$$

With the re-identification of $\hat{X}(K, Q')$, one reproduces Eq. (3.148)

$$\hat{\mathcal{R}}_{\alpha\beta\gamma}\hat{X}(K, Q)\hat{\mathcal{R}}_{\alpha\beta\gamma}^{-1} = \sum_{Q'} \mathcal{D}_{Q' Q}^{K}(\alpha\beta\gamma)\hat{X}(K, Q')$$

Chapter 4

Scattering Theory

Problem 4.1 (a) Write out the $n = 2$ term in Eq. (4.15). There will be two contributions, one for $t_1 > t_2$ and one for $t_1 < t_2$. Change dummy integration variables in the second term, and show that it is identical to the first. Hence show that the $n = 2$ term in Eq. (4.15) explicitly reproduces the $n = 2$ contribution in Eq. (4.14);

(b) Demonstrate the relation in Eq. (4.22) for $n = 2$.

Solution to Problem 4.1

(a) The term for $n = 2$ in Eq. (4.15) reads

$$
\left(-\frac{i}{\hbar}\right)^2 \frac{1}{2!} \int_{t_0}^{t} e^{-\epsilon|t_1|} dt_1 \int_{t_0}^{t} e^{-\epsilon|t_2|} dt_2 \, T\left[\hat{H}_I(t_1)\hat{H}_I(t_2)\right] =
$$
$$
\left(-\frac{i}{\hbar}\right)^2 \frac{1}{2!} \int_{t_0}^{t} e^{-\epsilon|t_1|} dt_1 \int_{t_0}^{t} e^{-\epsilon|t_2|} dt_2 \, \Big[\theta(t_1 - t_2)\hat{H}_I(t_1)\hat{H}_I(t_2)+
$$
$$
\theta(t_2 - t_1)\hat{H}_I(t_2)\hat{H}_I(t_1)\Big]
$$

where $\theta(t)$ is the step function. Consider now the second term in square brackets. Under the change of dummy variables $t_1 \leftrightharpoons t_2$, it transforms to

$$
\left(-\frac{i}{\hbar}\right)^2 \frac{1}{2!} \int_{t_0}^{t} e^{-\epsilon|t_1|} dt_1 \int_{t_0}^{t} e^{-\epsilon|t_2|} dt_2 \, \theta(t_1 - t_2)\hat{H}_I(t_1)\hat{H}_I(t_2)
$$

which is identical to the first term. Therefore the $n = 2$ term of Eq. (4.15) reduces to

$$\left(-\frac{i}{\hbar}\right)^2 \frac{1}{2!} \int_{t_0}^t e^{-\epsilon|t_1|} dt_1 \int_{t_0}^t e^{-\epsilon|t_2|} dt_2 \, T\left[\hat{H}_I(t_1)\hat{H}_I(t_2)\right] =$$
$$\left(-\frac{i}{\hbar}\right)^2 \int_{t_0}^t e^{-\epsilon|t_1|} dt_1 \int_{t_0}^{t_1} e^{-\epsilon|t_2|} dt_2 \, \hat{H}_I(t_1)\hat{H}_I(t_2)$$

which coincides with the $n = 2$ term of Eq. (4.14).

(b) Assume that $t > t_0$, then

$$\hat{U}_\epsilon(t_0, t)\hat{U}_\epsilon(t, t_0) =$$
$$\sum_{n=0}^\infty \left(\frac{i}{\hbar}\right)^n \frac{1}{n!} \int_{t_0}^t e^{-\epsilon|t_1|} dt_1 \cdots \int_{t_0}^t e^{-\epsilon|t_n|} dt_n \, \overline{T}\left[\hat{H}_I(t_1) \ldots \hat{H}_I(t_n)\right] \times$$
$$\sum_{m=0}^\infty \left(-\frac{i}{\hbar}\right)^m \frac{1}{m!} \int_{t_0}^t e^{-\epsilon|t_1'|} dt_1' \cdots \int_{t_0}^t e^{-\epsilon|t_m'|} dt_m' \, T\left[\hat{H}_I(t_1') \ldots \hat{H}_I(t_m')\right]$$
$$= 1 - \left(\frac{i}{\hbar}\right)^2 \int_{t_0}^t e^{-\epsilon|t_1|} dt_1 \int_{t_0}^t e^{-\epsilon|t_2|} dt_2 \, \hat{H}_I(t_1)\hat{H}_I(t_2) + \left(\frac{i}{\hbar}\right)^2 \frac{1}{2!} \times$$
$$\int_{t_0}^t e^{-\epsilon|t_1|} dt_1 \int_{t_0}^t e^{-\epsilon|t_2|} dt_2 \left\{\overline{T}\left[\hat{H}_I(t_1)\hat{H}_I(t_2)\right] + T\left[\hat{H}_I(t_1)\hat{H}_I(t_2)\right]\right\} + \cdots$$

where we have written the terms up through second order, using Eqs. (4.19) and (4.15).[1] Consider now the term containing the sum of the time-ordered and anti-time-ordered terms

$$\overline{T}\left[\hat{H}_I(t_1)\hat{H}_I(t_2)\right] + T\left[\hat{H}_I(t_1)\hat{H}_I(t_2)\right] =$$
$$\theta(t_1 - t_2)\left[\hat{H}_I(t_1)\hat{H}_I(t_2) + \hat{H}_I(t_2)\hat{H}_I(t_1)\right] +$$
$$\theta(t_2 - t_1)\left[\hat{H}_I(t_2)\hat{H}_I(t_1) + \hat{H}_I(t_1)\hat{H}_I(t_2)\right]$$
$$= \hat{H}_I(t_1)\hat{H}_I(t_2) + \hat{H}_I(t_2)\hat{H}_I(t_1)$$

where in the last line we have used the property $\theta(x) + \theta(-x) = 1$. Now notice that with a simple change of variable $t_1 \leftrightharpoons t_2$, the integral containing the product $\hat{H}_I(t_2)\hat{H}_I(t_1)$ transforms into the same integral containing the product $\hat{H}_I(t_1)\hat{H}_I(t_2)$, thus providing a factor of 2 which cancels the 2! in the denominator of the third term in the expression of $\hat{U}_\epsilon(t_0, t)\hat{U}_\epsilon(t, t_0)$.

---

[1]Recall that $\overline{T}[\cdots]$ is the *anti*-time-ordered product.

Therefore

$$\hat{U}_\epsilon(t_0, t)\hat{U}_\epsilon(t, t_0) = 1 - \left(\frac{i}{\hbar}\right)^2 \int_{t_0}^t e^{-\epsilon|t_1|} dt_1 \int_{t_0}^t e^{-\epsilon|t_2|} dt_2 \, \hat{H}_I(t_1)\hat{H}_I(t_2)$$

$$+ \left(\frac{i}{\hbar}\right)^2 \int_{t_0}^t e^{-\epsilon|t_1|} dt_1 \int_{t_0}^t e^{-\epsilon|t_2|} dt_2 \, \hat{H}_I(t_1)\hat{H}_I(t_2) + \cdots$$

$$= 1 + O\left(\hat{H}_I^3\right)$$

This proves Eq. (4.22) up through second order.

Problem 4.2 Write out the proof of the group property in Eq. (4.26) in detail for $n = 3$.

Solution to Problem 4.2

Let $t_0 < t_2 < t_1$ and consider the product

$$\hat{U}_\epsilon(t_2, t_1)\hat{U}_\epsilon(t_1, t_0) =$$

$$\sum_{n=0}^\infty \left(\frac{i}{\hbar}\right)^n \frac{1}{n!} \int_{t_2}^{t_1} e^{-\epsilon|u_1|} du_1 \cdots \int_{t_2}^{t_1} e^{-\epsilon|u_n|} du_n \, \overline{T}\left[\hat{H}_I(u_1)\ldots\hat{H}_I(u_n)\right] \times$$

$$\sum_{m=0}^\infty \left(-\frac{i}{\hbar}\right)^m \frac{1}{m!} \int_{t_0}^{t_1} e^{-\epsilon|v_1|} dv_1 \cdots \int_{t_0}^{t_1} e^{-\epsilon|v_m|} dv_m \, T\left[\hat{H}_I(v_1)\ldots\hat{H}_I(v_m)\right]$$

Let us collect the terms of order three in the interaction

$$\left[\hat{U}_\epsilon(t_2, t_1)\hat{U}_\epsilon(t_1, t_0)\right]^{(3)} = \left(-\frac{i}{\hbar}\right)^3 \times$$

$$\left\{-\frac{1}{3!} \int_{t_2}^{t_1} e^{-\epsilon|u_1|} du_1 \int_{t_2}^{t_1} e^{-\epsilon|u_2|} du_2 \int_{t_2}^{t_1} e^{-\epsilon|u_3|} du_3 \, \overline{T}\left[\hat{H}_I(u_1)\hat{H}_I(u_2)\hat{H}_I(u_3)\right]\right.$$

$$+\frac{1}{2!} \int_{t_2}^{t_1} e^{-\epsilon|u_1|} du_1 \int_{t_2}^{t_1} e^{-\epsilon|u_2|} du_2 \int_{t_0}^{t_1} e^{-\epsilon|v_1|} dv_1 \, \overline{T}\left[\hat{H}_I(u_1)\hat{H}_I(u_2)\right] \hat{H}_I(v_1)$$

$$-\frac{1}{2!} \int_{t_2}^{t_1} e^{-\epsilon|u_1|} du_1 \int_{t_0}^{t_1} e^{-\epsilon|v_1|} dv_1 \int_{t_0}^{t_1} e^{-\epsilon|v_2|} dv_2 \, \hat{H}_I(u_1)T\left[\hat{H}_I(v_1)\hat{H}_I(v_2)\right]$$

$$\left.+\frac{1}{3!} \int_{t_0}^{t_1} e^{-\epsilon|v_1|} dv_1 \int_{t_0}^{t_1} e^{-\epsilon|v_2|} dv_2 \int_{t_0}^{t_1} e^{-\epsilon|v_3|} dv_3 \, T\left[\hat{H}_I(v_1)\hat{H}_I(v_2)\hat{H}_I(v_3)\right]\right\}$$

Since $t_0 < t_2 < t_1$, it is convenient to split the region of integration (t_0, t_1) into the regions (t_0, t_2) and (t_2, t_1). In this way, the expression above can be written in terms of the following:

(1) Terms containing only integrals between t_2 and t_1

$$\left(-\frac{i}{\hbar}\right)^3 \int_{t_2}^{t_1} e^{-\epsilon|u_1|} du_1 \int_{t_2}^{t_1} e^{-\epsilon|u_2|} du_2 \int_{t_2}^{t_1} e^{-\epsilon|u_3|} du_3 \times$$

$$\left\{ -\frac{1}{3!} T\left[\hat{H}_I(u_1)\hat{H}_I(u_2)\hat{H}_I(u_3)\right] + \frac{1}{2!} \overline{T}\left[\hat{H}_I(u_1)\hat{H}_I(u_2)\right]\hat{H}_I(u_3) \right.$$

$$\left. -\frac{1}{2!}\hat{H}_I(u_1)\, T\left[\hat{H}_I(u_2)\hat{H}_I(u_3)\right] + \frac{1}{3!} T\left[\hat{H}_I(u_1)\hat{H}_I(u_2)\hat{H}_I(u_3)\right] \right\}$$

Here a change of dummy variable of integration $v_i \leftrightarrows u_i$ has been made where needed.

Notice that, because of the integration over (u_1, u_2, u_3), one can perform a suitable re-labeling of the variable of integration, transforming the integrand inside the curly brackets into a symmetrized expression in the variables of integration; for instance

$$\overline{T}\left[\hat{H}_I(u_1)\hat{H}_I(u_2)\right]\hat{H}_I(u_3) \rightarrow \frac{1}{3}\left\{\overline{T}\left[\hat{H}_I(u_1)\hat{H}_I(u_2)\right]\hat{H}_I(u_3) + \right.$$

$$\left. \overline{T}\left[\hat{H}_I(u_1)\hat{H}_I(u_3)\right]\hat{H}_I(u_2) + \overline{T}\left[\hat{H}_I(u_2)\hat{H}_I(u_3)\right]\hat{H}_I(u_1) \right\}$$

Also

$$\hat{H}_I(u_1)\, T\left[\hat{H}_I(u_2)\hat{H}_I(u_3)\right] \rightarrow \frac{1}{3}\left\{\hat{H}_I(u_3)T\left[\hat{H}_I(u_1)\hat{H}_I(u_2)\right] + \right.$$

$$\left. \hat{H}_I(u_2)T\left[\hat{H}_I(u_1)\hat{H}_I(u_3)\right] + \hat{H}_I(u_1)T\left[\hat{H}_I(u_2)\hat{H}_I(u_3)\right] \right\}$$

Notice that the T-products or $\overline{T}$-products of three operators are already symmetrized.

Once these expressions are substituted in the previous expression, we have for the terms containing only integrals between t_2 and t_1

$$\left[\hat{U}_\epsilon(t_2,t_1)\hat{U}_\epsilon(t_1,t_0)\right]^{(3)} = \int_{t_2}^{t_1} e^{-\epsilon|u_1|} du_1 \int_{t_2}^{t_1} e^{-\epsilon|u_2|} du_2 \int_{t_2}^{t_1} e^{-\epsilon|u_3|} du_3 \times$$

$$\frac{1}{3!}\left(-\frac{i}{\hbar}\right)^3 \left\{ -\overline{T}\left[\hat{H}_I(u_1)\hat{H}_I(u_2)\hat{H}_I(u_3)\right] + \overline{T}\left[\hat{H}_I(u_1)\hat{H}_I(u_2)\right]\hat{H}_I(u_3) + \right.$$

$$\overline{T}\left[\hat{H}_I(u_1)\hat{H}_I(u_3)\right]\hat{H}_I(u_2) + \overline{T}\left[\hat{H}_I(u_2)\hat{H}_I(u_3)\right]\hat{H}_I(u_1) -$$

$$\hat{H}_I(u_1)\, T\left[\hat{H}_I(u_2)\hat{H}_I(u_3)\right] - \hat{H}_I(u_2)\, T\left[\hat{H}_I(u_1)\hat{H}_I(u_3)\right] -$$

$$\left. \hat{H}_I(u_3)\, T\left[\hat{H}_I(u_1)\hat{H}_I(u_2)\right] + T\left[\hat{H}_I(u_1)\hat{H}_I(u_2)\hat{H}_I(u_3)\right] \right\}$$

We will now verify that the expression in the curly brackets vanishes. For instance consider the case $u_1 < u_2 < u_3$ and write the expression in the curly brackets explicitly as

$$
-\hat{H}_I(u_1)\hat{H}_I(u_2)\hat{H}_I(u_3) + \hat{H}_I(u_1)\hat{H}_I(u_2)\hat{H}_I(u_3) + \hat{H}_I(u_1)\hat{H}_I(u_3)\hat{H}_I(u_2)
$$
$$
+ \hat{H}_I(u_2)\hat{H}_I(u_3)\hat{H}_I(u_1) - \hat{H}_I(u_1)\hat{H}_I(u_3)\hat{H}_I(u_2) - \hat{H}_I(u_2)\,\hat{H}_I(u_3)\hat{H}_I(u_1)
$$
$$
-\hat{H}_I(u_3)\hat{H}_I(u_2)\hat{H}_I(u_1) + \hat{H}_I(u_3)\hat{H}_I(u_2)\hat{H}_I(u_1) = 0
$$

Since the expression in the curly brackets is completely symmetric in the integration variable, any other ordering is equivalent, and the sum always vanishes.

(2) Terms in $\left[\hat{U}_\epsilon(t_2, t_1)\hat{U}_\epsilon(t_1, t_0)\right]^{(3)}$ containing two integrals between t_2 and t_1 and one integral between t_0 and t_2 (recall $t_0 < t_2 < t_1$)

$$
\left(-\frac{i}{\hbar}\right)^3 \int_{t_2}^{t_1} e^{-\epsilon|u_1|}du_1 \int_{t_2}^{t_1} e^{-\epsilon|u_2|}du_2 \int_{t_0}^{t_2} e^{-\epsilon|u_3|}du_3 \times
$$
$$
\left\{ \frac{1}{2!}\overline{T}\left[\hat{H}_I(u_1)\hat{H}_I(u_2)\right]\hat{H}_I(u_3) - \frac{1}{2!}\hat{H}_I(u_1)\,T\left[\hat{H}_I(u_2)\hat{H}_I(u_3)\right] \right.
$$
$$
\left. - \frac{1}{2!}\hat{H}_I(u_2)\,T\left[\hat{H}_I(u_1)\hat{H}_I(u_3)\right] + \frac{1}{2!}T\left[\hat{H}_I(u_1)\hat{H}_I(u_2)\hat{H}_I(u_3)\right] \right\}
$$

Since $u_3 \le \min(u_1, u_2)$, one can simplify the expression inside the curly brackets

$$
\frac{1}{2!}\overline{T}\left[\hat{H}_I(u_1)\hat{H}_I(u_2)\right]\hat{H}_I(u_3) - \frac{1}{2!}\hat{H}_I(u_1)\hat{H}_I(u_2)\hat{H}_I(u_3)
$$
$$
- \frac{1}{2!}\hat{H}_I(u_2)\hat{H}_I(u_1)\hat{H}_I(u_3) + \frac{1}{2!}T\left[\hat{H}_I(u_1)\hat{H}_I(u_2)\right]\hat{H}_I(u_3)
$$

However, as we have seen in Prob. (4.1), one has

$$
\overline{T}\left[\hat{H}_I(u_1)\hat{H}_I(u_2)\right] + T\left[\hat{H}_I(u_1)\hat{H}_I(u_2)\right] = \hat{H}_I(u_1)\hat{H}_I(u_2) + \hat{H}_I(u_2)\hat{H}_I(u_1)
$$

Therefore this expression also vanishes

$$
\frac{1}{2!}\overline{T}\left[\hat{H}_I(u_1)\hat{H}_I(u_2)\right]\hat{H}_I(u_3) - \frac{1}{2!}\hat{H}_I(u_1)\hat{H}_I(u_2)\hat{H}_I(u_3)
$$
$$
- \frac{1}{2!}\hat{H}_I(u_2)\hat{H}_I(u_1)\hat{H}_I(u_3) + \frac{1}{2!}T\left[\hat{H}_I(u_1)\hat{H}_I(u_2)\right]\hat{H}_I(u_3) = 0
$$

(3) Terms in $\left[\hat{U}_\epsilon(t_2, t_1)\hat{U}_\epsilon(t_1, t_0)\right]^{(3)}$ containing one integral between t_2

and t_1 and two integrals between t_0 and t_2

$$\left(-\frac{i}{\hbar}\right)^3 \int_{t_2}^{t_1} e^{-\epsilon|u_1|} du_1 \int_{t_0}^{t_2} e^{-\epsilon|u_2|} du_2 \int_{t_0}^{t_2} e^{-\epsilon|u_3|} du_3 \times$$

$$\left\{-\frac{1}{2!}\hat{H}_I(u_1)\, T\left[\hat{H}_I(u_2)\hat{H}_I(u_3)\right] + \frac{1}{2!}T\left[\hat{H}_I(u_1)\hat{H}_I(u_2)\hat{H}_I(u_3)\right]\right\}$$

Since $u_1 > \max(u_2, u_3)$, we have

$$T\left[\hat{H}_I(u_1)\hat{H}_I(u_2)\hat{H}_I(u_3)\right] = \hat{H}_I(u_1)T\left[\hat{H}_I(u_2)\hat{H}_I(u_3)\right]$$

Therefore, this particular contribution also vanishes

$$\left(-\frac{i}{\hbar}\right)^3 \int_{t_2}^{t_1} e^{-\epsilon|u_1|} du_1 \int_{t_0}^{t_2} e^{-\epsilon|u_2|} du_2 \int_{t_0}^{t_2} e^{-\epsilon|u_3|} du_3 \times$$

$$\left\{-\frac{1}{2!}\hat{H}_I(u_1)\, T\left[\hat{H}_I(u_2)\hat{H}_I(u_3)\right] + \frac{1}{2!}\hat{H}_I(u_1)T\left[\hat{H}_I(u_2)\hat{H}_I(u_3)\right]\right\} = 0$$

(4) We are left with the remaining contribution containing three integrals between t_0 and t_2, arising from the last term in curly brackets in $\left[\hat{U}_\epsilon(t_2,t_1)\hat{U}_\epsilon(t_1,t_0)\right]^{(3)}$, and we conclude that

$$\left[\hat{U}_\epsilon(t_2,t_1)\hat{U}_\epsilon(t_1,t_0)\right]^{(3)} = \left(-\frac{i}{\hbar}\right)^3 \frac{1}{3!} \times$$

$$\int_{t_0}^{t_2} e^{-\epsilon|u_1|} du_1 \int_{t_0}^{t_2} e^{-\epsilon|u_2|} du_2 \int_{t_0}^{t_2} e^{-\epsilon|u_3|} du_3\, T\left[\hat{H}_I(u_1)\hat{H}_I(u_2)\hat{H}_I(u_3)\right]$$

Let us now write out the contribution corresponding to $n = 3$ for $\hat{U}_\epsilon(t_2,t_0)$ with $t_2 > t_0$ from Eq. (4.15)

$$\left[\hat{U}_\epsilon(t_2,t_0)\right]^{(3)} = \left(-\frac{i}{\hbar}\right)^3 \frac{1}{3!} \int_{t_0}^{t_2} e^{-\epsilon|u_1|} du_1 \int_{t_0}^{t_2} e^{-\epsilon|u_2|} du_2 \int_{t_0}^{t_2} e^{-\epsilon|u_3|} du_3 \times$$

$$T\left[\hat{H}_I(u_1)\hat{H}_I(u_2)\hat{H}_I(u_3)\right]$$

A simple comparison shows that

$$\left[\hat{U}_\epsilon(t_2,t_1)\hat{U}_\epsilon(t_1,t_0)\right]^{(3)} = \left[\hat{U}_\epsilon(t_2,t_0)\right]^{(3)}$$

which proves Eq. (4.26) for $n = 3$.

Problem 4.3 Carry out an argument parallel to that in the text for $|\psi_i^{(+)}\rangle$, and derive Eq. (4.69) for the incoming scattering state $|\psi_f^{(-)}\rangle$.

Solution to Problem 4.3

The analog of Eq. (4.64), using the time-ordered form of Eq. (4.19), is

$$\hat{U}_\epsilon^{(n)}(0,+\infty)|\psi_f\rangle = \left(\frac{i}{\hbar}\right)^n \int_0^\infty e^{-\epsilon t_1}\, dt_1 \int_{t_1}^\infty e^{-\epsilon t_2}\, dt_2 \cdots \int_{t_{n-1}}^\infty e^{-\epsilon t_n}\, dt_n \times$$

$$e^{\frac{i}{\hbar}\hat{H}_0 t_1}\hat{H}_1 e^{-\frac{i}{\hbar}\hat{H}_0(t_1-t_2)}\hat{H}_1 e^{-\frac{i}{\hbar}\hat{H}_0(t_2-t_3)}\cdots e^{-\frac{i}{\hbar}\hat{H}_0(t_{n-1}-t_n)}\hat{H}_1 e^{-\frac{i}{\hbar}\hat{H}_0 t_n}|\psi_f\rangle$$

Here the operators $\hat{H}_I(t_i)$ appear in anti-time-ordered form, with the operator with the *earliest* time on the left.

The analysis then proceeds as in Eqs. (4.65)–(4.67) in the text. The only differences are:

- The hamiltonian $\hat{H}_0$ acting on the state on the right gives

$$\hat{H}_0|\psi_f\rangle = E_f|\psi_f\rangle$$

- The integrals over dx_i now give[2]

$$\frac{i}{\hbar}\int_0^\infty dx_i\, e^{-\epsilon x_i} e^{-\frac{i}{\hbar}(E_f-\hat{H}_0)x_i} = \frac{i/\hbar}{i(E_f-\hat{H}_0)/\hbar+\epsilon}$$

$$= \frac{1}{E_f-\hat{H}_0-i\varepsilon}$$

Then, exactly as in Eqs. (4.65)–(4.67), one arrives at Eq. (4.69)

$$|\psi_f^{(-)}\rangle \equiv \hat{U}_\varepsilon(0,+\infty)|\psi_f\rangle$$

$$= \sum_{n=0}^\infty \left(\frac{1}{E_f-\hat{H}_0-i\varepsilon}\hat{H}_1\right)^n |\psi_f\rangle$$

Problem 4.4 Consider the integral for the Green's function $G_0(\mathbf{x}-\mathbf{y})$ in Eq. (4.104). Show that the contribution around the large semi-circle in the complex t-plane in Fig. 4.1 in the text vanishes as $R \to \infty$, and hence conclude that in that limit, the integral is identical to the integral around the contour C.

Solution to Problem 4.4

Equation (4.104) reads

$$G_0(\mathbf{x}-\mathbf{y}) = \frac{2m}{\hbar^2}\frac{4\pi}{(2\pi)^3}\frac{1}{2ir}\int_{-\infty}^\infty t\,dt\, e^{itr}\frac{1}{t^2-k^2-i\varepsilon} \qquad ; \mathbf{r} \equiv \mathbf{x}-\mathbf{y}$$

---

[2]Again, $\varepsilon = \hbar\epsilon$.

The integral is converted to a contour integral by adding the contribution from the semi-circle in the upper-half plane in Fig. 4.1 in the text. Consider the contribution from that semi-circle when R is very large. For t on that semi-circle, write

$$t = Re^{i\phi}$$
$$dt = iRe^{i\phi}\, d\phi$$

The contribution to the integral from the semi-circle then takes the form[3]

$$\int_R \frac{t\, dt}{t^2 - (k + i\varepsilon)^2} e^{it|\mathbf{x}-\mathbf{y}|} = i \int_0^\pi d\phi\, \frac{R^2 e^{2i\phi}}{R^2 e^{2i\phi} - (k + i\varepsilon)^2} e^{iR(\cos\phi + i\sin\phi)|\mathbf{x}-\mathbf{y}|}$$

If $R \gg k$, then the $(k + i\varepsilon)^2$ in the denominator can be neglected, and

$$\int_R \frac{t\, dt}{t^2 - (k + i\varepsilon)^2} e^{it|\mathbf{x}-\mathbf{y}|} \approx i \int_0^\pi d\phi\, e^{iR(\cos\phi + i\sin\phi)}\, |\mathbf{x}-\mathbf{y}|$$

This expression is bounded by

$$\left| \int_R \frac{t\, dt}{t^2 - (k + i\varepsilon)^2} e^{it|\mathbf{x}-\mathbf{y}|} \right| \le \int_0^\pi d\phi\, e^{-R\sin\phi\, |\mathbf{x}-\mathbf{y}|}$$

The final integral is convergent and vanishes as $R \to \infty$ for any $|\mathbf{x} - \mathbf{y}| \ne 0$.

Problem 4.5 (a) Prove the following operator relations

$$\frac{1}{\hat{A} - \hat{B}} = \frac{1}{\hat{A}} + \frac{1}{\hat{A} - \hat{B}} \hat{B} \frac{1}{\hat{A}} = \frac{1}{\hat{A}} + \frac{1}{\hat{A}} \hat{B} \frac{1}{\hat{A} - \hat{B}}$$

(b) Now iterate to obtain a power series in $\hat{B}$

$$\frac{1}{\hat{A} - \hat{B}} = \frac{1}{\hat{A}} + \frac{1}{\hat{A}} \hat{B} \frac{1}{\hat{A}} + \cdots = \sum_{n=0}^\infty \left(\frac{1}{\hat{A}} \hat{B} \right)^n \frac{1}{\hat{A}} = \sum_{n=0}^\infty \frac{1}{\hat{A}} \left(\hat{B} \frac{1}{\hat{A}} \right)^n$$

This produces, for example, a power series in the coupling constant, and when the l.h.s. can be iterated, it must be equal to the r.h.s.

Solution to Problem 4.5

(a) Write

$$\hat{A} = \hat{A} - \hat{B} + \hat{B}$$

---

[3] Recall the limit $\varepsilon \to 0$ is implied.

Multiply on the left by the inverse operator $(\hat{A} - \hat{B})^{-1}$

$$\frac{1}{\hat{A} - \hat{B}}\hat{A} = \hat{1} + \frac{1}{\hat{A} - \hat{B}}\hat{B}$$

Now multiply on the right by the inverse $\hat{A}^{-1}$

$$\frac{1}{\hat{A} - \hat{B}} = \frac{1}{\hat{A}} + \frac{1}{\hat{A} - \hat{B}}\hat{B}\frac{1}{\hat{A}}$$

Multiplication of the starting relation $\hat{A} = \hat{A} - \hat{B} + \hat{B}$ first on the right by $(\hat{A} - \hat{B})^{-1}$ and then on the left by $\hat{A}^{-1}$ leads to

$$\frac{1}{\hat{A} - \hat{B}} = \frac{1}{\hat{A}} + \frac{1}{\hat{A}}\hat{B}\frac{1}{\hat{A} - \hat{B}}$$

These are the desired operator relations.

(b) Take the last relation, for example, and substitute the expression for $(\hat{A} - \hat{B})^{-1}$ into the r.h.s.

$$\frac{1}{\hat{A} - \hat{B}} = \frac{1}{\hat{A}} + \frac{1}{\hat{A}}\hat{B}\frac{1}{\hat{A}} + \frac{1}{\hat{A}}\hat{B}\frac{1}{\hat{A}}\hat{B}\frac{1}{\hat{A} - \hat{B}}$$

This is still an exact expression. Assume the operator $\hat{B}$ is characterized by some small parameter. Repeated application of this process then leads to the following power series in $\hat{B}$

$$\frac{1}{\hat{A} - \hat{B}} = \frac{1}{\hat{A}} + \frac{1}{\hat{A}}\hat{B}\frac{1}{\hat{A}} + \cdots = \sum_{n=0}^{\infty}\left(\frac{1}{\hat{A}}\hat{B}\right)^{n}\frac{1}{\hat{A}}$$

A similar result is obtained from the iteration of the first relation

$$\frac{1}{\hat{A} - \hat{B}} = \frac{1}{\hat{A}} + \frac{1}{\hat{A}}\hat{B}\frac{1}{\hat{A}} + \cdots = \sum_{n=0}^{\infty}\frac{1}{\hat{A}}\left(\hat{B}\frac{1}{\hat{A}}\right)^{n}$$

Problem 4.6 (a) Use the results in Prob. 4.5 to prove the following relation

$$\frac{1}{E_0 - \hat{H}_0 + i\varepsilon - \hat{H}_1} = \sum_{n=0}^{\infty}\left(\frac{1}{E_0 - \hat{H}_0 + i\varepsilon}\hat{H}_1\right)^{n}\frac{1}{E_0 - \hat{H}_0 + i\varepsilon}$$

(b) Hence conclude that

$$\frac{1}{E_0 - \hat{H} + i\varepsilon}\hat{H}_1 = \sum_{n=1}^{\infty}\left(\frac{1}{E_0 - \hat{H}_0 + i\varepsilon}\hat{H}_1\right)^{n}$$

where $\hat{H} = \hat{H}_0 + \hat{H}_1$ is the full hamiltonian, and the sum starts with $n = 1$.

Solution to Problem 4.6

(a) Identify

$$\hat{A} = E_0 - \hat{H}_0 + i\varepsilon \qquad ; \ \hat{B} = \hat{H}_1$$

Then from the results in Prob. 4.5

$$\frac{1}{E_0 - \hat{H}_0 + i\varepsilon - \hat{H}_1} = \sum_{n=0}^{\infty} \left(\frac{1}{E_0 - \hat{H}_0 + i\varepsilon} \hat{H}_1 \right)^n \frac{1}{E_0 - \hat{H}_0 + i\varepsilon}$$

(b) Multiply the result in part (a) on the right by $\hat{H}_1$, and note that

$$\hat{H} = \hat{H}_0 + \hat{H}_1 \qquad \qquad ; \ \text{full hamiltonian}$$

is the full hamiltonian. It follows that

$$\frac{1}{E_0 - \hat{H} + i\varepsilon} \hat{H}_1 = \sum_{n=1}^{\infty} \left(\frac{1}{E_0 - \hat{H}_0 + i\varepsilon} \hat{H}_1 \right)^n$$

where the sum on the right now starts with $n = 1$. This expression provides a very useful perturbation expansion of the exact scattering Green's function $(E_0 - \hat{H} + i\varepsilon)^{-1}$ multiplied by the interaction $\hat{H}_1$.

Problem 4.7 Use the results in Prob. 4.6 to obtain *explicit* expressions for the scattering states $|\psi_i^{(+)}\rangle$ and $|\psi_f^{(-)}\rangle$ in terms of the full hamiltonian $\hat{H}$

$$|\psi_i^{(+)}\rangle = \left\{ 1 + \frac{1}{E_0 - \hat{H} + i\varepsilon} \hat{H}_1 \right\} |\psi_i\rangle$$

$$|\psi_f^{(-)}\rangle = \left\{ 1 + \frac{1}{E_f - \hat{H} - i\varepsilon} \hat{H}_1 \right\} |\psi_f\rangle$$

Solution to Problem 4.7

The scattering states with outgoing- and incoming-wave boundary conditions are given in terms of a power series in $\hat{H}_1$ in Eqs. (4.67) and (4.69) as[4]

$$|\psi_i^{(+)}\rangle = \sum_{n=0}^{\infty} \left(\frac{1}{E_0 - \hat{H}_0 + i\varepsilon} \hat{H}_1 \right)^n |\psi_i\rangle$$

$$|\psi_f^{(-)}\rangle = \sum_{n=0}^{\infty} \left(\frac{1}{E_f - \hat{H}_0 - i\varepsilon} \hat{H}_1 \right)^n |\psi_f\rangle$$

The results in Prob. 4.6 can then be used to obtain an explicit expression for these states in terms of the full scattering Green's functions

$$|\psi_i^{(+)}\rangle = \left\{ 1 + \frac{1}{E_0 - \hat{H} + i\varepsilon} \hat{H}_1 \right\} |\psi_i\rangle$$

$$|\psi_f^{(-)}\rangle = \left\{ 1 + \frac{1}{E_f - \hat{H} - i\varepsilon} \hat{H}_1 \right\} |\psi_f\rangle$$

Many useful results follow from these relations (see the next problem).

Problem 4.8 Derive the following relations from the results in Prob. 4.7.

(a) Show the incoming and outgoing scattering states $|\psi_f^{(-)}\rangle$ and $|\psi_f^{(+)}\rangle$ are explicitly related by

$$|\psi_f^{(-)}\rangle = |\psi_f^{(+)}\rangle + \left(\frac{1}{E_f - \hat{H} - i\varepsilon} - \frac{1}{E_f - \hat{H} + i\varepsilon} \right) \hat{H}_1 |\psi_f\rangle$$

(b) Show the S-matrix, as expressed in Eq. (4.74), can be written as

$$\langle \psi_f | \hat{S} | \psi_i \rangle = \langle \psi_f^{(-)} | \psi_i^{(+)} \rangle$$

$$= \langle \psi_f^{(+)} | \psi_i^{(+)} \rangle + \langle \psi_f | \hat{H}_1 \left(\frac{1}{E_f - \hat{H} + i\varepsilon} - \frac{1}{E_f - \hat{H} - i\varepsilon} \right) | \psi_i^{(+)} \rangle$$

(c) Make use of the limiting relation in Eq. (4.58), and then prove the following [compare Eq. (4.90)]

$$\text{Lim}_{\varepsilon \to 0} \left(\frac{1}{E_f - E_0 + i\varepsilon} - \frac{1}{E_f - E_0 - i\varepsilon} \right) = \text{Lim}_{\varepsilon \to 0} \frac{-2i\varepsilon}{(E_f - E_0)^2 + \varepsilon^2}$$

$$= -2\pi i \delta(E_f - E_0)$$

---

[4]The $n = 0$ term in the sum is again just unity; we suppress the hat on the unit operator.

(d) Now use the results in part (c) to reproduce the general expression for the S-matrix

$$\langle\psi_f|\hat{S}|\psi_i\rangle = \langle\psi_f|\psi_i\rangle - 2\pi i\delta(E_f - E_0)\langle\psi_f|\hat{H}_1|\psi_i^{(+)}\rangle$$

Solution to Problem 4.8

(a) Take the difference of the two states in Prob. 4.7

$$|\psi_f^{(-)}\rangle - |\psi_f^{(+)}\rangle = \left(\frac{1}{E_f - \hat{H} - i\varepsilon} - \frac{1}{E_f - \hat{H} + i\varepsilon}\right)\hat{H}_1|\psi_f\rangle$$

Hence

$$|\psi_f^{(-)}\rangle = |\psi_f^{(+)}\rangle + \left(\frac{1}{E_f - \hat{H} - i\varepsilon} - \frac{1}{E_f - \hat{H} + i\varepsilon}\right)\hat{H}_1|\psi_f\rangle$$

(b) The orthonormality of the states with outgoing- and incoming-wave boundary conditions is exhibited in Eqs. (4.71)–(4.72)

$$\langle\psi_{i'}^{(+)}|\psi_i^{(+)}\rangle = \langle\psi_{i'}|\psi_i\rangle = \delta_{i'i}$$
$$\langle\psi_{f'}^{(-)}|\psi_f^{(-)}\rangle = \langle\psi_{f'}|\psi_f\rangle = \delta_{f'f}$$

According to Eq. (4.74), their inner product produces the S-matrix

$$\langle\psi_f|\hat{S}|\psi_i\rangle = \langle\psi_f^{(-)}|\psi_i^{(+)}\rangle$$

The matrix element of the state $|\psi_f^{(-)}\rangle$ in part (a) with $|\psi_i^{(+)}\rangle$ then yields

$$\langle\psi_f|\hat{S}|\psi_i\rangle = \langle\psi_f|\psi_i\rangle + \langle\psi_f|\hat{H}_1\left(\frac{1}{E_f - \hat{H} + i\varepsilon} - \frac{1}{E_f - \hat{H} - i\varepsilon}\right)|\psi_i^{(+)}\rangle$$

(c) Equations (4.58) tell us that in the limit as the quantization volume gets very large $\Omega \to \infty$, and the adiabatic damping factor goes to zero $\varepsilon \to 0$, the state $|\psi_i^{(+)}\rangle$ is an eigenstate of the total hamiltonian with the initial energy

$$\hat{H}|\psi_i^{(+)}\rangle = E_0|\psi_i^{(+)}\rangle \qquad ; \Omega \to \infty$$
$$\varepsilon \to 0$$

In this limit, the last factor in the S-matrix becomes

$$\text{Lim}_{\varepsilon \to 0} \left(\frac{1}{E_f - E_0 + i\varepsilon} - \frac{1}{E_f - E_0 - i\varepsilon} \right)$$

$$= -2i \, \text{Lim}_{\varepsilon \to 0} \frac{\varepsilon}{(E_f - E_0)^2 + \varepsilon^2}$$

$$= -2\pi i \, \delta(E_f - E_0)$$

where $\delta(E_f - E_0)$ is a Dirac delta-function.[5]

(d) Hence we derive the general expression for the S-matrix

$$\langle \psi_f | \hat{S} | \psi_i \rangle = \langle \psi_f | \psi_i \rangle - 2\pi i \, \delta(E_f - E_0) \langle \psi_f | \hat{H}_1 | \psi_i^{(+)} \rangle$$

Problem 4.9 Consider the *incoming* scattering wave function $\psi_f^{(-)}(\mathbf{x})$ with $E_f = E_0$ in potential scattering. For this, one needs the Green's function $G_0^{(-)}(\mathbf{x} - \mathbf{y})$ given by

$$G_0^{(-)}(\mathbf{x} - \mathbf{y}) \equiv \langle \mathbf{x} | \frac{1}{\hat{H}_0 - E_0 + i\varepsilon} | \mathbf{y} \rangle$$

(a) Follow the arguments in the text for $G_0(\mathbf{x} - \mathbf{y}) \equiv G_0^{(+)}(\mathbf{x} - \mathbf{y})$. Convert to a contour integral, locate the singularities as in Fig. 4.1 in the text, and use the method of residues to show

$$G_0^{(-)}(\mathbf{x} - \mathbf{y}) = \frac{2m}{\hbar^2} \frac{e^{-ikr}}{4\pi r} \qquad ; \mathbf{r} \equiv \mathbf{x} - \mathbf{y}$$

(b) Show the wave function $\psi_f^{(-)}(\mathbf{x})$ satisfies the integral equation

$$\psi_f^{(-)}(\mathbf{x}) = e^{i\mathbf{k}_f \cdot \mathbf{x}} - \frac{2m}{\hbar^2} \int d^3 y \, \frac{e^{-ik|\mathbf{x}-\mathbf{y}|}}{4\pi |\mathbf{x} - \mathbf{y}|} V(y) \, \psi_f^{(-)}(\mathbf{y})$$

(c) Show the T-matrix can be written

$$\tilde{T}_{fi} = \frac{1}{\Omega} \int d^3 y \, \psi_f^{(-)}(\mathbf{y})^\star \, V(y) \, e^{i\mathbf{k}\cdot\mathbf{y}}$$

Solution to Problem 4.9

(a) The argument for obtaining the scattering Green's function $G_0^{(-)}(\mathbf{x} - \mathbf{y})$ follows exactly as in Eqs. (4.95)–(4.104), where the contour C is that

---

[5]See the solution to Prob. 7.14 in [Amore and Walecka (2014)].

shown in Fig. 4.1 in the text. The only difference is that now the denominator in Eq. (4.105) reads[6]

$$\frac{1}{t^2 - k^2 + i\varepsilon} = \frac{1}{(t - k + i\varepsilon)(t + k - i\varepsilon)}$$

As a consequence, the position of the singularities in Fig. 4.1 in the text is reflected with respect to the $y = 0$ axis; it is the pole on the *left* that now lies inside the contour.

The method of residues then gives

$$G_0^{(-)}(\mathbf{x} - \mathbf{y}) = \frac{2m}{\hbar^2} \frac{4\pi}{(2\pi)^3} \frac{2\pi i}{2ir} \text{Res} \left[\frac{te^{itr}}{t^2 - k^2 + i\varepsilon} \right]_{t=-k+i\varepsilon}$$

$$= \frac{2m}{\hbar^2} \frac{1}{2\pi r} \left[\frac{-ke^{-ikr}}{-2k} \right]$$

$$= \frac{2m}{\hbar^2} \frac{e^{-ikr}}{4\pi r}$$

(b) The scattering state $|\psi_f^{(-)}\rangle$ is then projected onto eigenstates of position exactly as in Eqs. (4.108)–(4.109)

$$\psi_f^{(-)}(\mathbf{x}) = e^{i\mathbf{k}_f \cdot \mathbf{x}} - \int d^3y\, G_0^{(-)}(\mathbf{x} - \mathbf{y})\, V(y)\, \psi_f^{(-)}(\mathbf{y})$$

$$= e^{i\mathbf{k}_f \cdot \mathbf{x}} - \frac{2m}{\hbar^2} \int d^3y\, \frac{e^{-ik|\mathbf{x} - \mathbf{y}|}}{4\pi|\mathbf{x} - \mathbf{y}|}\, V(y)\, \psi_f^{(-)}(\mathbf{y})$$

(c) The T-matrix is obtained exactly as in Eqs. (4.110)–(4.111), starting from the relation in Eq. (4.75)[7]

$$\tilde{T}_{fi} = \int d^3y\, \langle \psi_f^{(-)} | \mathbf{y} \rangle V(y) \langle \mathbf{y} | \mathbf{k} \rangle$$

$$= \frac{1}{\Omega} \int d^3y\, \psi_f^{(-)}(\mathbf{y})^*\, V(y)\, e^{i\mathbf{k} \cdot \mathbf{y}}$$

Problem 4.10 (a) Consider the asymptotic form of the Green's function $G_0(\mathbf{x} - \mathbf{y})$ in Eq. (4.107), where $x = |\mathbf{x}| \to \infty$ while $\mathbf{y}$ is confined to the region of the potential. Show

$$G_0(\mathbf{x} - \mathbf{y}) \to \frac{2m}{\hbar^2} \frac{e^{ikx}}{4\pi x} e^{-i\mathbf{k}_f \cdot \mathbf{y}} \qquad ; \ x = |\mathbf{x}| \to \infty$$

$$\mathbf{y} \text{ in potential}$$

---

[6] Again, the limit $\varepsilon \to 0$ is implied.

[7] Here $\tilde{T}_{fi} = \langle \psi_f | \hat{T} | \psi_i \rangle = \langle \psi_f^{(-)} | \hat{H}_1 | \psi_i \rangle$ with $E_f = E_0$ [compare Eq. (4.87)].

Here $\mathbf{k}_f = k\,(\mathbf{x}/|\mathbf{x}|)$ is a vector of length k that points in the direction of observation.

(b) Show that the corresponding asymptotic form of the scattering wave function in Eq. (4.118) is

$$\psi_i^{(+)}(\mathbf{x}) \to e^{i\mathbf{k}\cdot\mathbf{x}} + f(k,\theta)\frac{e^{ikx}}{x}$$

where the scattering amplitude is defined in Eq. (4.117).

Solution to Problem 4.10

(a) Suppose the distance from the scatterer $x = |\mathbf{x}|$ is very large, while $\mathbf{y}$ is confined to the region of the potential as illustrated in Fig. 4.1 below.

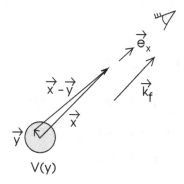

Fig. 4.1   Vector configuration where $|\mathbf{x}| \to \infty$ while $\mathbf{y}$ is confined to the potential. Here $\mathbf{x} = x\mathbf{e}_x$ where $\mathbf{e}_x$ is a unit vector in the direction of observation, and the final wave vector is defined by $\mathbf{k}_f \equiv k\mathbf{e}_x$. Also here $|\mathbf{x}| = x \equiv r$ is the usual radial coordinate in three dimensions.

Then as $x \to \infty$

$$\begin{aligned}
|\mathbf{x} - \mathbf{y}| &= \left(x^2 + y^2 - 2xy\cos\theta_{xy}\right)^{1/2} \\
&\approx x - y\cos\theta_{xy} \\
&= x - \mathbf{e}_x \cdot \mathbf{y}
\end{aligned}$$

Here $\mathbf{x} = x\mathbf{e}_x$ where $\mathbf{e}_x$ is a unit vector in the direction of observation. Then, while one can replace $|\mathbf{x} - \mathbf{y}| \approx x$ in the *amplitude* of the Green's function, in the *phase* one must use

$$e^{ik|\mathbf{x}-\mathbf{y}|} \approx e^{ikx}e^{-i\mathbf{k}_f\cdot\mathbf{y}} \qquad ; \ \mathbf{k}_f \equiv k\mathbf{e}_x$$

where the final wave vector is defined by $\mathbf{k}_f \equiv k\mathbf{e}_x$.

Thus the asymptotic form of the Green's function $G_0(\mathbf{x} - \mathbf{y})$ in Eq. (4.107), where $x = |\mathbf{x}| \to \infty$ while $\mathbf{y}$ is confined to the region of the potential, is

$$G_0(\mathbf{x} - \mathbf{y}) \to \frac{2m}{\hbar^2} \frac{e^{ikx}}{4\pi x} e^{-i\mathbf{k}_f \cdot \mathbf{y}} \qquad ; \; x = |\mathbf{x}| \to \infty$$

$$\mathbf{y} \text{ in potential}$$

(b) It follows from Eq. (4.118) that the asymptotic form of the scattering wave function is

$$\psi_i^{(+)}(\mathbf{x}) = e^{i\mathbf{k}\cdot\mathbf{x}} - \frac{2m}{\hbar^2} \int d^3y \, \frac{e^{ik|\mathbf{x}-\mathbf{y}|}}{4\pi|\mathbf{x} - \mathbf{y}|} V(y) \, \psi_i^{(+)}(\mathbf{y})$$

$$\to e^{i\mathbf{k}\cdot\mathbf{x}} - \frac{2m}{\hbar^2} \frac{e^{ikx}}{4\pi x} \int d^3y \, e^{-i\mathbf{k}_f \cdot \mathbf{y}} \, V(y) \, \psi_i^{(+)}(\mathbf{y}) \qquad ; \; x = |\mathbf{x}| \to \infty$$

Hence

$$\psi_i^{(+)}(\mathbf{x}) \to e^{i\mathbf{k}\cdot\mathbf{x}} + f(k, \theta) \frac{e^{ikx}}{x} \qquad\qquad ; \; x = |\mathbf{x}| \to \infty$$

$$f(k, \theta) \equiv -\frac{1}{4\pi} \frac{2m}{\hbar^2} \int d^3y \, e^{-i\mathbf{k}_f \cdot \mathbf{y}} \, V(y) \, \psi_i^{(+)}(\mathbf{y})$$

where $f(k, \theta)$ is the scattering amplitude in Eq. (4.117).

Problem 4.11 Verify Eq. (4.115).

Solution to Problem 4.11

Consider

$$\int \delta(E_f - E_0) k_f^2 dk_f \qquad\qquad ; \; E_f = \frac{\hbar^2 k_f^2}{2m}$$

Write this as

$$\frac{1}{2} \int k_f \delta(E_f - E_0) dk_f^2 = \frac{m}{\hbar^2} \int k_f \delta(E_f - E_0) dE_f$$

$$= \frac{m k_f}{\hbar^2} \qquad\qquad ; \; E_f = E_0$$

This is Eq. (4.115)

$$\int \delta(E_f - E_0) k_f^2 dk_f = \frac{2m}{\hbar^2} \frac{k_f}{2} \qquad\qquad ; \; |\mathbf{k}_f| = |\mathbf{k}|$$

Chapter 5

Lagrangian Field Theory

Problem 5.1 Consider lagrangian particle mechanics with a set of generalized coordinates $(q^1, \cdots, q^n)$. Assume the kinetic energy is a quadratic form in the generalized velocities $\dot{q}^i \equiv dq^i(t)/dt$ so that

$$L(q, \dot{q}) = T - V = \sum_i \sum_j t_{ij}(q)\dot{q}^i\dot{q}^j - V(q)$$

(a) Show that one can assume that $t_{ij}(q)$ is symmetric in (i, j);

(b) Prove that the hamiltonian is now the energy $H = T + V = E$.

Solution to Problem 5.1

(a) We can assume that $t_{ij}(q)$ is symmetric under the interchange of the indices (i, j). If not, introduce

$$t_{ij}(q) \equiv \frac{1}{2}\left[t_{ij}(q) + t_{ji}(q)\right] + \frac{1}{2}\left[t_{ij}(q) - t_{ji}(q)\right]$$

The first term is symmetric, and the second term makes a vanishing contribution to the sum $\sum_i \sum_j \dot{q}^i\dot{q}^j$ with a symmetric summand.

(b) Then with a symmetric $t_{ij}(q)$, the canonical momentum is given by

$$p^i = \frac{\partial L}{\partial \dot{q}^i} = \sum_j t_{ij}(q)\dot{q}^j + \sum_j t_{ji}(q)\dot{q}^j = 2\sum_j t_{ij}(q)\dot{q}^j$$

The hamiltonian follows as

$$H = \sum_i p^i\dot{q}^i - L$$

$$= \sum_i \sum_j t_{ij}(q)\dot{q}^i\dot{q}^j + V(q)$$

This is the sum of the kinetic and potential energies

$$H = T + V = E$$

Problem 5.2 Work in the coordinate representation. Assume a generalized coordinate q and its conjugate momentum p satisfying the canonical commutation relations $[p, q] = \hbar/i$.

(a) Prove $[p, F(q)] = -i\hbar \, \partial F(q)/\partial q$;

(b) Prove $[q, G(p)] = i\hbar \, \partial G(p)/\partial p$;

For parts (a,b) you may assume that the functions (F, G) have power series expansions in their respective arguments [*Hint*: try a few terms].

(c) Extend these results to the case where the series contain additional inverse powers of the arguments.

Solution to Problem 5.2

(a) We assume that $F(q)$ has a series representation[1]

$$F(q) = \sum_{n=0}^{\infty} \frac{1}{n!} \left[\frac{\partial^n F}{\partial q^n} \right]_{q=0} q^n$$

Therefore

$$[p, F(q)] = \sum_{n=0}^{\infty} \frac{1}{n!} \left[\frac{\partial^n F}{\partial q^n} \right]_{q=0} [p, q^n]$$

We will prove by induction that

$$[p, q^n] = n \frac{\hbar}{i} q^{n-1}$$

The proof goes as follows: first we explicitly calculate the case $n = 2$

$$[p, q^2] = [p, q]q + q[p, q] = 2 \frac{\hbar}{i} q$$

where we have used the commutator $[p, q] = \hbar/i$. We then assume that this relation holds to order $n - 1$

$$[p, q^{n-1}] = (n-1) \frac{\hbar}{i} q^{n-2}$$

and explicitly calculate the order n

$$[p, q^n] = [p, q^{n-1}]q + q^{n-1}[p, q] = n \frac{\hbar}{i} q^{n-1}$$

---

[1] Here the coefficients $[\partial^n F/\partial q^n]_{q=0}$ and $[\partial^n G/\partial p^n]_{p=0}$ are assumed to be c-numbers.

This completes the proof.

Now we can go back to our original commutator, substituting the result that we just obtained

$$[p, F(q)] = \frac{\hbar}{i} \sum_{n=1}^{\infty} \frac{1}{(n-1)!} \left[\frac{\partial^n F}{\partial q^n} \right]_{q=0} q^{n-1} = \frac{\hbar}{i} \frac{\partial F}{\partial q}$$

(b) We assume that $G(p)$ has a similar series representation

$$G(p) = \sum_{n=0}^{\infty} \frac{1}{n!} \left[\frac{\partial^n G}{\partial p^n} \right]_{p=0} p^n$$

Therefore

$$[q, G(p)] = \sum_{n=1}^{\infty} \frac{1}{n!} \left[\frac{\partial^n G}{\partial p^n} \right]_{p=0} [q, p^n]$$

It is straightforward to prove by induction, as in part (a), that

$$[q, p^n] = -n \frac{\hbar}{i} p^{n-1}$$

and therefore

$$[q, G(p)] = -\frac{\hbar}{i} \sum_{n=1}^{\infty} \frac{1}{(n-1)!} \left[\frac{\partial^n G}{\partial p^n} \right]_{p=0} p^{n-1} = -\frac{\hbar}{i} \frac{\partial G}{\partial p}$$

(c) We will now extend these results to the case where the series contain inverse powers of the arguments. For example, we may consider the case in which $F(q)$ contains a term q^{-r} where r is a positive integer. In this case

$$[p, q^{-r}] = pq^{-r} - q^{-r}p = q^{-r} \left(q^r p - p q^r \right) q^{-r} = -q^{-r}[p, q^r]q^{-r}$$

Now the commutator has the general form considered in part (a), and thus

$$[p, q^{-r}] = -\frac{\hbar}{i} r q^{-r-1}$$

As a result

$$[p, q^{-r}] = \frac{\hbar}{i} \frac{\partial}{\partial q} q^{-r}$$

and

$$[p, F(q)] = \frac{\hbar}{i} \frac{\partial F}{\partial q}$$

even when $F(q)$ contains negative powers.

The case in which $G(p)$ contains terms p^{-r}, with r a positive integer, is completely analogous.

Problem 5.3 Ehrenfest's theorem relates the time derivative of the expectation value of an operator $\hat{O}$ to the expectation value of the commutator of the operator with the hamiltonian. Show that in the abstract Hilbert space it reads (compare ProbI. 4.9)[2]

$$\frac{d}{dt}\langle\Psi(t)|\hat{O}|\Psi(t)\rangle = \langle\Psi(t)|\frac{i}{\hbar}[\hat{H},\,\hat{O}]\,|\Psi(t)\rangle \qquad ; \text{ Ehrenfest's theorem}$$

(a) Now work in the coordinate representation. Assume a hermitian hamiltonian $H(p,q)$ with a specified, ordered power-series expansion in (p,q), and write out Ehrenfest's theorem for both (p,q). Use the canonical commutation relations $[p,\,q] = \hbar/i$, and the results in Prob. 5.2, to show that in each case the operator on the r.h.s. is the appropriate derivative of $H(p,q)$ appearing in Hamilton's equations;

(b) Explain under what circumstances one now satisfies the *correspondence principle*.

Solution to Problem 5.3

We start with the Schrödinger equation satisfied by the state $|\Psi(t)\rangle$

$$i\hbar\frac{\partial}{\partial t}|\Psi(t)\rangle = \hat{H}|\Psi(t)\rangle$$

and by $\langle\Psi(t)|$ [3]

$$-i\hbar\frac{\partial}{\partial t}\langle\Psi(t)| = \langle\Psi(t)|\hat{H}$$

Now we can explicitly evaluate the time derivative of the expectation value of an operator $\hat{O}$ using the expressions that we have just obtained

$$\frac{d}{dt}\langle\Psi(t)|\hat{O}|\Psi(t)\rangle = \frac{i}{\hbar}\langle\Psi(t)|\hat{H}\hat{O} - \hat{O}\hat{H}|\Psi(t)\rangle$$

$$= \langle\Psi(t)|\frac{i}{\hbar}[\hat{H},\,\hat{O}]\,|\Psi(t)\rangle$$

This is Ehrenfest's theorem.

---

[2]This assumes that $\hat{O}$ has no *explicit* time dependence.

[3]See the solution to Prob. 2.3.

(a) For concreteness, we work in the coordinate representation where the coordinate $q = x$ is a c-number, and $p = (\hbar/i)\partial/\partial x$ satisfying

$$[p, x] = \frac{\hbar}{i}$$

With no explicit time dependence in O, Ehrenfest's theorem now reads

$$\frac{d}{dt}\int dx\,\Psi^\star(x,t)O(p,x)\Psi(x,t) = \int dx\,\Psi^\star(x,t)\frac{i}{\hbar}[H(p,x),O(p,x)]\,\Psi(x,t)$$

For generality, assume that the hamiltonian is some double power series in p and x

$$H(p,x) = \sum_m \sum_n a_{mn}(p^m x^n)_{\text{ord}}$$

where

- The coefficients a_{mn} are real numbers;
- The expression $(p^m x^n)_{\text{ord}}$ is a specified ordered product of m factors of p and n factors of x;
- The product $(p^m x^n)_{\text{ord}}$ is hermitian.

Just as in Prob. 5.2, it then follows that[4]

$$\frac{i}{\hbar}[H(p,x),\,x] = \frac{\partial}{\partial p}H(p,x)$$

$$\frac{i}{\hbar}[H(p,x),\,p] = -\frac{\partial}{\partial x}H(p,x)$$

Ehrenfest's theorem therefore gives

$$\frac{d}{dt}\int dx\,\Psi^\star(x,t)\,x\,\Psi(x,t) = \int dx\,\Psi^\star(x,t)\frac{\partial H(p,x)}{\partial p}\,\Psi(x,t)$$

$$\frac{d}{dt}\int dx\,\Psi^\star(x,t)\,p\,\Psi(x,t) = -\int dx\,\Psi^\star(x,t)\frac{\partial H(p,x)}{\partial x}\,\Psi(x,t)$$

These are the quantum form of *Hamilton's equations* of classical mechanics.

(b) Let us assume that

$$H(p,x) = \frac{p^2}{2m} + V(x)$$

---

[4]For example, $(\partial/\partial x)(px^2p^2x^2p) = 2(pxp^2x^2p + px^2p^2xp)$, etc.

and reduce the expressions calculated in part (a) to the form

$$\frac{d}{dt} \int dx\, \Psi^*(x,t)\, x\, \Psi(x,t) = \int dx\, \Psi^*(x,t) \frac{p}{m}\, \Psi(x,t)$$

$$\frac{d}{dt} \int dx\, \Psi^*(x,t)\, p\, \Psi(x,t) = -\int dx\, \Psi^*(x,t) \frac{\partial V(x)}{\partial x}\, \Psi(x,t)$$

Substitution of the second equation into the time derivative of the first gives

$$m\frac{d^2}{dt^2} \int dx\, \Psi^*(x,t)\, x\, \Psi(x,t) = -\int dx\, \Psi^*(x,t) \frac{\partial V(x)}{\partial x}\, \Psi(x,t)$$

For a well-localized wave packet we may approximate

$$\int dx\, \Psi^*(x,t) \frac{\partial V(x)}{\partial x}\, \Psi(x,t) \approx \frac{\partial V(\langle x \rangle)}{\partial \langle x \rangle}$$

where we have introduced the notation $\int dx\, \Psi^*(x,t)\, x\, \Psi(x,t) \to \langle x \rangle$. In this case $\langle x \rangle$ plays the role of the classical coordinate of a particle of mass m, moving under the effect of a potential V, and the correspondence principle is satisfied

$$m\frac{d^2 \langle x \rangle}{dt^2} = -\frac{\partial V(\langle x \rangle)}{\partial \langle x \rangle}$$

This is simply Newton's second law.

Problem 5.4 Consider the differential operator p_r^2 in polar coordinates (r, θ) defined in Eq. (5.43).

(a) Show it is hermitian with respect to the volume element in Eq. (5.41). State carefully any assumptions about the boundary contributions;

(b) Show it satisfies the commutation relation in the second line of Eq. (5.42).

Solution to Problem 5.4

(a) We have

$$p_r^2 = -\hbar^2 \frac{1}{\rho(r)} \frac{\partial}{\partial r} \rho(r) \frac{\partial}{\partial r} \qquad ;\ \rho(r) = r$$

Let $\psi_l(r)$ and $\psi_m(r)$ be functions normalizable on the interval $[0, \infty]$ and well-behaved at $r = 0$. Consider the matrix element

$$\int_0^\infty \psi_m^\star(r) \left[p_r^2 \, \psi_l(r) \right] \rho(r) \, dr = -\hbar^2 \int_0^\infty \rho(r) dr \, \psi_m^\star(r) \frac{1}{\rho(r)} \frac{\partial}{\partial r} \left[\rho(r) \frac{\partial}{\partial r} \psi_l(r) \right]$$

$$= -\hbar^2 \left[\psi_m^\star(r) \rho(r) \frac{\partial \psi_l(r)}{\partial r} \right]_0^\infty + \hbar^2 \int_0^\infty \frac{\partial \psi_m^\star(r)}{\partial r} \rho(r) \frac{\partial \psi_l(r)}{\partial r} dr$$

where the second line is obtained through an integration by parts. Since the wave functions are normalizable and well–behaved at the origin, the first term in the second line vanishes and we are free to integrate by parts a second time

$$\int_0^\infty \psi_m^\star(r) \left[p_r^2 \psi_l(r) \right] \rho(r) \, dr = \hbar^2 \left[\psi_l(r) \rho(r) \frac{\partial \psi_m^\star(r)}{\partial r} \right]_0^\infty -$$

$$\hbar^2 \int_0^\infty \psi_l(r) \frac{\partial}{\partial r} \left[\rho(r) \frac{\partial \psi_m^\star(r)}{\partial r} \right] dr$$

Once again the surface terms vanish, and we finally obtain

$$\int_0^\infty \psi_m^\star(r) \left[p_r^2 \psi_l(r) \right] \rho(r) \, dr = \int_0^\infty \left[p_r^2 \psi_m(r) \right]^\star \psi_l(r) \rho(r) \, dr$$

$$\equiv \int_0^\infty \left[p_r^{2\,\dagger} \psi_m(r) \right]^\star \psi_l(r) \rho(r) \, dr$$

which proves that $p_r^2 = p_r^{2\,\dagger}$ is hermitian with respect to $\rho(r) = r$.

(b) We first calculate the inner commutator

$$[p_r^2, r] f(r) = -\hbar^2 \frac{1}{r} \frac{\partial}{\partial r} r \frac{\partial}{\partial r} r f(r) + \hbar^2 \frac{\partial}{\partial r} r \frac{\partial f}{\partial r}$$

$$= -\hbar^2 \left[2 \frac{\partial}{\partial r} + \frac{1}{r} \right] f(r)$$

Now we can calculate the outer commutator

$$[[p_r^2, r], r] f(r) = -\hbar^2 [2 \frac{\partial}{\partial r} + \frac{1}{r}, r] f(r)$$

$$= -2\hbar^2 \left\{ \frac{\partial}{\partial r} [r f(r)] - r \frac{\partial f(r)}{\partial r} \right\} = -2\hbar^2 f(r)$$

which proves the second of Eqs. (5.42).

Problem 5.5 (a) Use the expression for p_r^2 in Eq. (5.43), and show that the operator p_r determined from the first of Eqs. (5.42) is

$$p_r = \frac{\hbar}{i} \left[\frac{\partial}{\partial r} + \frac{\rho'(r)}{2\rho(r)} \right]$$

(b) Show this p_r satisfies the commutation relation $[p_r, r] = \hbar/i$;

(c) Show this p_r is hermitian with respect to the volume element in Eq. (5.41).

Solution to Problem 5.5

(a) The comparison of the first of Eqs. (5.42) with the inner commutator calculated in part (a) of the previous problem allows one to identify

$$p_r = \frac{\hbar}{i} \left[\frac{\partial}{\partial r} + \frac{1}{2r} \right]$$

which reads

$$p_r = \frac{\hbar}{i} \left[\frac{\partial}{\partial r} + \frac{\rho'(r)}{2\rho(r)} \right]$$

once the identification $\rho(r) = r$ is made.

(b) The proof here is straightforward since the component of p_r which does not commute with r is just $(\hbar/i)(\partial/\partial r)$. Therefore, just as in the case of one-dimensional canonically conjugate variables p and q,

$$[p_r, r] = \frac{\hbar}{i}$$

(c) As in part (a) of the previous problem, we consider a pair of functions $\psi_l(r)$ and $\psi_m(r)$ which are normalizable for r on the positive semi-axis $[0, \infty]$, and well-behaved at the origin. We then evaluate the matrix element

$$\int_0^\infty \psi_m^\star(r) \left[p_r \psi_l(r) \right] \rho(r) \, dr = \frac{\hbar}{i} \int_0^\infty \psi_m^\star(r) \left[\frac{\partial \psi_l(r)}{\partial r} + \frac{1}{2r} \psi_l(r) \right] \rho(r) \, dr$$

$$= \frac{\hbar}{i} \left[\psi_m^\star(r) \psi_l(r) \rho(r) \right]_0^\infty - \frac{\hbar}{i} \int_0^\infty \left\{ \frac{\partial}{\partial r} [\rho(r) \psi_m^\star(r)] \right\} \psi_l(r) \, dr +$$

$$\frac{\hbar}{2i} \int_0^\infty \psi_m^\star(r) \psi_l(r) \, dr$$

The surface terms that appear after integrating by parts clearly vanish under the hypothesis that the functions are normalizable and well-behaved at the origin. Therefore

$$\int_0^\infty \psi_m^*(r)\left[p_r\psi_l(r)\right]\rho(r)\,dr = \int_0^\infty \psi_l(r)\left[\frac{\hbar}{i}\left(\frac{\partial}{\partial r}+\frac{1}{2r}\right)\psi_m(r)\right]^* \psi_l(r)\rho(r)\,dr$$

$$= \int_0^\infty \left[p_r\psi_m(r)\right]^* \psi_l(r)\rho(r)\,dr$$

$$\equiv \int_0^\infty \left[p_r^\dagger\psi_m(r)\right]^* \psi_l(r)\rho(r)\,dr$$

This proves that $p_r = p_r^\dagger$ is hermitian with respect to the volume element in Eq. (5.41).

Problem 5.6 Given the free-field hamiltonian $\hat{H}_0$ in Eq. (5.50), and the commutation relations for the creation and destruction operators in Eq. (5.48),
(a) Show

$$[\hat{H}_0,\, a_k] = -\hbar\omega_k a_k \qquad ; [\hat{H}_0,\, a_k^\dagger] = \hbar\omega_k a_k^\dagger$$

(b) Use the operator relation in Eq. (3.9) to then prove Eqs. (5.51);
(c) Take the time derivative of the l.h.s. of Eqs. (5.51) and show that

$$\frac{da_k(t)}{dt} = \frac{i}{\hbar}[\hat{H}_0,\, a_k(t)] = -i\omega_k a_k(t) \qquad ; a_k(0) = a_k$$

$$\frac{da_k^\dagger(t)}{dt} = \frac{i}{\hbar}[\hat{H}_0,\, a_k^\dagger(t)] = i\omega_k a_k^\dagger(t) \qquad ; a_k^\dagger(0) = a_k^\dagger$$

(d) Now integrate the results in part (c) with respect to time to re-derive the result in part (b).

Solution to Problem 5.6

(a) The hamiltonian in Eq. (5.50) is

$$\hat{H}_0 = \sum_k \hbar\omega_k \left(a_k^\dagger a_k + \frac{1}{2}\right)$$

and the commutation relations in Eqs. (5.48) are

$$[a_k,\, a_{k'}^\dagger] = \delta_{kk'}$$

$$[a_k^\dagger,\, a_{k'}^\dagger] = [a_k,\, a_{k'}] = 0$$

We can then compute

$$[\hat{H}_0, a_k] = \hbar\omega_k[a_k^\dagger a_k, a_k]$$
$$\equiv \hbar\omega_k\left(a_k^\dagger[a_k, a_k] + [a_k^\dagger, a_k]a_k\right)$$
$$= -\hbar\omega_k a_k$$

In an analogous fashion[5]

$$[\hat{H}_0, a_k^\dagger] = \hbar\omega_k a_k^\dagger$$

(b) The operator relation in Eq. (3.9) is

$$e^{i\hat{A}}\hat{B}e^{-i\hat{A}} = \hat{B} + i[\hat{A}, \hat{B}] + \frac{i^2}{2!}[\hat{A}, [\hat{A}, \hat{B}]] + \frac{i^3}{3!}[\hat{A}, [\hat{A}, [\hat{A}, \hat{B}]]] + \cdots$$

Identify

$$\hat{A} = \frac{\hat{H}_0 t}{\hbar} \qquad ; \hat{B} = a_k$$

It follows that

$$e^{i\hat{H}_0 t/\hbar}a_k e^{-i\hat{H}_0 t/\hbar} = a_k + (-i\omega_k t)\, a_k + \frac{(-i\omega_k t)^2}{2!}a_k + \frac{(-i\omega_k t)^3}{3!}a_k + \cdots$$
$$= a_k e^{-i\omega_k t}$$

In exactly the same manner

$$e^{i\hat{H}_0 t/\hbar}a_k^\dagger e^{-i\hat{H}_0 t/\hbar} = a_k^\dagger e^{i\omega_k t}$$

(c) Define the operators in the Heisenberg picture for the free fields as

$$a_k(t) \equiv e^{i\hat{H}_0 t/\hbar}a_k e^{-i\hat{H}_0 t/\hbar}$$
$$a_k^\dagger(t) \equiv e^{i\hat{H}_0 t/\hbar}a_k^\dagger e^{-i\hat{H}_0 t/\hbar}$$

With the use of the results in part (a), the time derivative and initial values of these quantities are[6]

$$\frac{da_k(t)}{dt} = \frac{i}{\hbar}[\hat{H}_0, a_k(t)] = -i\omega_k a_k(t) \qquad ; a_k(0) = a_k$$
$$\frac{da_k^\dagger(t)}{dt} = \frac{i}{\hbar}[\hat{H}_0, a_k^\dagger(t)] = i\omega_k a_k^\dagger(t) \qquad ; a_k^\dagger(0) = a_k^\dagger$$

---

[5]Or by simply taking the adjoint of the previous result (recall $[\hat{H}_0, a_k]^\dagger = -[\hat{H}_0, a_k^\dagger]$). In line with our usage in the text, we henceforth leave the hats off the creation and destruction operators.

[6]Remember that any operator commutes with itself.

(d) These relations are immediately integrated with respect to time to reproduce the results in part (b)

$$a_k(t) = a_k e^{-i\omega_k t}$$
$$a_k^\dagger(t) = a_k^\dagger e^{i\omega_k t}$$

Problem 5.7 (a) Explicitly convert the Schrödinger-picture hamiltonian in Eq. (5.49) to the interaction picture, and derive the expression in Eq. (5.54);

(b) Explain why the expressions in Eqs. (5.50) and (5.55) turn out to be identical.

Solution to Problem 5.7

(a) The hamiltonian for the string in the interaction picture, which is the same as the Heisenberg picture for free fields, is obtained from Eq. (5.49) as[7]

$$e^{i\hat{H}_0 t/\hbar} \int_0^l dx \left\{ \frac{1}{2\sigma} \hat{\pi}^2(x) + \frac{\sigma c^2}{2} \left[\frac{\partial \hat{q}(x)}{\partial x} \right]^2 \right\} e^{-i\hat{H}_0 t/\hbar} =$$
$$\int_0^l dx \left\{ \frac{1}{2\sigma} \hat{\pi}^2(x,t) + \frac{\sigma c^2}{2} \left[\frac{\partial \hat{q}(x,t)}{\partial x} \right]^2 \right\}$$

Here, with the use of the results in the previous problem, the fields in the interaction pictures are those given in Eqs. (5.52)

$$\hat{q}(x,t) = e^{i\hat{H}_0 t/\hbar} \hat{q}(x) e^{-i\hat{H}_0 t/\hbar}$$
$$= \sum_k \left(\frac{\hbar}{2\omega_k \sigma l} \right)^{1/2} \left[a_k e^{i(kx - \omega_k t)} + a_k^\dagger e^{-i(kx - \omega_k t)} \right]$$
$$\hat{\pi}(x,t) = e^{i\hat{H}_0 t/\hbar} \hat{\pi}(x) e^{-i\hat{H}_0 t/\hbar}$$
$$= \frac{1}{i} \sum_k \left(\frac{\hbar \omega_k \sigma}{2l} \right)^{1/2} \left[a_k e^{i(kx - \omega_k t)} - a_k^\dagger e^{-i(kx - \omega_k t)} \right]$$

This reproduces the results in Eq. (5.54).

(b) Since the hamiltonian is simply $\hat{H}_0$, one has

$$e^{i\hat{H}_0 t/\hbar} \hat{H}_0 e^{-i\hat{H}_0 t/\hbar} = \hat{H}_0$$

This is why the expressions in Eqs. (5.50) and (5.55) turn out to be identical.

---

[7] Use $e^{i\hat{H}_0 t/\hbar} \hat{\pi}^2(x) e^{-i\hat{H}_0 t/\hbar} = e^{i\hat{H}_0 t/\hbar} \hat{\pi}(x) e^{-i\hat{H}_0 t/\hbar} e^{i\hat{H}_0 t/\hbar} \hat{\pi}(x) e^{-i\hat{H}_0 t/\hbar}$, etc.

Problem 5.8 Start from the lagrangian for the massive scalar field in Eq. (5.63).

(a) Show that Lagrange's equation produces the Klein-Gordon equation $(\Box - m_s^2)\phi(\mathbf{x}, t) = 0$;

(b) Show the angular frequency appearing in the quantum fields and hamiltonian is $\omega_k = c\sqrt{\mathbf{k}^2 + m_s^2}$;

(c) Interpret the result in (b) in terms of the relativistic energy-momentum relation for the quanta (particles).

Solution to Problem 5.8

(a) The lagrangian density for the massive scalar field in Eq. (5.63) is

$$\mathcal{L}\left(\frac{\partial \phi}{\partial x_\mu}, \phi\right) = -\frac{c^2}{2}\left[\left(\frac{\partial \phi}{\partial x_\mu}\right)^2 + m_s^2 \phi^2\right]$$

where $m_s \equiv m_0 c/\hbar$ is the inverse Compton wavelength. Lagrange's Eq. (5.58) is

$$\frac{\partial}{\partial x_\mu}\left[\frac{\partial \mathcal{L}}{\partial(\partial \phi/\partial x_\mu)}\right] - \frac{\partial \mathcal{L}}{\partial \phi} = 0$$

The required derivatives are

$$\frac{\partial \mathcal{L}}{\partial(\partial \phi/\partial x_\mu)} = -c^2 \frac{\partial \phi}{\partial x_\mu}$$

$$\frac{\partial \mathcal{L}}{\partial \phi} = -c^2 m_s^2 \phi$$

Hence Lagrange's equation reads

$$-c^2 \frac{\partial}{\partial x_\mu}\frac{\partial \phi}{\partial x_\mu} + c^2 m_s^2 \phi = 0$$

This is re-written as

$$\left(\Box - m_s^2\right)\phi(\mathbf{x}, t) = 0$$

where

$$\Box = \frac{\partial^2}{\partial x_\mu \partial x_\mu} = \nabla^2 - \frac{1}{c^2}\frac{\partial^2}{\partial t^2}$$

(b) In analogy to Eqs. (5.60), the quantum fields for the massive scalar field are again written as

$$\hat{\phi}(\mathbf{x}, t) = \sum_{\mathbf{k}} \left(\frac{\hbar}{2\omega_k \Omega} \right)^{1/2} \left[a_{\mathbf{k}} e^{i(\mathbf{k}\cdot\mathbf{x} - \omega_k t)} + a_{\mathbf{k}}^{\dagger} e^{-i(\mathbf{k}\cdot\mathbf{x} - \omega_k t)} \right]$$

$$\hat{\pi}(\mathbf{x}, t) = \frac{1}{i} \sum_{\mathbf{k}} \left(\frac{\hbar\omega_k}{2\Omega} \right)^{1/2} \left[a_{\mathbf{k}} e^{i(\mathbf{k}\cdot\mathbf{x} - \omega_k t)} - a_{\mathbf{k}}^{\dagger} e^{-i(\mathbf{k}\cdot\mathbf{x} - \omega_k t)} \right]$$

Now, however, in order to satisfy the equation of motion in part (a), one must use

$$\omega_k = c\sqrt{\mathbf{k}^2 + m_s^2}$$

instead of the previous massless expression $\omega_k = |\mathbf{k}|c$.

Since the additional m_s^2 term ultimately enters in the same fashion as $\mathbf{k}^2$, the derivation of the hamiltonian goes through as before[8]

$$\hat{H}_0 = \sum_{\mathbf{k}} \hbar\omega_k \left(a_{\mathbf{k}}^{\dagger} a_{\mathbf{k}} + \frac{1}{2} \right)$$

(c) The energy of each quantum is now

$$E_k = \hbar\omega_k = \sqrt{(\hbar k c)^2 + (m_0 c^2)^2}$$

This is the relativistic dispersion relation for a particle of momentum $\mathbf{p} = \hbar\mathbf{k}$ and rest mass m_0.

Problem 5.9 Despite the fact that the creation and destruction operators for the Dirac (fermion) field obey the *anticommutation* relations of Eq. (5.103), all of the results in Prob. 5.6 continue to hold for them when the hamiltonian $\hat{H}_0$ of Eq. (5.105) is employed. Prove this statement.

Solution to Problem 5.9

The free Dirac hamiltonian in Eq. (5.105) is

$$\hat{H}_0 = \sum_{\mathbf{k}} \sum_{\lambda} \hbar\omega_k \left(a_{\mathbf{k}\lambda}^{\dagger} a_{\mathbf{k}\lambda} + b_{\mathbf{k}\lambda}^{\dagger} b_{\mathbf{k}\lambda} - 1 \right)$$

---

[8]Compare the solution to Prob. 5.16.

where the creation and destruction operators obey the canonical *anticommutation* relations

$$\{a_{\mathbf{k}\lambda}, a^\dagger_{\mathbf{k}'\lambda'}\} = \{b_{\mathbf{k}\lambda}, b^\dagger_{\mathbf{k}'\lambda'}\} = \delta_{\mathbf{k}\mathbf{k}'}\delta_{\lambda\lambda'}$$
; all other anticommutators vanish

Use the following *essential identities* to re-write the commutators required in Prob. 5.6 in terms of anticommutators

$$[a^\dagger_{\mathbf{k}\lambda}a_{\mathbf{k}\lambda}, a_{\mathbf{k}'\lambda'}] \equiv a^\dagger_{\mathbf{k}\lambda}\{a_{\mathbf{k}\lambda}, a_{\mathbf{k}'\lambda'}\} - \left\{a^\dagger_{\mathbf{k}\lambda}, a_{\mathbf{k}'\lambda'}\right\}a_{\mathbf{k}\lambda}$$

$$[a^\dagger_{\mathbf{k}\lambda}a_{\mathbf{k}\lambda}, a^\dagger_{\mathbf{k}'\lambda'}] \equiv a^\dagger_{\mathbf{k}\lambda}\left\{a_{\mathbf{k}\lambda}, a^\dagger_{\mathbf{k}'\lambda'}\right\} - \left\{a^\dagger_{\mathbf{k}\lambda}, a^\dagger_{\mathbf{k}'\lambda'}\right\}a_{\mathbf{k}\lambda}$$

Then, with the aid of the C.A.R. we have

$$[\hat{H}_0, a_{\mathbf{k}\lambda}] = -\hbar\omega_k a_{\mathbf{k}\lambda} \qquad ; [\hat{H}_0, a^\dagger_{\mathbf{k}\lambda}] = \hbar\omega_k a^\dagger_{\mathbf{k}\lambda}$$

In the same way

$$[\hat{H}_0, b_{\mathbf{k}\lambda}] = -\hbar\omega_k b_{\mathbf{k}\lambda} \qquad ; [\hat{H}_0, b^\dagger_{\mathbf{k}\lambda}] = \hbar\omega_k b^\dagger_{\mathbf{k}\lambda}$$

The remainder of Prob. 5.6 now follows exactly as before.

Problem 5.10 Start from the quantum field $\hat{\psi}(\mathbf{x})$ in the Schrödinger picture, and use the results in Prob. 5.9 to derive the field $\hat{\psi}(\mathbf{x}, t)$ in the interaction picture given in Eq. (5.102).

Solution to Problem 5.10

The Dirac field in the Schrödinger picture is given by

$$\hat{\psi}(\mathbf{x}) = \frac{1}{\sqrt{\Omega}}\sum_{\mathbf{k}}\sum_{\lambda}\left[a_{\mathbf{k}\lambda}u_\lambda(\mathbf{k})e^{i\mathbf{k}\cdot\mathbf{x}} + b^\dagger_{\mathbf{k}\lambda}v_\lambda(-\mathbf{k})e^{-i\mathbf{k}\cdot\mathbf{x}}\right]$$

$$\omega_k = c\sqrt{\mathbf{k}^2 + M^2} \qquad ; \text{Schrödinger picture}$$

where $\lambda = (\uparrow, \downarrow)$ denotes the helicity. It was established in Prob. 5.9 that

$$e^{\frac{i}{\hbar}\hat{H}_0 t}a_{\mathbf{k}\lambda}e^{-\frac{i}{\hbar}\hat{H}_0 t} = a_{\mathbf{k}\lambda}e^{-i\omega_k t}$$

$$e^{\frac{i}{\hbar}\hat{H}_0 t}b^\dagger_{\mathbf{k}\lambda}e^{-\frac{i}{\hbar}\hat{H}_0 t} = b^\dagger_{\mathbf{k}\lambda}e^{i\omega_k t}$$

Hence the Dirac field in the interaction picture is that given in Eq. (5.102)

$$\hat{\psi}(\mathbf{x}, t) = \frac{1}{\sqrt{\Omega}} \sum_{\mathbf{k}} \sum_{\lambda} \left[a_{\mathbf{k}\lambda} u_\lambda(\mathbf{k}) e^{i(\mathbf{k}\cdot\mathbf{x} - \omega_k t)} + b^\dagger_{\mathbf{k}\lambda} v_\lambda(-\mathbf{k}) e^{-i(\mathbf{k}\cdot\mathbf{x} - \omega_k t)} \right]$$

$$\omega_k = c\sqrt{\mathbf{k}^2 + M^2} \qquad\qquad ; \text{ interaction picture}$$

Problem 5.11 Prove that the second of Eqs. (5.94) is the proper adjoint of the Dirac equation in the first of Eqs. (5.94).

Solution to Problem 5.11

We start by writing the first of Eqs. (5.94), which reads

$$\left(\gamma_\mu \frac{\partial}{\partial x_\mu} + M \right) \psi = 0$$

We now take the adjoint of this equation, by taking its complex conjugate and then transposing it, obtaining

$$\psi^\dagger \left(\gamma^\dagger_\mu \frac{\overleftarrow{\partial}}{\partial x^\star_\mu} + M \right) = 0$$

where $x_\mu = (\mathbf{x}, ict)$ and $x^\star_\mu = (\mathbf{x}, -ict)$.

It is useful to recall that $\overline{\psi} = \psi^\dagger \gamma_4$. Since $\gamma_4^2 = 1$, we also have $\psi^\dagger = \overline{\psi} \gamma_4$. If we substitute this identity in the above equation, and multiply on the right by γ_4, we obtain

$$\overline{\psi} \left(\gamma_4 \gamma^\dagger_\mu \gamma_4 \frac{\overleftarrow{\partial}}{\partial x^\star_\mu} + M \right) = 0$$

In the solution to Prob. 9.1 in [Amore and Walecka (2013)], it is proven that the γ_μ are hermitian and they satisfy the Clifford algebra

$$\gamma_\mu \gamma_\nu + \gamma_\nu \gamma_\mu = 2\delta_{\mu\nu}$$

If we use these results, we obtain

$$\gamma_4 \gamma^\dagger_\mu \gamma_4 = \gamma_4 \gamma_\mu \gamma_4 = \begin{cases} \gamma_4 & ; \mu = 4 \\ -\gamma_\mu & ; \mu = 1, 2, 3 \end{cases}$$

Therefore

$$\gamma_4 \gamma^\dagger_\mu \gamma_4 \frac{\overleftarrow{\partial}}{\partial x^\star_\mu} = -\gamma_\mu \frac{\overleftarrow{\partial}}{\partial x_\mu}$$

Thus we finally reach the second of Eqs. (5.94)

$$\bar{\psi}\left(\gamma_\mu \frac{\overleftarrow{\partial}}{\partial x_\mu} - M\right) = 0$$

Problem 5.12 (a) Derive the second of Eqs. (5.79) for the momentum $\hat{\mathbf{P}}$ in the massive scalar field;

(b) Derive Eq. (5.106) for the momentum $\hat{\mathbf{P}}$ in the Dirac field.

Solution to Problem 5.12

(a) Our starting point is the second of Eqs. (5.78), which provides an expression for the total momentum of the scalar field as

$$\hat{\mathbf{P}} = -\int_\Omega d^3x \, \hat{\pi}(\mathbf{x}, t) \nabla \hat{\phi}(\mathbf{x}, t)$$

Here the quantum fields are given in Eqs. (5.60) and (5.65)

$$\hat{\phi}(\mathbf{x}, t) = \sum_k \left(\frac{\hbar}{2\omega_k \Omega}\right)^{1/2} \left[a_{\mathbf{k}} e^{i(\mathbf{k}\cdot\mathbf{x} - \omega_k t)} + a_{\mathbf{k}}^\dagger e^{-i(\mathbf{k}\cdot\mathbf{x} - \omega_k t)}\right]$$

$$\hat{\pi}(\mathbf{x}, t) = \frac{1}{i}\sum_k \left(\frac{\hbar\omega_k}{2\Omega}\right)^{1/2} \left[a_{\mathbf{k}} e^{i(\mathbf{k}\cdot\mathbf{x} - \omega_k t)} - a_{\mathbf{k}}^\dagger e^{-i(\mathbf{k}\cdot\mathbf{x} - \omega_k t)}\right]$$

with $\omega_k = c\sqrt{\mathbf{k}^2 + m_s^2}$.

The direct substitution of the expressions for the quantum fields inside the first equation provides

$$\hat{\mathbf{P}} = -\frac{1}{i}\int_\Omega d^3x \sum_{k'} \left(\frac{\hbar\omega_{k'}}{2\Omega}\right)^{1/2} \left[a_{\mathbf{k}'} e^{i(\mathbf{k}'\cdot\mathbf{x} - \omega_{k'} t)} - a_{\mathbf{k}'}^\dagger e^{-i(\mathbf{k}'\cdot\mathbf{x} - \omega_{k'} t)}\right] \times$$

$$\sum_k \left(\frac{\hbar}{2\omega_k \Omega}\right)^{1/2} i\mathbf{k} \left[a_{\mathbf{k}} e^{i(\mathbf{k}\cdot\mathbf{x} - \omega_k t)} - a_{\mathbf{k}}^\dagger e^{-i(\mathbf{k}\cdot\mathbf{x} - \omega_k t)}\right]$$

$$= -\frac{1}{2}\sum_{k,k'} \hbar\mathbf{k} \left(a_{\mathbf{k}'} a_{\mathbf{k}} e^{-2i\omega_k t}\delta_{\mathbf{k}, -\mathbf{k}'} - a_{\mathbf{k}'} a_{\mathbf{k}}^\dagger \delta_{\mathbf{k}\mathbf{k}'} - \right.$$

$$\left. a_{\mathbf{k}'}^\dagger a_{\mathbf{k}}\delta_{\mathbf{k}\mathbf{k}'} + a_{\mathbf{k}'}^\dagger a_{\mathbf{k}}^\dagger e^{2i\omega_k t}\delta_{\mathbf{k}, -\mathbf{k}'}\right)$$

$$= -\sum_k \frac{\hbar\mathbf{k}}{2}\left(a_{-\mathbf{k}} a_{\mathbf{k}} e^{-2i\omega_k t} - a_{\mathbf{k}} a_{\mathbf{k}}^\dagger - a_{\mathbf{k}}^\dagger a_{\mathbf{k}} + a_{-\mathbf{k}}^\dagger a_{\mathbf{k}}^\dagger e^{2i\omega_k t}\right)$$

Let us concentrate on the contribution

$$-\sum_{\mathbf{k}} \frac{\hbar \mathbf{k}}{2} \left(a_{-\mathbf{k}} a_{\mathbf{k}} e^{-2i\omega_k t} + a_{-\mathbf{k}}^\dagger a_{\mathbf{k}}^\dagger e^{2i\omega_k t} \right)$$

Given that like operators commute, the summand changes sign under the re-labeling $\mathbf{k} \to -\mathbf{k}$, and therefore the sum (which is done over all wave numbers) must vanish. Thus

$$\hat{\mathbf{P}} = \sum_{\mathbf{k}} \frac{\hbar \mathbf{k}}{2} \left(2a_{\mathbf{k}}^\dagger a_{\mathbf{k}} + 1 \right)$$

where we have used the commutator $[a_{\mathbf{k}}, a_{\mathbf{k}}^\dagger] = 1$. It is easy to see that the contribution $\sum_{\mathbf{k}} \hbar \mathbf{k}/2$ vanishes, and therefore we reach the final expression

$$\hat{\mathbf{P}} = \sum_{\mathbf{k}} \hbar \mathbf{k}\, a_{\mathbf{k}}^\dagger a_{\mathbf{k}}$$

(b) We now consider the second of Eqs. (5.99)

$$\hat{\mathbf{P}} = \int d^3x\; \hat{\psi}^\dagger \mathbf{p}\, \hat{\psi}$$

where $\mathbf{p} = (\hbar/i)\boldsymbol{\nabla}$, and the Dirac field is given in Eq. (5.102)

$$\hat{\psi}(\mathbf{x}, t) = \frac{1}{\sqrt{\Omega}} \sum_{\mathbf{k},\lambda} \left[a_{\mathbf{k}\lambda} u_\lambda(\mathbf{k}) e^{i(\mathbf{k}\cdot\mathbf{x} - \omega_k t)} + b_{\mathbf{k}\lambda}^\dagger v_\lambda(-\mathbf{k}) e^{-i(\mathbf{k}\cdot\mathbf{x} - \omega_k t)} \right]$$

with $\omega_k = c\sqrt{\mathbf{k}^2 + M^2}$. The appropriate expression for $\hat{\psi}^\dagger$ is obtained taking the adjoint of this expression.

If we substitute the fields inside the expression for $\hat{\mathbf{P}}$, we obtain

$$\hat{\mathbf{P}} = \frac{1}{\Omega} \int d^3x \sum_{k'\lambda'} \left[a_{\mathbf{k}'\lambda'}^\dagger u_{\lambda'}^\dagger(\mathbf{k}') e^{-i(\mathbf{k}'\cdot\mathbf{x} - \omega_{k'}t)} + b_{\mathbf{k}'\lambda'} v_{\lambda'}^\dagger(-\mathbf{k}') e^{i(\mathbf{k}'\cdot\mathbf{x} - \omega_{k'}t)} \right] \times$$

$$\frac{\hbar}{i} \sum_{k\lambda} i\mathbf{k} \left[a_{\mathbf{k}\lambda} u_\lambda(\mathbf{k}) e^{i(\mathbf{k}\cdot\mathbf{x} - \omega_k t)} - b_{\mathbf{k}\lambda}^\dagger v_\lambda(-\mathbf{k}) e^{-i(\mathbf{k}\cdot\mathbf{x} - \omega_k t)} \right]$$

$$= \sum_{k\lambda} \sum_{k'\lambda'} \hbar \mathbf{k} \left[a_{\mathbf{k}'\lambda'}^\dagger a_{\mathbf{k}\lambda} u_{\lambda'}^\dagger(\mathbf{k}') u_\lambda(\mathbf{k}) \delta_{\mathbf{k}\mathbf{k}'} + b_{\mathbf{k}'\lambda'} a_{\mathbf{k}\lambda} v_{\lambda'}^\dagger(-\mathbf{k}') u_\lambda(\mathbf{k}) e^{-2i\omega_k t} \delta_{\mathbf{k},-\mathbf{k}'} \right.$$

$$\left. - a_{\mathbf{k}'\lambda'}^\dagger b_{\mathbf{k}\lambda}^\dagger u_{\lambda'}^\dagger(\mathbf{k}') v_\lambda(-\mathbf{k}) e^{2i\omega_k t} \delta_{\mathbf{k},-\mathbf{k}'} - b_{\mathbf{k}'\lambda'} b_{\mathbf{k}\lambda}^\dagger v_{\lambda'}^\dagger(-\mathbf{k}') v_\lambda(-\mathbf{k}) \delta_{\mathbf{k}\mathbf{k}'} \right]$$

After using the Kronecker-delta's to eliminate the second wave number, we are left with a simpler expression

$$\hat{\mathbf{P}} = \sum_{\mathbf{k}} \sum_{\lambda,\lambda'} \hbar\mathbf{k} \ \left[a^{\dagger}_{\mathbf{k}\lambda'} a_{\mathbf{k},\lambda} u^{\dagger}_{\lambda'}(\mathbf{k}) u_{\lambda}(\mathbf{k}) + b_{-\mathbf{k}\lambda'} a_{\mathbf{k}\lambda} v^{\dagger}_{\lambda'}(\mathbf{k}) u_{\lambda}(\mathbf{k}) e^{-2i\omega_k t} - \right.$$

$$\left. a^{\dagger}_{-\mathbf{k}\lambda'} b^{\dagger}_{\mathbf{k}\lambda} u^{\dagger}_{\lambda'}(-\mathbf{k}) v_{\lambda}(-\mathbf{k}) e^{2i\omega_k t} - b_{\mathbf{k}\lambda'} b^{\dagger}_{\mathbf{k}\lambda} v^{\dagger}_{\lambda'}(-\mathbf{k}) v_{\lambda}(-\mathbf{k}) \right]$$

In the solution to Prob. 9.7 in [Amore and Walecka (2013)], it is proven that the Dirac spinors have the following properties

$$u^{\dagger}_{\lambda'}(\mathbf{k}) u_{\lambda}(\mathbf{k}) = v^{\dagger}_{\lambda'}(-\mathbf{k}) v_{\lambda}(-\mathbf{k}) = \delta_{\lambda\lambda'}$$
$$u^{\dagger}_{\lambda'}(\mathbf{k}) v_{\lambda}(\mathbf{k}) = v^{\dagger}_{\lambda'}(\mathbf{k}) u_{\lambda}(\mathbf{k}) = 0$$

With the use of these properties we have

$$\hat{\mathbf{P}} = \sum_{\mathbf{k}} \sum_{\lambda} \hbar\mathbf{k} \left(a^{\dagger}_{\mathbf{k}\lambda} a_{\mathbf{k}\lambda} - b_{\mathbf{k}\lambda} b^{\dagger}_{\mathbf{k}\lambda} \right)$$

$$= \sum_{\mathbf{k}} \sum_{\lambda} \hbar\mathbf{k} \left(a^{\dagger}_{\mathbf{k}\lambda} a_{\mathbf{k}\lambda} + b^{\dagger}_{\mathbf{k}\lambda} b_{\mathbf{k}\lambda} - \left\{ b_{\mathbf{k}\lambda}, b^{\dagger}_{\mathbf{k}\lambda} \right\} \right)$$

However

$$\sum_{\mathbf{k}} \sum_{\lambda} \hbar\mathbf{k} \left\{ b_{\mathbf{k}\lambda}, b^{\dagger}_{\mathbf{k}\lambda} \right\} = \sum_{\mathbf{k}} \sum_{\lambda} \hbar\mathbf{k} = 0$$

Hence we obtain the final expression

$$\hat{\mathbf{P}} = \sum_{\mathbf{k}} \sum_{\lambda} \hbar\mathbf{k} \left(a^{\dagger}_{\mathbf{k}\lambda} a_{\mathbf{k}\lambda} + b^{\dagger}_{\mathbf{k}\lambda} b_{\mathbf{k}\lambda} \right)$$

Problem 5.13 Show that if there are n generalized coordinates $q^i(\mathbf{x}, t)$ with $i = 1, 2, \cdots, n$, then the stress tensor in Eq. (5.69) gets generalized to

$$T_{\mu\nu} \equiv \mathcal{L} \delta_{\mu\nu} - \sum_{i=1}^{n} \frac{\partial \mathcal{L}}{\partial(\partial q^i / \partial x_{\mu})} \frac{\partial q^i}{\partial x_{\nu}}$$

Solution to Problem 5.13

We go through the same demonstration as presented in the text for a single generalized coordinate. Consider the four-divergence of this new stress tensor[9]

$$\frac{\partial}{\partial x_\mu} T_{\mu\nu} = \frac{\partial \mathcal{L}}{\partial x_\nu} - \sum_{i=1}^{n} \frac{\partial \mathcal{L}}{\partial(\partial q^i/\partial x_\mu)} \frac{\partial^2 q^i}{\partial x_\mu \partial x_\nu} - \sum_{i=1}^{n} \left[\frac{\partial}{\partial x_\mu} \frac{\partial \mathcal{L}}{\partial(\partial q^i/\partial x_\mu)} \right] \frac{\partial q^i}{\partial x_\nu}$$

The first term on the r.h.s. is now given through the chain rule as

$$\frac{\partial \mathcal{L}}{\partial x_\nu} = \sum_{i=1}^{n} \frac{\partial \mathcal{L}}{\partial q^i} \frac{\partial q^i}{\partial x_\nu} + \sum_{i=1}^{n} \frac{\partial \mathcal{L}}{\partial(\partial q^i/\partial x_\mu)} \frac{\partial^2 q^i}{\partial x_\nu \partial x_\mu} + \left[\frac{\partial \mathcal{L}}{\partial x_\nu} \right]_{(q^i,\, \partial q^i/\partial x_\mu)}$$

The second term in $\partial T_{\mu\nu}/\partial x_\mu$ is cancelled by the second term above. The last term in $\partial T_{\mu\nu}/\partial x_\mu$ cancels the first term in the above through the use of Lagrange's equation for each generalized coordinate

$$\frac{\partial}{\partial x_\mu} \frac{\partial \mathcal{L}}{\partial(\partial q^i/\partial x_\mu)} = \frac{\partial \mathcal{L}}{\partial q^i} \qquad ; i = 1, \cdots, n$$

Thus one is left with

$$\frac{\partial T_{\mu\nu}}{\partial x_\mu} = \left[\frac{\partial \mathcal{L}}{\partial x_\nu} \right]_{(q^i,\, \partial q^i/\partial x_\mu)}$$

It follows that if the lagrangian density has no *explicit* dependence on the space-time coordinate x_ν, then the new stress tensor has a vanishing divergence in Minkowski space.

The four-momentum of the system is defined by

$$P_\mu = \left(\mathbf{P}, \frac{i}{c} H \right) \equiv \frac{1}{ic} \int_\Omega d^3x\, T_{4\mu}$$

Consider the fourth component of this relation, and substitute the new definition of the stress tensor

$$H = - \int_\Omega d^3x\, T_{44}$$

$$= - \int_\Omega d^3x \left[\mathcal{L} - \sum_{i=1}^{n} \frac{\partial \mathcal{L}}{\partial(\partial q^i/\partial t)} \frac{\partial q^i}{\partial t} \right]$$

$$= \int_\Omega d^3x \left[\sum_{i=1}^{n} \pi^i \frac{\partial q^i}{\partial t} - \mathcal{L} \right]$$

Here the canonical momentum densities have been identified through

$$\pi^i = \frac{\partial \mathcal{L}}{\partial(\partial q^i/\partial t)} \qquad ; i = 1, \cdots, n$$

---

[9]Recall that repeated Greek indices are summed from 1 to 4.

The integrand in H now indeed reproduces the correct definition of the hamiltonian density.

Problem 5.14 (a) Construct the stress tensor for the string. Use the two-vector (x, ict) where c is the wave velocity in the string;

(b) Construct the two-vector $P_\mu = (P, iH/c)$;

(c) Insert the interaction-picture expansion of the string field in Eq. (5.52) and rederive the previous expression for the hamiltonian $\hat{H}_0$ in Eq. (5.55);

(d) Repeat part (c), and determine the momentum $\hat{P}$.

Solution to Problem 5.14

(a) The lagrangian density for the string is given in Eq. (5.30)

$$\mathcal{L} = -\frac{\tau}{2} \left(\frac{\partial q}{\partial x_\mu} \right)^2$$

Here the "two-vector" is $x_\mu = (x_1, x_2) = (x, ict)$, with $c^2 = \tau/\sigma$, and repeated Greek indices are here summed from 1 to 2. The stress tensor is given by Eq. (5.69) as

$$T_{\mu\nu} \equiv \mathcal{L}\, \delta_{\mu\nu} - \frac{\partial \mathcal{L}}{\partial(\partial q/\partial x_\mu)} \frac{\partial q}{\partial x_\nu}$$

Hence the stress tensor for the string is[10]

$$T_{\mu\nu} = \mathcal{L}\, \delta_{\mu\nu} + \tau \frac{\partial q}{\partial x_\mu} \frac{\partial q}{\partial x_\nu}$$

(b) The two-momentum follows from the analog of Eq. (5.74)

$$P_\mu = \left(P, \frac{i}{c} H \right) \equiv \frac{1}{ic} \int_l d^3x\, T_{2\mu}$$

The canonical momentum density for the string is given by Eq. (5.34)

$$\pi(x, t) = \sigma \frac{\partial q(x, t)}{\partial t}$$

---

[10]Note that $T_{\mu\nu} = T_{\nu\mu}$ is symmetric.

The hamiltonian then follows from the above as

$$H = -\int_l dx \left[\mathcal{L} - \frac{\tau}{c^2} \frac{\partial q}{\partial t} \frac{\partial q}{\partial t} \right]$$

$$= \int_l dx \left[\frac{1}{2\sigma} \pi^2 + \frac{\tau}{2} \left(\frac{\partial q}{\partial x} \right)^2 \right]$$

The hamiltonian density now coincides with the expression in Eq. (5.36). The momentum is given by

$$P = -\frac{\tau}{c^2} \int_l dx \left(\frac{\partial q}{\partial t} \right) \left(\frac{\partial q}{\partial x} \right)$$

$$= -\int_l dx \left[\pi \frac{\partial q}{\partial x} \right]$$

(c) The normal-mode expansion of (π, q) in terms of creation and destruction operators is given in the interaction picture in Eqs. (5.52)

$$\hat{q}(x,t) = \sum_k \left(\frac{\hbar}{2\omega_k \sigma l} \right)^{1/2} \left[a_k e^{i(kx - \omega_k t)} + a_k^\dagger e^{-i(kx - \omega_k t)} \right]$$

$$\hat{\pi}(x,t) = \frac{1}{i} \sum_k \left(\frac{\hbar \omega_k \sigma}{2l} \right)^{1/2} \left[a_k e^{i(kx - \omega_k t)} - a_k^\dagger e^{-i(kx - \omega_k t)} \right]$$

We apply periodic boundary conditions, and the sum goes over the modes in Eqs. (5.47). The creation and destruction operators satisfy Eqs. (5.48)

$$[a_k, a_{k'}^\dagger] = \delta_{kk'}$$

with all other commutators vanishing.

Substitute these expansions in the above expression for the hamiltonian

$$\hat{H} = \frac{\hbar}{4l} \int_l dx \sum_k \sum_{k'} \left\{ -\sqrt{\omega_k \omega_{k'}} \left[a_k e^{i(kx - \omega_k t)} - a_k^\dagger e^{-i(kx - \omega_k t)} \right] \times \right.$$

$$\left[a_{k'} e^{i(k'x - \omega_{k'} t)} - a_{k'}^\dagger e^{-i(k'x - \omega_{k'} t)} \right] - \frac{c^2 kk'}{\sqrt{\omega_k \omega_{k'}}} \times$$

$$\left[a_k e^{i(kx - \omega_k t)} - a_k^\dagger e^{-i(kx - \omega_k t)} \right] \left[a_{k'} e^{i(k'x - \omega_{k'} t)} - a_{k'}^\dagger e^{-i(k'x - \omega_{k'} t)} \right] \right\}$$

Use the orthonormality of the wave functions to do the spatial integral

$$\frac{1}{l} \int_l dx \, e^{i(k - k')x} = \delta_{kk'}$$

as well as the dispersion relation

$$\omega_k = c|k|$$

This leads to

$$\hat{H} = \frac{1}{4}\sum_k \hbar\omega_k \left[a_k a_k^\dagger + a_k^\dagger a_k - a_k a_{-k} e^{-2i\omega_k t} - a_k^\dagger a_{-k}^\dagger e^{2i\omega_k t} \right] +$$

$$\frac{1}{4}\sum_k \hbar\omega_k \left[a_k a_k^\dagger + a_k^\dagger a_k + a_k a_{-k} e^{-2i\omega_k t} + a_k^\dagger a_{-k}^\dagger e^{2i\omega_k t} \right]$$

The time-dependent terms cancel, and the use of the previous commutation relation gives Eq. (5.50)

$$\hat{H}_0 = \sum_k \hbar\omega_k \frac{1}{2}\left(a_k a_k^\dagger + a_k^\dagger a_k \right)$$

$$= \sum_k \hbar\omega_k \left(a_k^\dagger a_k + \frac{1}{2} \right)$$

(d) Substitution of the normal-mode expansions in the above expression for the momentum gives

$$\hat{P} = -\frac{\hbar}{2l}\int_l dx \sum_k \sum_{k'} \left\{ \left(\frac{\omega_k}{\omega_{k'}} \right)^{1/2} \left[a_k e^{i(kx-\omega_k t)} - a_k^\dagger e^{-i(kx-\omega_k t)} \right] \times \right.$$

$$\left. k' \left[a_{k'} e^{i(k'x-\omega_{k'}t)} - a_{k'}^\dagger e^{-i(k'x-\omega_{k'}t)} \right] \right\}$$

As above, this leads to

$$\hat{P} = \frac{\hbar}{2}\sum_k k \left[a_k a_k^\dagger + a_k^\dagger a_k + a_k a_{-k} e^{-2i\omega_k t} + a_k^\dagger a_{-k}^\dagger e^{2i\omega_k t} \right]$$

The time-dependent summands are each odd under the substitution $k \to -k$, and hence they cancel identically in the sum. Thus[11]

$$\hat{P} = \sum_k \hbar k \frac{1}{2}\left(a_k a_k^\dagger + a_k^\dagger a_k \right)$$

$$= \sum_k \hbar k \, a_k^\dagger a_k$$

Problem 5.15 Problems 5.15–5.17 concern the complex, or charged, scalar field (one can only get a current with a complex field). Assume

---

[11]Note $\sum_k \hbar k = 0$.

equal mass scalar fields (ϕ, ϕ^*). These are independent fields, since both the real and imaginary parts must be specified to specify both of them [compare part (a) below]. Assume a lagrangian density of the form

$$\mathcal{L}\left(\phi, \frac{\partial \phi}{\partial x_\mu}, \phi^*, \frac{\partial \phi^*}{\partial x_\mu}\right) = -c^2 \left[\frac{\partial \phi^*}{\partial x_\mu}\frac{\partial \phi}{\partial x_\mu} + m_s^2 \phi^* \phi\right]$$

a) Write out Hamilton's principle. Show that if $A\delta\phi + B\delta\phi^* = 0$ then $A = B = 0$;[12]

b) Show that Lagrange's equations are just the Klein-Gordon equations

$$(\Box - m_s^2)\phi = 0 \qquad ; \phi^* \text{ eqn.}$$
$$(\Box - m_s^2)\phi^* = 0 \qquad ; \phi \text{ eqn.}$$

c) Show the stress tensor is [note Prob. 5.13]

$$T_{\mu\nu} = \mathcal{L}\,\delta_{\mu\nu} + c^2 \frac{\partial \phi^*}{\partial x_\mu}\frac{\partial \phi}{\partial x_\nu} + c^2 \frac{\partial \phi^*}{\partial x_\nu}\frac{\partial \phi}{\partial x_\mu}$$

d) Show the momentum densities are[13]

$$\pi_\phi = \frac{\partial \phi^*}{\partial t} \qquad ; \pi_{\phi^*} = \frac{\partial \phi}{\partial t}$$

e) Show that the four-momentum gives

$$H = \int_\Omega d^3x \left(\pi_{\phi^*}\pi_\phi + c^2 \nabla\phi^* \cdot \nabla\phi + m_s^2 c^2 \phi^*\phi\right)$$

$$\mathbf{P} = -\int_\Omega d^3x \left(\pi_{\phi^*}\nabla\phi^* + \pi_\phi\nabla\phi\right)$$

Solution to Problem 5.15

(a) Hamilton's principle states that

$$\delta \int d^4x\, \mathcal{L}\left(\phi, \frac{\partial \phi}{\partial x_\mu}, \phi^*, \frac{\partial \phi^*}{\partial x_\mu}\right) = 0$$

with fixed endpoints in time. As in Eqs. (5.27)–(5.28), this gives

$$\int d^4x \left\{\left[\frac{\partial \mathcal{L}}{\partial \phi} - \frac{\partial}{\partial x_\mu}\frac{\partial \mathcal{L}}{\partial(\partial\phi/\partial x_\mu)}\right]\delta\phi + \left[\frac{\partial \mathcal{L}}{\partial \phi^*} - \frac{\partial}{\partial x_\mu}\frac{\partial \mathcal{L}}{\partial(\partial\phi^*/\partial x_\mu)}\right]\delta\phi^*\right\} = 0$$

---

[12] *Hint:* First take $\delta\phi$ to be a real variation, and then an imaginary one.
[13] Note carefully where the ϕ and ϕ^* appear in these equations.

The above is of the form

$$\int d^4x \, (A\delta\phi + B\delta\phi^\star) = 0$$

with $\delta\phi$ an arbitrary complex variation. We can first take it to be real, and then imaginary, to conclude that

$$A + B = 0 \qquad\qquad ; \, \delta\phi \text{ real}$$
$$A - B = 0 \qquad\qquad ; \, \delta\phi \text{ imaginary}$$
$$\implies \qquad A = B = 0$$

As a consequence, we can treat $(\delta\phi, \delta\phi^\star)$ as *arbitrary, independent* variations. The result is Lagrange's equation for each independent field $(\phi, \phi^\star)$

$$\frac{\partial}{\partial x_\mu} \frac{\partial\mathcal{L}}{\partial(\partial\phi/\partial x_\mu)} - \frac{\partial\mathcal{L}}{\partial\phi} = 0 \qquad ; \, \phi \text{ eqn}$$

$$\frac{\partial}{\partial x_\mu} \frac{\partial\mathcal{L}}{\partial(\partial\phi^\star/\partial x_\mu)} - \frac{\partial\mathcal{L}}{\partial\phi^\star} = 0 \qquad ; \, \phi^\star \text{ eqn}$$

(b) Consider the given lagrangian density

$$\mathcal{L}\left(\phi, \frac{\partial\phi}{\partial x_\mu}, \phi^\star, \frac{\partial\phi^\star}{\partial x_\mu}\right) = -c^2\left[\frac{\partial\phi^\star}{\partial x_\mu} \frac{\partial\phi}{\partial x_\mu} + m_s^2\phi^\star\phi\right]$$

Lagrange's equations are then just the Klein-Gordon equations

$$\frac{\partial}{\partial x_\mu} \frac{\partial\phi}{\partial x_\mu} - m_s^2\phi = (\Box - m_s^2)\phi = 0 \qquad ; \, \phi^\star \text{ eqn.}$$

$$\frac{\partial}{\partial x_\mu} \frac{\partial\phi^\star}{\partial x_\mu} - m_s^2\phi^\star = (\Box - m_s^2)\phi^\star = 0 \qquad ; \, \phi \text{ eqn.}$$

c) From Prob. 5.13, the stress tensor gets an additive contribution from each of the independent fields, and thus[14]

$$T_{\mu\nu} = \mathcal{L}\,\delta_{\mu\nu} - \frac{\partial\mathcal{L}}{\partial(\partial\phi/\partial x_\mu)} \frac{\partial\phi}{\partial x_\nu} - \frac{\partial\mathcal{L}}{\partial(\partial\phi^\star/\partial x_\mu)} \frac{\partial\phi^\star}{\partial x_\nu}$$

$$= \mathcal{L}\,\delta_{\mu\nu} + c^2\frac{\partial\phi^\star}{\partial x_\mu} \frac{\partial\phi}{\partial x_\nu} + c^2\frac{\partial\phi^\star}{\partial x_\nu} \frac{\partial\phi}{\partial x_\mu}$$

(d) The canonical momentum densities are given by

$$\pi_\phi = \frac{\partial\mathcal{L}}{\partial(\partial\phi/\partial t)} \qquad ; \, \pi_{\phi^\star} = \frac{\partial\mathcal{L}}{\partial(\partial\phi^\star/\partial t)}$$

---

[14]Note that the stress tensor is symmetric, with $T_{\mu\nu} = T_{\nu\mu}$.

Hence

$$\pi_\phi = \frac{\partial \phi^\star}{\partial t} \qquad ; \pi_{\phi^\star} = \frac{\partial \phi}{\partial t}$$

(e) The four-momentum of the system is defined by

$$P_\mu = \left(\mathbf{P}, \frac{i}{c}H\right) \equiv \frac{1}{ic} \int_\Omega d^3x\, T_{4\mu}$$

This gives the hamiltonian

$$H = -\int_\Omega d^3x\, T_{44} = \int_\Omega d^3x \left[2\frac{\partial \phi^\star}{\partial t}\frac{\partial \phi}{\partial t} - \mathcal{L}\right]$$

Therefore

$$H = \int_\Omega d^3x \left(\pi_{\phi^\star}\pi_\phi + c^2 \boldsymbol{\nabla}\phi^\star \cdot \boldsymbol{\nabla}\phi + m_s^2 c^2 \phi^\star \phi\right)$$

The j-th component of the momentum is

$$P_j = \frac{1}{ic}\int_\Omega d^3x\, T_{4j} = -\int_\Omega d^3x \left[\frac{\partial \phi^\star}{\partial t}\frac{\partial \phi}{\partial x_j} + \frac{\partial \phi}{\partial t}\frac{\partial \phi^\star}{\partial x_j}\right]$$

Hence

$$\mathbf{P} = -\int_\Omega d^3x \left(\pi_{\phi^\star}\boldsymbol{\nabla}\phi^\star + \pi_\phi \boldsymbol{\nabla}\phi\right)$$

Problem 5.16 Now quantize the continuum problem formulated in Prob. 5.15.[15]

(a) Write out the canonical equal-time commutation relations in the interaction picture for $(\pi_{\phi^\star}, \phi^\star)$ and (π_ϕ, ϕ);

(b) Show the following expansions of the fields in the interaction picture satisfy these commutation relations

$$\hat{\phi}(\mathbf{x}, t) = \sum_{\mathbf{k}} \left(\frac{\hbar}{2\omega_k \Omega}\right)^{1/2} \left[a_{\mathbf{k}} e^{i(\mathbf{k}\cdot\mathbf{x} - \omega_k t)} + b_{\mathbf{k}}^\dagger e^{-i(\mathbf{k}\cdot\mathbf{x} - \omega_k t)}\right]$$

$$\hat{\pi}_\phi(\mathbf{x}, t) = i\sum_{\mathbf{k}} \left(\frac{\hbar\omega_k}{2\Omega}\right)^{1/2} \left[a_{\mathbf{k}}^\dagger e^{-i(\mathbf{k}\cdot\mathbf{x} - \omega_k t)} - b_{\mathbf{k}} e^{i(\mathbf{k}\cdot\mathbf{x} - \omega_k t)}\right]$$

---

[15]The interpretation of the Klein-Gordon equation as the field equation in quantum field theory is due to [Pauli and Weisskopf (1934)].

Here p.b.c. are assumed, and $(a^\dagger, b^\dagger, a, b)$ are the usual boson creation and destruction operators. The operators $(\hat{\phi}^*, \hat{\pi}_{\phi^*})$ are just the adjoints of these expressions;

(c) Substitute these expressions in part (e) of Prob. 5.15, and show the hamiltonian and momentum become

$$\hat{H}_0 = \sum_k \hbar\omega_k \left(a^\dagger_\mathbf{k} a_\mathbf{k} + b^\dagger_\mathbf{k} b_\mathbf{k} + 1\right)$$

$$\hat{\mathbf{P}} = \sum_k \hbar\mathbf{k} \left(a^\dagger_\mathbf{k} a_\mathbf{k} + b^\dagger_\mathbf{k} b_\mathbf{k}\right)$$

Solution to Problem 5.16

(a) The theory is quantized by imposing canonical commutation relations, which in the interaction picture are

$$[\hat{\phi}(\mathbf{x}, t), \hat{\pi}_\phi(\mathbf{x}', t')]_{t=t'} = i\hbar\, \delta^{(3)}(\mathbf{x} - \mathbf{x}')$$
$$[\hat{\phi}^*(\mathbf{x}, t), \hat{\pi}_{\phi^*}(\mathbf{x}', t')]_{t=t'} = i\hbar\, \delta^{(3)}(\mathbf{x} - \mathbf{x}')$$

(b) A representation satisfying these commutation relations is that given in the statement of the problem

$$\hat{\phi}(\mathbf{x}, t) = \sum_k \left(\frac{\hbar}{2\omega_k \Omega}\right)^{1/2} \left[a_\mathbf{k} e^{i(\mathbf{k}\cdot\mathbf{x} - \omega_k t)} + b^\dagger_\mathbf{k} e^{-i(\mathbf{k}\cdot\mathbf{x} - \omega_k t)}\right]$$

$$\hat{\pi}_\phi(\mathbf{x}, t) = i \sum_k \left(\frac{\hbar\omega_k}{2\Omega}\right)^{1/2} \left[a^\dagger_\mathbf{k} e^{-i(\mathbf{k}\cdot\mathbf{x} - \omega_k t)} - b_\mathbf{k} e^{i(\mathbf{k}\cdot\mathbf{x} - \omega_k t)}\right]$$

Here p.b.c. are assumed, and $(a^\dagger, b^\dagger, a, b)$ are the usual boson creation and destruction operators, satisfying

$$[a_\mathbf{k}, a^\dagger_{\mathbf{k}'}] = \delta_{\mathbf{kk}'} \qquad ; \ [b_\mathbf{k}, b^\dagger_{\mathbf{k}'}] = \delta_{\mathbf{kk}'}$$

with all other commutators vanishing. The operators $(\hat{\phi}^*, \hat{\pi}_{\phi^*})$ are just the adjoints of these expressions;[16]

---

[16] Note the other equal-time commutators of the fields vanish

$$[\hat{\phi}(\mathbf{x}, t), \hat{\phi}^*(\mathbf{x}', t)] = [\hat{\pi}_\phi(\mathbf{x}, t), \hat{\pi}_{\phi^*}(\mathbf{x}', t)] = [\hat{\phi}(\mathbf{x}, t), \hat{\pi}_{\phi^*}(\mathbf{x}', t)] = [\hat{\phi}^*(\mathbf{x}, t), \hat{\pi}_\phi(\mathbf{x}', t)] = 0$$

(c) Insert the above expansions in the hamiltonian in Prob. 5.16(e)

$$\hat{H} = \frac{\hbar}{2\Omega} \int_\Omega d^3x \sum_{\mathbf{k}} \sum_{\mathbf{k}'} \left\{ \sqrt{\omega_k \omega_{k'}} \left[a_{\mathbf{k}}^\dagger e^{-i(\mathbf{k}\cdot\mathbf{x}-\omega_k t)} - b_{\mathbf{k}} e^{i(\mathbf{k}\cdot\mathbf{x}-\omega_k t)} \right] \times \right.$$

$$\left[a_{\mathbf{k}'} e^{i(\mathbf{k}'\cdot\mathbf{x}-\omega_{k'}t)} - b_{\mathbf{k}'}^\dagger e^{-i(\mathbf{k}'\cdot\mathbf{x}-\omega_{k'}t)} \right] +$$

$$\frac{c^2 \mathbf{k}\cdot\mathbf{k}'}{\sqrt{\omega_k \omega_{k'}}} \left[a_{\mathbf{k}} e^{i(\mathbf{k}\cdot\mathbf{x}-\omega_k t)} - b_{\mathbf{k}}^\dagger e^{-i(\mathbf{k}\cdot\mathbf{x}-\omega_k t)} \right] \left[a_{\mathbf{k}'}^\dagger e^{-i(\mathbf{k}'\cdot\mathbf{x}-\omega_{k'}t)} - b_{\mathbf{k}'} e^{i(\mathbf{k}'\cdot\mathbf{x}-\omega_{k'}t)} \right] +$$

$$\left. \frac{m_s^2 c^2}{\sqrt{\omega_k \omega_{k'}}} \left[a_{\mathbf{k}} e^{i(\mathbf{k}\cdot\mathbf{x}-\omega_k t)} + b_{\mathbf{k}}^\dagger e^{-i(\mathbf{k}\cdot\mathbf{x}-\omega_k t)} \right] \left[a_{\mathbf{k}'}^\dagger e^{-i(\mathbf{k}'\cdot\mathbf{x}-\omega_{k'}t)} + b_{\mathbf{k}'} e^{i(\mathbf{k}'\cdot\mathbf{x}-\omega_{k'}t)} \right] \right\}$$

Use the orthonormality relation to do the spatial integrals

$$\frac{1}{\Omega} \int_\Omega d^3x \, e^{i(\mathbf{k}-\mathbf{k}')\cdot\mathbf{x}} = \delta_{\mathbf{k}\mathbf{k}'}$$

as well as the dispersion relation

$$\omega_k = c\sqrt{\mathbf{k}^2 + m_s^2}$$

Then

$$\hat{H} = \frac{\hbar}{2} \sum_{\mathbf{k}} \omega_k \left[a_{\mathbf{k}}^\dagger a_{\mathbf{k}} + b_{\mathbf{k}} b_{\mathbf{k}}^\dagger - a_{\mathbf{k}}^\dagger b_{-\mathbf{k}}^\dagger e^{2i\omega_k t} - b_{\mathbf{k}} a_{-\mathbf{k}} e^{-2i\omega_k t} \right] +$$

$$\frac{\hbar}{2} \sum_{\mathbf{k}} \omega_k \left[a_{\mathbf{k}} a_{\mathbf{k}}^\dagger + b_{\mathbf{k}}^\dagger b_{\mathbf{k}} + b_{\mathbf{k}}^\dagger a_{-\mathbf{k}}^\dagger e^{2i\omega_k t} + a_{\mathbf{k}} b_{-\mathbf{k}} e^{-2i\omega_k t} \right]$$

Use of the previous commutation relations, and a change of dummy summation variable $\mathbf{k} \to -\mathbf{k}$ in the time-dependent terms in the second line, reduces this to the given expression

$$\hat{H}_0 = \sum_{\mathbf{k}} \hbar\omega_k \left(a_{\mathbf{k}}^\dagger a_{\mathbf{k}} + b_{\mathbf{k}}^\dagger b_{\mathbf{k}} + 1 \right)$$

The momentum is obtained in a similar fashion as

$$\hat{\mathbf{P}} = \frac{\hbar}{2\Omega} \int_\Omega d^3x \sum_{\mathbf{k}} \sum_{\mathbf{k}'} \left(\frac{\omega_k}{\omega_{k'}} \right)^{1/2} \mathbf{k}' \left\{ \left[a_{\mathbf{k}} e^{i(\mathbf{k}\cdot\mathbf{x}-\omega_k t)} - b_{\mathbf{k}}^\dagger e^{-i(\mathbf{k}\cdot\mathbf{x}-\omega_k t)} \right] \times \right.$$

$$\left[a_{\mathbf{k}'}^\dagger e^{-i(\mathbf{k}'\cdot\mathbf{x}-\omega_{k'}t)} - b_{\mathbf{k}'} e^{i(\mathbf{k}'\cdot\mathbf{x}-\omega_{k'}t)} \right] +$$

$$\left. \left[a_{\mathbf{k}}^\dagger e^{-i(\mathbf{k}\cdot\mathbf{x}-\omega_k t)} - b_{\mathbf{k}} e^{i(\mathbf{k}\cdot\mathbf{x}-\omega_k t)} \right] \left[a_{\mathbf{k}'} e^{i(\mathbf{k}'\cdot\mathbf{x}-\omega_{k'}t)} - b_{\mathbf{k}'}^\dagger e^{-i(\mathbf{k}'\cdot\mathbf{x}-\omega_{k'}t)} \right] \right\}$$

Integration over the spatial coordinates leads to

$$\hat{\mathbf{P}} = \frac{\hbar}{2}\sum_{\mathbf{k}}\mathbf{k}\left[a_{\mathbf{k}}a_{\mathbf{k}}^{\dagger} + b_{\mathbf{k}}^{\dagger}b_{\mathbf{k}} + a_{\mathbf{k}}b_{-\mathbf{k}}e^{-2i\omega_k t} + b_{\mathbf{k}}^{\dagger}a_{-\mathbf{k}}^{\dagger}e^{2i\omega_k t}\right] +$$
$$\frac{\hbar}{2}\sum_{\mathbf{k}}\mathbf{k}\left[a_{\mathbf{k}}^{\dagger}a_{\mathbf{k}} + b_{\mathbf{k}}b_{\mathbf{k}}^{\dagger} + b_{\mathbf{k}}a_{-\mathbf{k}}e^{-2i\omega_k t} + a_{\mathbf{k}}^{\dagger}b_{-\mathbf{k}}^{\dagger}e^{2i\omega_k t}\right]$$

Use of the previous commutation relations, and a change of dummy summation variable $\mathbf{k} \to -\mathbf{k}$ in the time-dependent terms in the second line, give

$$\hat{\mathbf{P}} = \sum_{\mathbf{k}}\hbar\mathbf{k}\left(a_{\mathbf{k}}^{\dagger}a_{\mathbf{k}} + b_{\mathbf{k}}^{\dagger}b_{\mathbf{k}} + 1\right)$$

The final term vanishes, and hence we arrive at the given result

$$\hat{\mathbf{P}} = \sum_{\mathbf{k}}\hbar\mathbf{k}\left(a_{\mathbf{k}}^{\dagger}a_{\mathbf{k}} + b_{\mathbf{k}}^{\dagger}b_{\mathbf{k}}\right)$$

Problem 5.17 (a) Show that the lagrangian density in Prob. 5.15 has an invariance under global phase transformations;

(b) Use Noether's theorem to construct the corresponding conserved current

$$j_\mu(x) = \frac{c}{i\hbar}\left[\phi^\star\frac{\partial\phi}{\partial x_\mu} - \frac{\partial\phi^\star}{\partial x_\mu}\phi\right]$$

(c) Show that the integral over all space of the fourth component of this current is a constant of the motion, and show that when quantized, this charge is given by[17]

$$\hat{Q} = \sum_{\mathbf{k}}\left(a_{\mathbf{k}}^{\dagger}a_{\mathbf{k}} - b_{\mathbf{k}}^{\dagger}b_{\mathbf{k}}\right)$$

(d) Hence conclude that this quantum field theory contains both *particles and antiparticles* with the opposite sign of this charge.

Solution to Problem 5.17

(a) The lagrangian density in Prob. 5.15 is

$$\mathcal{L}\left(\phi, \frac{\partial\phi}{\partial x_\mu}, \phi^\star, \frac{\partial\phi^\star}{\partial x_\mu}\right) = -c^2\left[\frac{\partial\phi^\star}{\partial x_\mu}\frac{\partial\phi}{\partial x_\mu} + m_s^2\phi^\star\phi\right]$$

---

[17]Here we write $j_\mu = (\mathbf{j}/c, i\rho)$, where ρ is the charge density.

This is invariant under a global phase transformation

$$\phi \to e^{i\theta}\phi$$
$$\phi^\star \to e^{-i\theta}\phi^\star$$

If $\theta \equiv \varepsilon \to 0$, then the change in the fields under this infinitesimal transformation takes the form [compare Eqs. (5.112)–(5.113)]

$$\delta\phi = i\varepsilon\phi$$
$$\delta\phi^\star = -i\varepsilon\phi^\star$$

(b) Noether's theorem says the invariance of $\mathcal{L}$ leads to the conserved current in Eq. (5.111), where there is one contribution for each independent field $(\phi, \phi^\star)$[18]

$$J_\mu(x) = \frac{\partial \mathcal{L}}{\partial(\partial\phi/\partial x_\mu)}\delta\phi + \frac{\partial \mathcal{L}}{\partial(\partial\phi^\star/\partial x_\mu)}\delta\phi^\star$$

If we remove a factor of $-\varepsilon\hbar c$, this current is given by

$$j_\mu(x) = \frac{c}{i\hbar}\left[\frac{\partial\phi}{\partial x_\mu}\phi^\star - \frac{\partial\phi^\star}{\partial x_\mu}\phi\right]$$

The fourth component of the current is $i\rho(x)$, and this gives a charge density of

$$\rho(x) = \frac{i}{\hbar}\left[\frac{\partial\phi}{\partial t}\phi^\star - \frac{\partial\phi^\star}{\partial t}\phi\right]$$
$$= \frac{i}{\hbar}\left[\pi_{\phi^\star}\phi^\star - \pi_\phi\phi\right]$$

(c) Now substitute the normal-mode expansions in Prob. 5.16, and compute the total charge[19]

$$\hat{Q} = \frac{1}{2\Omega}\int_\Omega d^3x \sum_{\mathbf{k}}\sum_{\mathbf{k}'}\left(\frac{\omega_k}{\omega_{k'}}\right)^{1/2}\left\{\left[a_\mathbf{k}e^{i(\mathbf{k}\cdot\mathbf{x}-\omega_k t)} - b_\mathbf{k}^\dagger e^{-i(\mathbf{k}\cdot\mathbf{x}-\omega_k t)}\right] \times \right.$$
$$\left[a_{\mathbf{k}'}^\dagger e^{-i(\mathbf{k}'\cdot\mathbf{x}-\omega_{k'} t)} + b_{\mathbf{k}'}e^{i(\mathbf{k}'\cdot\mathbf{x}-\omega_{k'} t)}\right] +$$
$$\left.\left[a_\mathbf{k}^\dagger e^{-i(\mathbf{k}\cdot\mathbf{x}-\omega_k t)} - b_\mathbf{k}e^{i(\mathbf{k}\cdot\mathbf{x}-\omega_k t)}\right]\left[a_{\mathbf{k}'}e^{i(\mathbf{k}'\cdot\mathbf{x}-\omega_{k'} t)} + b_{\mathbf{k}'}^\dagger e^{-i(\mathbf{k}'\cdot\mathbf{x}-\omega_{k'} t)}\right]\right\}$$

---

[18]Compare the argument in Eqs. (5.112)–(5.115).
[19]In units of e.

The integral is performed with the aid of the orthonormality relation, and the result is

$$\hat{Q} = \frac{1}{2} \sum_{k} \left[a_k a_k^\dagger - b_k^\dagger b_k + a_k b_{-k} e^{-2i\omega_k t} - b_k^\dagger a_{-k}^\dagger e^{2i\omega_k t} \right] + $$

$$\frac{1}{2} \sum_{k} \left[a_k^\dagger a_k - b_k b_k^\dagger - b_k a_{-k} e^{-2i\omega_k t} + a_k^\dagger b_{-k}^\dagger e^{2i\omega_k t} \right]$$

Use of the previous commutation relations, and a change of dummy summation variable $k \to -k$ in the time-dependent terms in the second line, produces the given result

$$\hat{Q} = \sum_{k} \left(a_k^\dagger a_k - b_k^\dagger b_k \right)$$

This operator is indeed independent of time,[20] and since $a_k^\dagger a_k$ and $b_k^\dagger b_k$ are number operators, we know the eigenvalues of $\hat{Q}$.

(d) The operator $a_k^\dagger$ evidently creates a particle with a positive charge, while $b_k^\dagger$ evidently creates a particle with a negative charge. Hence we conclude that this quantum field theory contains both *particles and antiparticles* with the opposite sign of this charge.

Problem 5.18 One would like to have a *symmetric* stress tensor $T_{\mu\nu} = T_{\nu\mu}$ in order to construct a conserved angular momentum density

$$M_{\mu\nu\nu'} = T_{\mu\nu} x_{\nu'} - T_{\mu\nu'} x_\nu \qquad ; \text{ angular momentum density}$$

(a) Show that $\partial M_{\mu\nu\nu'} / \partial x_\mu = 0$ if $T_{\mu\nu}$ is symmetric;

It is always possible to symmetrize $T_{\mu\nu}$, while not changing the integrated four-momentum in Eq. (5.74), and the general procedure for doing this is in an appendix in [Wentzel (1949)]. Here we simply note that the scalar stress tensors in Eq. (5.76) and Prob. 5.15(c) are already symmetric, and we develop a procedure for symmetrizing the Dirac stress tensor.

(b) The Dirac $\mathcal{L}$ vanishes along the dynamical trajectory, and $T_{\mu\nu}$ in

---

[20]Note that current conservation implies the total charge is a constant of the motion

$$\frac{d}{dt} \int_\Omega \rho(\mathbf{x}, t) \, d^3x = - \int_\Omega \boldsymbol{\nabla} \cdot \mathbf{j}(\mathbf{x}, t) \, d^3x = - \int_S \mathbf{j}(\mathbf{x}, t) \cdot d\mathbf{S} = 0$$

Eq. (5.96) can equally well be written as[21]

$$\tilde{T}_{\mu\nu} = \frac{\hbar c}{2}\left[\bar{\psi}\gamma_\mu\frac{\partial\psi}{\partial x_\nu} - \frac{\partial\bar{\psi}}{\partial x_\nu}\gamma_\mu\psi\right]$$

Show this $\tilde{T}_{\mu\nu}$ has the following properties:

$$\frac{\partial\tilde{T}_{\mu\nu}}{\partial x_\mu} = \frac{\partial\tilde{T}_{\nu\mu}}{\partial x_\mu} = 0$$

$$\int_\Omega d^3x\,\tilde{T}_{4\nu} = \int_\Omega d^3x\,T_{4\nu}$$

(c) Now take $\theta_{\mu\nu} \equiv \tilde{T}_{\mu\nu} + (\tilde{T}_{\nu\mu} - \tilde{T}_{\mu\nu})/2$. Show[22]

$$\frac{\partial\theta_{\mu\nu}}{\partial x_\mu} = 0 \qquad\qquad ; \theta_{\mu\nu} = \theta_{\nu\mu}$$

$$\int_\Omega d^3x\,\theta_{4\nu} = \int_\Omega d^3x\,T_{4\nu}$$

Hence conclude that $\theta_{\mu\nu}$ has all the desired properties for the stress tensor.

Solution to Problem 5.18

(a) Suppose one has a conserved stress tensor $\partial T_{\mu\nu}/\partial x_\mu = 0$. Define the angular momentum tensor by

$$M_{\mu\nu\nu'} \equiv T_{\mu\nu}x_{\nu'} - T_{\mu\nu'}x_\nu$$

The four-divergence of this tensor is

$$\frac{\partial}{\partial x_\mu}M_{\mu\nu\nu'} = T_{\nu'\nu} - T_{\nu\nu'}$$

which vanishes if $T_{\mu\nu} = T_{\nu\mu}$ is *symmetric*.

So far, for the Dirac field we have been using the conserved stress tensor in Eq. (5.96) and corresponding four-momentum in Eq. (5.97)

$$T_{\mu\nu} = \hbar c\bar{\psi}\gamma_\mu\frac{\partial\psi}{\partial x_\nu}$$

$$P_\nu = \left(\mathbf{P}, \frac{i}{c}H\right) = \frac{1}{ic}\int_\Omega d^3x\,T_{4\nu}$$

---

[21]This comes from the equivalent lagrangian density $\tilde{\mathcal{L}} = \mathcal{L}_{\text{Dirac}} + (\hbar c/2)\partial(\bar{\psi}\gamma_\mu\psi)/\partial x_\mu$.
[22]*Hint*: Show that the extra term $(\tilde{T}_{\nu\mu} - \tilde{T}_{\mu\nu})/2$ does not contribute in the integral.

Here we have used the Dirac Eq. (5.94) along the dynamical trajectory

$$\left(\gamma_\mu \frac{\partial}{\partial x_\mu} + M\right)\psi = 0 \qquad ; \ \bar\psi\left(\gamma_\mu \frac{\overleftarrow{\partial}}{\partial x_\mu} - M\right) = 0$$

We now describe a procedure for obtaining a symmetric stress tensor in the Dirac case, while leaving the integrated four-momentum unaffected.

(b) The addition of a total divergence to the lagrangian density does not change Lagrange's equations, and hence the following is a perfectly equivalent starting lagrangian density

$$\tilde{\mathcal{L}} = \mathcal{L}_{\text{Dirac}} + \frac{\hbar c}{2}\frac{\partial}{\partial x_\mu}(\bar\psi\gamma_\mu\psi)$$

$$\mathcal{L}_{\text{Dirac}} = -\hbar c\,\bar\psi\left(\gamma_\mu\frac{\partial}{\partial x_\mu} + M\right)\psi$$

where we have identified $\mathcal{L}_{\text{Dirac}}$ from Eq. (5.92). Since the Dirac current is also conserved, this leads to the following stress tensor in Eq. (5.95)

$$\tilde{T}_{\mu\nu} = \frac{\hbar c}{2}\left[\bar\psi\gamma_\mu\frac{\partial\psi}{\partial x_\nu} - \frac{\partial\bar\psi}{\partial x_\nu}\gamma_\mu\psi\right]$$

The use of the Dirac equation gives

$$\frac{\partial}{\partial x_\mu}\tilde{T}_{\mu\nu} = \frac{\hbar c}{2}(M - M)\left[\bar\psi\frac{\partial\psi}{\partial x_\nu} - \frac{\partial\bar\psi}{\partial x_\nu}\psi\right] = 0$$

It follows directly from the Dirac equation that each component of the Dirac field satisfied the Klein-Gordon equation

$$\left(\Box - M^2\right)\psi = \left(\Box - M^2\right)\bar\psi = 0$$

Therefore $\tilde{T}_{\mu\nu}$ is also divergenceless with respect to the second index

$$\frac{\partial}{\partial x_\nu}\tilde{T}_{\mu\nu} = \frac{\hbar c}{2}M^2\left[\bar\psi\gamma_\mu\psi - \bar\psi\gamma_\mu\psi\right] = 0$$

Hence we establish the first property that the new stress tensor is divergenceless with respect to both indices

$$\frac{\partial\tilde{T}_{\mu\nu}}{\partial x_\mu} = \frac{\partial\tilde{T}_{\mu\nu}}{\partial x_\nu} = 0$$

Consider, for k a spatial index,

$$\frac{1}{ic}\int_\Omega d^3x\,\tilde{T}_{4k} = \frac{1}{ic}\int_\Omega d^3x\,\frac{\hbar c}{2}\left[\bar\psi\gamma_4\frac{\partial\psi}{\partial x_k} - \frac{\partial\bar\psi}{\partial x_k}\gamma_4\psi\right]$$

A partial integration, discarding the boundary terms, gives

$$\frac{1}{ic}\int_\Omega d^3x\,\tilde{T}_{4k} = \int_\Omega d^3x\,\psi^\dagger(x)\mathbf{p}\,\psi(x) \qquad ; \mathbf{p} = \frac{\hbar}{i}\boldsymbol{\nabla}$$

which is our previous result for the momentum.

A similar calculation for $\nu = 4$ gives, with the use of the Dirac equation,

$$\frac{1}{ic}\int_\Omega d^3x\,\tilde{T}_{44} = \frac{1}{ic}\int_\Omega d^3x\,\frac{\hbar c}{2}\left[\bar\psi\gamma_4\frac{\partial\psi}{\partial x_4} - \frac{\partial\bar\psi}{\partial x_4}\gamma_4\psi\right]$$

$$= \frac{1}{ic}\int_\Omega d^3x\,\frac{\hbar c}{2}\left[\bar\psi(-M-\boldsymbol{\gamma}\cdot\boldsymbol{\nabla})\psi - \bar\psi\left(M - \boldsymbol{\gamma}\cdot\overleftarrow{\boldsymbol{\nabla}}\right)\psi\right]$$

Another partial integration, and the use of $\boldsymbol{\gamma} = i\boldsymbol{\alpha}\beta$ and $\gamma_4 = \beta$, gives

$$-\int_\Omega d^3x\,\tilde{T}_{44} = \int_\Omega d^3x\,\psi^\dagger(x)\left[c\boldsymbol{\alpha}\cdot\mathbf{p} + \beta m_0 c^2\right]\psi(x) \qquad ; \mathbf{p} = \frac{\hbar}{i}\boldsymbol{\nabla}$$

This is our previous result for the hamiltonian. Hence we verify the second property that the four-momentum is unaltered

$$\int_\Omega d^3x\,\tilde{T}_{4\nu} = \int_\Omega d^3x\,T_{4\nu}$$

(c) Now take as the new stress tensor

$$\theta_{\mu\nu} \equiv \tilde{T}_{\mu\nu} + \frac{1}{2}(\tilde{T}_{\nu\mu} - \tilde{T}_{\mu\nu}) = \frac{1}{2}(\tilde{T}_{\mu\nu} + \tilde{T}_{\nu\mu})$$

It is evident that $\theta_{\mu\nu}$ is both symmetric and divergenceless

$$\theta_{\mu\nu} = \theta_{\nu\mu} \qquad ; \frac{\partial\theta_{\mu\nu}}{\partial x_\mu} = 0$$

Since $(\tilde{T}_{44} - \tilde{T}_{44})/2 \equiv 0$, the integral over all space of $\tilde{T}_{44}$ is unchanged

$$\int_\Omega d^3x\,\theta_{44} = \int_\Omega d^3x\,\tilde{T}_{44}$$

This is also true for the integral over $\tilde{T}_{4k}$, since

$$\int_\Omega d^3x\,\frac{1}{2}(\tilde{T}_{k4} - \tilde{T}_{4k}) = \int_\Omega d^3x\,\frac{\hbar c}{4}\left[\bar\psi\gamma_k\frac{\partial\psi}{\partial x_4} - \frac{\partial\bar\psi}{\partial x_4}\gamma_k\psi - \bar\psi\gamma_4\frac{\partial\psi}{\partial x_k} + \frac{\partial\bar\psi}{\partial x_k}\gamma_4\psi\right]$$

From the Dirac equation

$$\left(\frac{\partial}{\partial x_4} + \gamma_4\boldsymbol{\gamma}\cdot\boldsymbol{\nabla} + \gamma_4 M\right)\psi = 0 \qquad ; \bar\psi\left(\frac{\overleftarrow{\partial}}{\partial x_4} + \boldsymbol{\gamma}\cdot\overleftarrow{\boldsymbol{\nabla}}\gamma_4 - M\gamma_4\right) = 0$$

Thus

$$\int_\Omega d^3x \, \frac{1}{2}(\tilde{T}_{k4} - \tilde{T}_{4k}) = \int_\Omega d^3x \, \frac{\hbar c}{4} \left[-\bar{\psi}\gamma_4 \frac{\partial \psi}{\partial x_k} + \frac{\partial \bar{\psi}}{\partial x_k}\gamma_4\psi \right.$$
$$\left. -\bar{\psi}\gamma_k \left(\gamma_4 \boldsymbol{\gamma} \cdot \boldsymbol{\nabla} + \gamma_4 M \right) \psi + \bar{\psi} \left(\boldsymbol{\gamma} \cdot \overleftarrow{\boldsymbol{\nabla}} \gamma_4 - M\gamma_4 \right) \gamma_k\psi \right]$$

Partial integration gives[23]

$$\int_\Omega d^3x \, \frac{1}{2}(\tilde{T}_{k4} - \tilde{T}_{4k}) = \int_\Omega d^3x \, \frac{\hbar c}{4} \left[-\bar{\psi}\gamma_4 \frac{\partial \psi}{\partial x_k} - \bar{\psi}\gamma_4 \frac{\partial \psi}{\partial x_k} \right.$$
$$\left. -\bar{\psi}\gamma_k\gamma_4 \boldsymbol{\gamma} \cdot \boldsymbol{\nabla}\psi - \bar{\psi}\boldsymbol{\gamma}\gamma_4\gamma_k \cdot \boldsymbol{\nabla}\psi \right]$$
$$= \int_\Omega d^3x \, \frac{\hbar c}{2} \left[-\bar{\psi}\gamma_4 \frac{\partial \psi}{\partial x_k} + \bar{\psi}\gamma_4 \frac{\partial \psi}{\partial x_k} \right] = 0$$

Hence we have the desired result that the four-momentum remains unaltered

$$\int_\Omega d^3x \, \theta_{4\nu} = \int_\Omega d^3x \, \tilde{T}_{4\nu} = \int_\Omega d^3x \, T_{4\nu}$$

Problem 5.19 Assume that under a Lorentz transformation the Dirac field transforms as $\psi(x) \to S\psi(x')$ where S is a 4×4 matrix with the properties in Prob. E.4.

(a) Show the lagrangian density in Eq. (5.92) is then a Lorentz scalar;

(b) Show the resulting Dirac equation is then Lorentz invariant.

Solution to Problem 5.19

(a) We assume that the Dirac field transforms according to

$$\hat{U}\hat{\psi}(x_\mu)\hat{U}^{-1} = \mathcal{S}(\Omega)\,\hat{\psi}(x'_\mu) \qquad ; \; x'_\mu = a_{\mu\nu}(-v)x_\nu$$

where[24]

$$\gamma_4 S^\dagger \gamma_4 = S^{-1}$$
$$S\gamma_\mu S^{-1} = a_{\mu\nu}(-v)\gamma_\nu$$

---

[23] Use $\gamma_k\gamma_4 + \gamma_4\gamma_k = 0$, and $\gamma_k\gamma_4\gamma_j + \gamma_j\gamma_4\gamma_k = -2\delta_{kj}\gamma_4$.

[24] These expressions for the spinor transformation differ from those in Prob. E.4 through the sign of v (see [Walecka (2013)]); it follows from the argument in Eqs. (E.34)–(E.41) that to produce $x'_\mu = a_{\mu\nu}(-v)x_\nu$, one must use $\tan i\Omega = +iv/c$. This requirement was overlooked in the formulation of Prob. E.4, which, nevertheless, remains an essential exercise. Note that $a_{\mu\nu}(v) = a_{\nu\mu}(-v)$.

The Lorentz transformation is orthogonal, with

$$a_{\mu\nu}a_{\mu\sigma} = a_{\nu\mu}a_{\sigma\mu} = \delta_{\nu\sigma}$$

Therefore

$$S^{-1}\gamma_\mu S = a_{\nu\mu}(-v)\gamma_\nu$$

Furthermore, the gradient transforms according to

$$\frac{\partial}{\partial x_\mu} = \frac{\partial x_\nu'}{\partial x_\mu}\frac{\partial}{\partial x_\nu'} = a_{\nu\mu}(-v)\frac{\partial}{\partial x_\nu'}$$

$$\frac{\partial}{\partial x_\mu'} = a_{\mu\nu}(-v)\frac{\partial}{\partial x_\nu}$$

The lagrangian density in Eq. (5.92) then transforms as

$$\hat{U}\hat{\mathcal{L}}(x)\,\hat{U}^{-1} = -\hbar c\,\hat{\psi}^\dagger(x')S^\dagger\gamma_4\left(\gamma_\mu\frac{\partial}{\partial x_\mu} + M\right)S\hat{\psi}(x')$$

$$= -\hbar c\,\hat{\bar{\psi}}(x')S^{-1}\left(\gamma_\mu\frac{\partial}{\partial x_\mu} + M\right)S\hat{\psi}(x')$$

$$= -\hbar c\,\hat{\bar{\psi}}(x')\left[a_{\nu\mu}(-v)\gamma_\nu\frac{\partial}{\partial x_\mu} + M\right]\hat{\psi}(x')$$

$$= -\hbar c\,\hat{\bar{\psi}}(x')\left[\gamma_\nu\frac{\partial}{\partial x_\nu'} + M\right]\hat{\psi}(x') = \hat{\mathcal{L}}(x')$$

This lagrangian density is a scalar under the Lorentz transformation.

(b) Under a Lorentz transformation, the Dirac wave function transforms according to

$$\psi(x) = S\psi'(x') \qquad ; \quad x_\mu' = a_{\mu\nu}(-v)x_\nu$$

The Dirac equation is therefore changed to

$$\left(\gamma_\mu\frac{\partial}{\partial x_\mu} + M\right)S\psi'(x') = 0$$

Multiply this equation on the left by S^{-1}, and use the results from part (a)

$$S^{-1}\left(\gamma_\mu\frac{\partial}{\partial x_\mu} + M\right)S\psi'(x') = \left[a_{\nu\mu}(-v)\gamma_\nu\frac{\partial}{\partial x_\mu} + M\right]\psi'(x')$$

$$= \left[\gamma_\nu\frac{\partial}{\partial x_\nu'} + M\right]\psi'(x')$$

Hence the Dirac equation is Lorentz invariant[25]

$$\left[\gamma_\nu \frac{\partial}{\partial x'_\nu} + M \right] \psi'(x') = 0$$

Problem 5.20 Show that the term that is subtracted off in the normal-ordered current in Eq. (5.116) simply counts the number of negative energy-states.

Solution to Problem 5.20

When the Dirac field expansion in Eq. (5.102) is substituted into the current in Eq. (5.115), the antiparticle term is normal-ordered according to

$$[J_\mu(x)]_{\text{a.p.}} \doteq \frac{i}{\Omega} \sum_{\mathbf{k}\lambda} \sum_{\mathbf{k}'\lambda'} \left[\bar{v}_\lambda(-\mathbf{k}) e^{i(\mathbf{k}\cdot\mathbf{x} - \omega_k t)} \right] \gamma_\mu \left[v_{\lambda'}(-\mathbf{k}') e^{-i(\mathbf{k}'\cdot\mathbf{x} - \omega_{k'} t)} \right] \times$$
$$\left[\delta_{\mathbf{k}\mathbf{k}'} \delta_{\lambda\lambda'} - b^\dagger_{\mathbf{k}'\lambda'} b_{\mathbf{k}\lambda} \right]$$

The remaining c-number contribution $\delta J_\mu(x)$ is therefore given by

$$\delta J_\mu(x) = \frac{i}{\Omega} \sum_{\mathbf{k}\lambda} \bar{v}_\lambda(-\mathbf{k}) \gamma_\mu v_\lambda(-\mathbf{k})$$

Now observe that the diagonal current has only a fourth component

$$\bar{v}_\lambda(-\mathbf{k}) \gamma_\mu v_\lambda(-\mathbf{k}) = \delta_{\mu 4} \, v^\dagger_\lambda(-\mathbf{k}) v_\lambda(-\mathbf{k})$$

and make use of the normalization of the Dirac spinors

$$v^\dagger_\lambda(-\mathbf{k}) v_\lambda(-\mathbf{k}) = 1$$

It follows that

$$\delta J_\mu(x) = i\delta_{\mu 4} \frac{1}{\Omega} \sum_{\mathbf{k}\lambda} 1$$

This term is independent of the dynamics, and simply counts the number of negative-energy states in computing the density.[26]

Problem 5.21 Show that $g^2/4\pi\hbar c^3$ is *dimensionless* in Eq. (5.119).

---

[25] Under the spinor transformation law, the quantity $\gamma_\mu \, \partial/\partial x_\mu$ behaves as the scalar product of two four-vectors. Note the transformed Dirac wave function is $\psi'(x') = S^{-1}\psi(x)$.

[26] In Dirac hole theory, each one of these negative-energy states would actually be filled by a particle.

Solution to Problem 5.21

The Dirac-scalar interaction in Eq. (5.119) is

$$\mathcal{L}_1 = g\bar{\psi}\psi\phi$$

We make use of the relation between the scalar field and inverse Compton wavelength of the Dirac particle in Eqs. (5.117)–(5.118)

$$M \to M - \frac{g}{\hbar c}\phi$$

as well as the lagrangian density for the scalar field in Eq. (5.117)

$$\mathcal{L}_0^{\text{scalar}} = -\frac{c^2}{2}\left[\left(\frac{\partial\phi}{\partial x_\mu}\right)^2 + m_s^2\phi^2\right]$$

Here m_s is the inverse Compton wavelength for the scalar field.

The dimension, indicated with square brackets, of the lagrangian density is the same as that of the hamiltonian density, which is energy/volume

$$[\mathcal{L}] = [\mathcal{H}] = [EL^{-3}] = [\hbar cL^{-4}]$$

The dimension of the scalar field follows from the above mass relation

$$[\phi] = [\hbar cMg^{-1}]$$

From the mass term for the scalar field, we have

$$[c^2m_s^2\phi^2] = [\hbar c\, m_s^2M^2(\hbar c^3g^{-2})] = [\hbar cL^{-4}]$$

We conclude from this that $(\hbar c^3g^{-2})$ is dimensionless

$$\frac{g^2}{4\pi\hbar c^3} \text{ is dimensionless}$$

Problem 5.22 Insert the field expansions, and discuss the processes described by the interaction hamiltonian $\hat{H}_I(t)$ for the Dirac-scalar theory in Eq. (5.121).

Solution to Problem 5.22

We will insert the interaction-picture field expansions

$$\hat{\phi}(\mathbf{x}, t) = \sum_{\mathbf{k}} \left(\frac{\hbar}{2\omega_k^s \Omega} \right)^{1/2} \left[c_{\mathbf{k}} e^{i(\mathbf{k}\cdot\mathbf{x} - \omega_k^s t)} + c_{\mathbf{k}}^\dagger e^{-i(\mathbf{k}\cdot\mathbf{x} - \omega_k^s t)} \right]$$

$$\hat{\psi}(\mathbf{x}, t) = \frac{1}{\sqrt{\Omega}} \sum_{\mathbf{k}} \sum_{\lambda} \left[a_{\mathbf{k}\lambda} u_\lambda(\mathbf{k}) e^{i(\mathbf{k}\cdot\mathbf{x} - \omega_k t)} + b_{\mathbf{k}\lambda}^\dagger v_\lambda(-\mathbf{k}) e^{-i(\mathbf{k}\cdot\mathbf{x} - \omega_k t)} \right]$$

in the interaction hamiltonian

$$\hat{H}_I(t) = -g \int_\Omega d^3x \, \hat{\bar{\psi}}(\mathbf{x}, t) \hat{\psi}(\mathbf{x}, t) \hat{\phi}(\mathbf{x}, t) = \sum_{i=1}^8 \hat{H}_I^{(i)}(t)$$

where $\hat{H}_I^{(i)}(t)$, with $i = 1, \ldots, 8$, correspond to the different contributions obtained from the product of the fields.[27] For instance, the first term is

$$\hat{H}_I^{(1)}(t) \equiv -g \sum_{\mathbf{k},\mathbf{k}',\mathbf{q}} \sum_{\lambda,\lambda'} \left(\frac{\hbar}{2\omega_q^s \Omega} \right)^{1/2} a_{\mathbf{k}'\lambda'}^\dagger a_{\mathbf{k}\lambda} c_{\mathbf{q}} \, \bar{u}_{\lambda'}(\mathbf{k}') u_\lambda(\mathbf{k}) \times$$

$$\frac{1}{\Omega} \int_\Omega d^3x \, e^{-i(\mathbf{k}'\cdot\mathbf{x} - \omega_{k'} t)} e^{i(\mathbf{k}\cdot\mathbf{x} - \omega_k t)} e^{i(\mathbf{q}\cdot\mathbf{x} - \omega_q^s t)}$$

$$= -g \sum_{\mathbf{k},\mathbf{q}} \sum_{\lambda,\lambda'} \left(\frac{\hbar}{2\omega_q^s \Omega} \right)^{1/2} a_{\mathbf{k}+\mathbf{q},\lambda'}^\dagger a_{\mathbf{k}\lambda} c_{\mathbf{q}} \, \bar{u}_{\lambda'}(\mathbf{k} + \mathbf{q}) u_\lambda(\mathbf{k}) \times$$

$$e^{i(\omega_{|\mathbf{q}+\mathbf{k}|} - \omega_k - \omega_q^s)t}$$

It describes all the physical processes where a fermion can absorb a scalar, conserving the total three-momentum and possibly changing helicity. Notice, in fact, that the operator $a_{\mathbf{k}\lambda}$ destroys a fermion of momentum $\mathbf{k}$ and helicity λ, while $c_{\mathbf{q}}$ destroys a scalar of momentum $\mathbf{q}$, and $a_{\mathbf{k}+\mathbf{q},\lambda'}^\dagger$ creates a fermion of momentum $\mathbf{k} + \mathbf{q}$ and helicity λ'.

Similarly, we have

$$\hat{H}_I^{(2)}(t) \equiv -g \sum_{\mathbf{k},\mathbf{k}',\mathbf{q}} \sum_{\lambda,\lambda'} \left(\frac{\hbar}{2\omega_q^s \Omega} \right)^{1/2} a_{\mathbf{k}'\lambda'}^\dagger a_{\mathbf{k}\lambda} c_{\mathbf{q}}^\dagger \, \bar{u}_{\lambda'}(\mathbf{k}') u_\lambda(\mathbf{k}) \times$$

$$\frac{1}{\Omega} \int_\Omega d^3x \, e^{-i(\mathbf{k}'\cdot\mathbf{x} - \omega_{k'} t)} e^{i(\mathbf{k}\cdot\mathbf{x} - \omega_k t)} e^{-i(\mathbf{q}\cdot\mathbf{x} - \omega_q^s t)}$$

$$= -g \sum_{\mathbf{k},\mathbf{q}} \sum_{\lambda,\lambda'} \left(\frac{\hbar}{2\omega_q^s \Omega} \right)^{1/2} a_{\mathbf{k}-\mathbf{q},\lambda'}^\dagger a_{\mathbf{k}\lambda} c_{\mathbf{q}}^\dagger \, \bar{u}_{\lambda'}(\mathbf{k} - \mathbf{q}) u_\lambda(\mathbf{k}) \times$$

$$e^{i(\omega_{|\mathbf{q}-\mathbf{k}|} - \omega_k + \omega_q^s)t}$$

---

[27]Here $\omega_k^s = c\sqrt{k^2 + m_s^2}$, and $\omega_k = c\sqrt{k^2 + M^2}$.

which describes all the physical processes where a fermion can emit a scalar, conserving the total three-momentum, and possibly changing its helicity. In this case the operators $a^\dagger_{\mathbf{k}-\mathbf{q},\lambda'} a_{\mathbf{k}\lambda} c^\dagger_{\mathbf{q}}$ destroy a fermion of momentum $\mathbf{k}$ and helicity λ, create a fermion of momentum $\mathbf{k} - \mathbf{q}$ and helicity λ', and create a scalar of momentum $\mathbf{q}$.

In a similar way we can obtain all the remaining contributions:

$$\hat{H}_I^{(3)}(t) = -g \sum_{\mathbf{k},\mathbf{q}} \sum_{\lambda,\lambda'} \left(\frac{\hbar}{2\omega_q^s \Omega} \right)^{1/2} a^\dagger_{\mathbf{q}-\mathbf{k},\lambda'} b^\dagger_{\mathbf{k}\lambda} c_{\mathbf{q}} \, \bar{u}_{\lambda'}(\mathbf{q} - \mathbf{k}) v_\lambda(-\mathbf{k}) \times$$
$$e^{i(\omega_{|\mathbf{q}-\mathbf{k}|}+\omega_k-\omega_q^s)t}$$

$$\hat{H}_I^{(4)}(t) = -g \sum_{\mathbf{k},\mathbf{q}} \sum_{\lambda,\lambda'} \left(\frac{\hbar}{2\omega_q^s \Omega} \right)^{1/2} a^\dagger_{-\mathbf{q}-\mathbf{k},\lambda'} b^\dagger_{\mathbf{k}\lambda} c^\dagger_{\mathbf{q}} \, \bar{u}_{\lambda'}(-\mathbf{q} - \mathbf{k}) v_\lambda(-\mathbf{k}) \times$$
$$e^{i(\omega_{|\mathbf{q}+\mathbf{k}|}+\omega_k+\omega_q^s)t}$$

$$\hat{H}_I^{(5)}(t) = -g \sum_{\mathbf{k},\mathbf{q}} \sum_{\lambda,\lambda'} \left(\frac{\hbar}{2\omega_q^s \Omega} \right)^{1/2} b_{-\mathbf{q}-\mathbf{k},\lambda'} a_{\mathbf{k}\lambda} c_{\mathbf{q}} \, \bar{v}_{\lambda'}(\mathbf{q} + \mathbf{k}) u_\lambda(\mathbf{k}) \times$$
$$e^{-i(\omega_{|\mathbf{q}+\mathbf{k}|}+\omega_k+\omega_q^s)t}$$

$$\hat{H}_I^{(6)}(t) = -g \sum_{\mathbf{k},\mathbf{q}} \sum_{\lambda,\lambda'} \left(\frac{\hbar}{2\omega_q^s \Omega} \right)^{1/2} b_{\mathbf{q}-\mathbf{k},\lambda'} a_{\mathbf{k}\lambda} c^\dagger_{\mathbf{q}} \, \bar{v}_{\lambda'}(\mathbf{k} - \mathbf{q}) u_\lambda(\mathbf{k}) \times$$
$$e^{-i(\omega_{|\mathbf{q}-\mathbf{k}|}+\omega_k-\omega_q^s)t}$$

$$\hat{H}_I^{(7)}(t) = -g \sum_{\mathbf{k},\mathbf{q}} \sum_{\lambda,\lambda'} \left(\frac{\hbar}{2\omega_q^s \Omega} \right)^{1/2} b_{\mathbf{k}-\mathbf{q},\lambda'} b^\dagger_{\mathbf{k}\lambda} c_{\mathbf{q}} \, \bar{v}_{\lambda'}(\mathbf{q} - \mathbf{k}) v_\lambda(-\mathbf{k}) \times$$
$$e^{-i(\omega_{|\mathbf{k}-\mathbf{q}|}-\omega_k+\omega_q^s)t}$$

$$\hat{H}_I^{(8)}(t) = -g \sum_{\mathbf{k},\mathbf{q}} \sum_{\lambda,\lambda'} \left(\frac{\hbar}{2\omega_q^s \Omega} \right)^{1/2} b_{\mathbf{k}+\mathbf{q},\lambda'} b^\dagger_{\mathbf{k}\lambda} c^\dagger_{\mathbf{q}} \, \bar{v}_{\lambda'}(-\mathbf{k} - \mathbf{q}) v_\lambda(-\mathbf{k}) \times$$
$$e^{-i(\omega_{|\mathbf{q}+\mathbf{k}|}-\omega_k-\omega_q^s)t}$$

where $\hat{H}_I^{(3)}(t)$ creates a fermion and an antifermion, destroying a scalar, $\hat{H}_I^{(4)}(t)$ creates a fermion, an antifermion and a scalar, $\hat{H}_I^{(5)}(t)$ destroys a fermion, an antifermion and a scalar, $\hat{H}_I^{(6)}(t)$ destroys a fermion and an antifermion, creating a scalar, $\hat{H}_I^{(7)}(t)$ destroys an antifermion and a scalar, and creates an antifermion, and $\hat{H}_I^{(8)}(t)$ destroys an antifermion, and creates a scalar and an antifermion. The three-momentum is conserved in each interaction.

Problem 5.23 Show that the canonical equal-time commutation relations in Eq. (5.53) are picture-independent.

Solution to Problem 5.23

The canonical equal-time commutation relations in the interaction picture for the string are given in Eqs. (5.53) as

$$[\hat{q}(x,t),\, \hat{\pi}(x',t')]_{t=t'} = i\hbar\delta(x - x')$$
$$[\hat{q}(x,t),\, \hat{q}(x',t')]_{t=t'} = [\hat{\pi}(x,t),\, \hat{\pi}(x',t')]_{t=t'} = 0$$

The relation between the time-dependent operators in the interaction picture and time-independent operators in the Schrödinger picture is

$$\hat{q}(x,t) = e^{i\hat{H}_0 t/\hbar}\hat{q}(x)e^{-i\hat{H}_0 t/\hbar} \qquad ;\ \hat{\pi}(x,t) = e^{i\hat{H}_0 t/\hbar}\hat{\pi}(x)e^{-i\hat{H}_0 t/\hbar}$$

Substitution into the above then gives, for example,

$$[\hat{q}(x,t),\, \hat{\pi}(x',t')]_{t=t'} = e^{i\hat{H}_0 t/\hbar}[\hat{q}(x),\, \hat{\pi}(x')]e^{-i\hat{H}_0 t/\hbar}$$
$$= [\hat{q}(x),\, \hat{\pi}(x')]$$

where the last result holds since the commutator is a c-number. Hence

$$[\hat{q}(x),\, \hat{\pi}(x')] = i\hbar\delta(x - x')$$

which is the canonical commutation relation in the Schrödinger picture.

Clearly the same demonstration holds if we replace $\hat{H}_0 \to \hat{H}$, with a time-independent $\hat{H}$, and proceed from the canonical equal-time commutation relations in the Heisenberg picture. Hence the canonical equal-time commutation relations in Eq. (5.53) are picture-independent.

Problem 5.24 Consider the interacting Dirac-scalar field theory of Eqs. (5.117) and (5.118).

(a) Write Lagrange's equations for each of the fields;

(b) In the Schrödinger picture, operators are independent of time, and their time derivative is defined through Ehrenfest's theorem[28]

$$\frac{d\hat{O}}{dt} \equiv \frac{i}{\hbar}[\hat{H},\, \hat{O}]$$

The expectation value of this operator taken with the time-dependent Schrödinger state vector gives the time derivative of the expectation value

---

[28]See Probs. 2.3 and 5.3. It is assumed here that $\hat{O}$ has no additional explicit time dependence.

of the operator. Take one time derivative of the scalar field, make use of the canonical commutation relations in the Schrödinger picture [the analogs of Eqs. (5.61)], and derive the operator form of Eq. (5.77);

(c) Compute the second time derivative, and derive the operator form of the scalar field equation in part (a);

(d) Discuss under what conditions the correspondence principle applies, that is, when can the quantum field be replaced by the corresponding classical field. (Compare Prob. 5.3.)

Solution to Problem 5.24

(a) From Eqs. (5.117)–(5.119), the lagrangian density for this system is

$$\mathcal{L} = -\hbar c\, \bar{\psi} \left[\gamma_\mu \frac{\partial}{\partial x_\mu} + \left(M - \frac{g\phi}{\hbar c} \right) \right] \psi - \frac{c^2}{2} \left[\left(\frac{\partial \phi}{\partial x_\mu} \right)^2 + m_s^2 \phi^2 \right]$$

Lagrange's equations for the Dirac fields are given in Eqs. (5.93). In analogy to Eqs. (5.94), they read

$$\left[\gamma_\mu \frac{\partial}{\partial x_\mu} + \left(M - \frac{g\phi}{\hbar c} \right) \right] \psi = 0 \qquad ; \bar{\psi} \text{ eqn}$$

$$\bar{\psi} \left[\gamma_\mu \frac{\overleftarrow{\partial}}{\partial x_\mu} - \left(M - \frac{g\phi}{\hbar c} \right) \right] = 0 \qquad ; \psi \text{ eqn}$$

Lagrange's equation for the scalar field is given in Eq. (5.58). In analogy to Eq. (5.64), it reads

$$(\Box - m_s^2)\phi = -\frac{g}{c^2}\bar{\psi}\psi \qquad ; \Box = \nabla^2 - \frac{1}{c^2}\frac{\partial^2}{\partial t^2}$$

(b) The hamiltonian for the quantized system is given in Eqs. (5.120)

$$\hat{H} = \hat{H}_0^{\text{Dirac}} + \hat{H}_0^{\text{scalar}} + \hat{H}_1$$

$$\hat{H}_1 = -g \int_\Omega d^3x\, \hat{\bar{\psi}}(\mathbf{x})\hat{\psi}(\mathbf{x})\hat{\phi}(\mathbf{x})$$

From Eqs. (5.61), the commutator of the scalar field at two distinct spatial points vanishes

$$[\hat{\phi}(\mathbf{x}),\, \hat{\phi}(\mathbf{x}')] = 0$$

Thus Ehrenfest's theorem gives for the first time derivative of the scalar field

$$\frac{d\hat{\phi}}{dt} = \frac{i}{\hbar}[\hat{H}, \hat{\phi}(\mathbf{x})] = \frac{i}{\hbar}[\hat{H}_0^{\text{scalar}}, \hat{\phi}(\mathbf{x})] = \hat{\pi}(\mathbf{x})$$

Here $\hat{\pi}(\mathbf{x})$ is the same as in the free-field case; it is given by $\hat{\pi}(\mathbf{x}, 0)$ in Eqs. (5.60)–(5.65). This is the operator form of Eq. (5.77).

(c) A second time derivative follows as

$$\frac{d^2\hat{\phi}}{dt^2} = \left(\frac{i}{\hbar}\right)^2 [\hat{H}, [\hat{H}, \hat{\phi}(\mathbf{x})]] = \frac{i}{\hbar}[\hat{H}, \hat{\pi}(\mathbf{x})]$$

$$= \frac{i}{\hbar}[\hat{H}_0^{\text{scalar}}, \hat{\pi}(\mathbf{x})] + \frac{i}{\hbar}[\hat{H}_1, \hat{\pi}(\mathbf{x})]$$

The first term is again exactly the same as in the free-field case

$$\frac{i}{\hbar}[\hat{H}_0^{\text{scalar}}, \hat{\pi}(\mathbf{x})] = \left(\frac{i}{\hbar}\right)^2 [\hat{H}_0^{\text{scalar}}, [\hat{H}_0^{\text{scalar}}, \hat{\phi}(\mathbf{x})]] = c^2(\nabla^2 - m_s^2)\hat{\phi}(\mathbf{x})$$

The second term in the above is evaluated with the aid of the canonical commutation relation in Eqs. (5.61)

$$[\hat{\phi}(\mathbf{x}), \hat{\pi}(\mathbf{x}')] = i\hbar\,\delta^{(3)}(\mathbf{x} - \mathbf{x}')$$

This gives

$$\frac{i}{\hbar}[\hat{H}_1, \hat{\pi}(\mathbf{x})] = g\hat{\bar{\psi}}(\mathbf{x})\hat{\psi}(\mathbf{x})$$

Ehrenfest's theorem therefore gives

$$(\nabla^2 - m_s^2)\hat{\phi}(\mathbf{x}) - \frac{1}{c^2}\left(\frac{i}{\hbar}\right)^2 [\hat{H}, [\hat{H}, \hat{\phi}(\mathbf{x})]] = -\frac{g}{c^2}\hat{\bar{\psi}}(\mathbf{x})\hat{\psi}(\mathbf{x})$$

This is the operator equation of motion in the Schrödinger picture corresponding to Lagrange's equation for ϕ.

(d) The classical field is defined by the expectation value

$$\langle\Psi(t)|\hat{\phi}(\mathbf{x})|\Psi(t)\rangle \equiv \phi_C(\mathbf{x}, t)$$

where $|\Psi(t)\rangle$ is the (normalized) solution to the Schrödinger equation. This quantity can be treated with classical field theory provided there are a sufficient number of quanta present in the modes of interest.[29]

---

[29] Recall there is an uncertainty relation between classical field strength and the number of quanta (see the solution to Prob. 7.7 in [Amore and Walecka (2014)]).

Problem 5.25 There is an alternate formulation of classical mechanics in terms of *Poisson brackets*, which bears a close analogy to quantum mechanics. Given a function $F(q^1, \cdots, q^n; p^1, \cdots p^n; t)$, and a similar function G, their Poisson bracket is defined by[30]

$$[F, G]_{\rm PB} \equiv \sum_{\sigma=1}^{n} \left(\frac{\partial F}{\partial q_\sigma} \frac{\partial G}{\partial p_\sigma} - \frac{\partial F}{\partial p_\sigma} \frac{\partial G}{\partial q_\sigma} \right) \quad ; \text{ Poisson bracket}$$

where the partial derivatives imply that all the other variables in the arguments of $F(q^1, \cdots, q^n; p^1, \cdots p^n; t)$ and G are to be held fixed.

(a) Show that Hamilton's equations take the form

$$\dot{q}_\sigma = -[H, q_\sigma]_{\rm PB} = \frac{\partial H}{\partial p_\sigma} \qquad ; \sigma = 1, \cdots, n$$

$$\dot{p}_\sigma = -[H, p_\sigma]_{\rm PB} = -\frac{\partial H}{\partial q_\sigma}$$

(b) Show that the total time derivative of F is given by

$$\frac{dF}{dt} = -[H, F]_{\rm PB} + \frac{\partial F}{\partial t}$$

(c) Show that

$$[p_\sigma, q_\rho]_{\rm PB} = -\delta_{\sigma\rho}$$

$$[p_\sigma, p_\rho]_{\rm PB} = [q_\sigma, q_\rho]_{\rm PB} = 0$$

(d) Compare with the expressions in Eqs. (5.37) and Prob. 5.24(b). Hence make the argument that the formal transition from classical to quantum mechanics is made through the replacement of a Poisson bracket by a commutator as follows[31]

$$[A, B]_{\rm PB} \to \frac{1}{i\hbar}[\hat{A}, \hat{B}]$$

Solution to Problem 5.25

(a) The Poisson bracket is defined in the statement of the problem. We proceed to calculate the Poisson brackets of the hamiltonian

---

[30] Here we do not distinguish between upper and lower indices on the coordinates and momenta.

[31] See [Fetter and Walecka (2003)].

$H(q^1, \cdots, q^n; p^1, \cdots p^n; t)$ with the coordinates (q_σ, p_σ) themselves

$$[H, q_\sigma]_{\text{PB}} = \sum_{\rho=1}^{n} \left[\frac{\partial H}{\partial q_\rho} \frac{\partial q_\sigma}{\partial p_\rho} - \frac{\partial H}{\partial p_\rho} \frac{\partial q_\sigma}{\partial q_\rho} \right] = -\frac{\partial H}{\partial p_\sigma}$$

$$[H, p_\sigma]_{\text{PB}} = \sum_{\rho=1}^{n} \left[\frac{\partial H}{\partial q_\rho} \frac{\partial p_\sigma}{\partial p_\rho} - \frac{\partial H}{\partial p_\rho} \frac{\partial p_\sigma}{\partial q_\rho} \right] = \frac{\partial H}{\partial q_\sigma}$$

where we have used the properties

$$\frac{\partial q_\sigma}{\partial q_\rho} = \frac{\partial p_\sigma}{\partial p_\rho} = \delta_{\rho\sigma} \qquad ; \quad \frac{\partial q_\sigma}{\partial p_\rho} = \frac{\partial p_\sigma}{\partial q_\rho} = 0$$

As a result, Hamilton's equations take the form[32]

$$\dot{q}_\sigma = -[H, q_\sigma]_{\text{PB}} = \frac{\partial H}{\partial p_\sigma} \qquad ; \quad \sigma = 1, \cdots, n$$

$$\dot{p}_\sigma = -[H, p_\sigma]_{\text{PB}} = -\frac{\partial H}{\partial q_\sigma}$$

(b) Let us explicitly calculate the Poisson bracket of the hamiltonian $H(q^1, \cdots, q^n; p^1, \cdots p^n; t)$ with a function $F(q^1, \cdots, q^n; p^1, \cdots p^n; t)$

$$[H, F]_{\text{PB}} = \sum_{\rho=1}^{n} \left[\frac{\partial H}{\partial q_\rho} \frac{\partial F}{\partial p_\rho} - \frac{\partial H}{\partial p_\rho} \frac{\partial F}{\partial q_\rho} \right]$$

$$= -\sum_{\rho=1}^{n} \left[\dot{p}_\rho \frac{\partial F}{\partial p_\rho} + \dot{q}_\rho \frac{\partial F}{\partial q_\rho} \right]$$

where we have used Hamilton's equations. Therefore, the total time derivative of F is given by

$$\frac{dF}{dt} = \sum_{\rho=1}^{n} \left[\dot{p}_\rho \frac{\partial F}{\partial p_\rho} + \dot{q}_\rho \frac{\partial F}{\partial q_\rho} \right] + \frac{\partial F}{\partial t}$$

$$= -[H, F]_{\text{PB}} + \frac{\partial F}{\partial t}$$

(c) The elementary Poisson brackets of p_σ and q_σ themselves are given

---

[32] Here $dq_\sigma/dt \equiv \dot{q}_\sigma$ and $dp_\sigma/dt \equiv \dot{p}_\sigma$.

by

$$[p_\sigma, q_\rho]_{\text{PB}} = \sum_{\lambda=1}^{n} \left[\frac{\partial p_\sigma}{\partial q_\lambda} \frac{\partial q_\rho}{\partial p_\lambda} - \frac{\partial p_\sigma}{\partial p_\lambda} \frac{\partial q_\rho}{\partial q_\lambda} \right] = -\delta_{\sigma\rho}$$

$$[p_\sigma, p_\rho]_{\text{PB}} = \sum_{\lambda=1}^{n} \left[\frac{\partial p_\sigma}{\partial q_\lambda} \frac{\partial p_\rho}{\partial p_\lambda} - \frac{\partial p_\sigma}{\partial p_\lambda} \frac{\partial p_\rho}{\partial q_\lambda} \right] = 0$$

$$[q_\sigma, q_\rho]_{\text{PB}} = \sum_{\lambda=1}^{n} \left[\frac{\partial q_\sigma}{\partial q_\lambda} \frac{\partial q_\rho}{\partial p_\lambda} - \frac{\partial q_\sigma}{\partial p_\lambda} \frac{\partial q_\rho}{\partial q_\lambda} \right] = 0$$

(d) Equations (5.37) are the quantum commutation relations obeyed by the coordinates and their conjugate momenta, which are both operators; the Poisson brackets evaluated in (c) are obeyed by the classical coordinates and their conjugate momenta, which are not operators. The passage from the Poisson brackets of part (c) to the quantum commutators in Eq. (5.37) can be formally obtained with the substitution

$$[p_\sigma, q_\rho]_{\text{PB}} \rightarrow \frac{1}{i\hbar} [\hat{p}_\sigma, \hat{q}_\rho]$$

In a classical description, the Poisson brackets of the hamiltonian with the coordinates and their conjugate momenta can be used to obtain the dynamics of the system since, as seen in part (a), Hamilton's equations can be expressed directly in terms of Poisson brackets. Given a function of the coordinates and momenta, $O(q_1, \ldots, q_n, p_1, \ldots, p_n; t)$, its time derivative is simply

$$\frac{dO}{dt} \equiv -[H, O]_{PB} + \frac{\partial O}{\partial t} \qquad ; \text{ Poisson bracket}$$

In a quantum description, on the other hand, the time derivative of an operator $\hat{O}$ in the Schrödinger picture can be expressed in terms of a commutator through Ehrenfest's theorem

$$\frac{d\hat{O}}{dt} \equiv \frac{i}{\hbar} [\hat{H}, \hat{O}] + \frac{\partial \hat{O}}{\partial t} \qquad ; \text{ Ehrenfest's theorem}$$

Therefore we see that the passage from the classical to the corresponding quantum equation is obtained with the replacement

$$[H, O]_{\text{PB}} \rightarrow \frac{1}{i\hbar} [\hat{H}, \hat{O}]$$

****

(*Aside*) The replacement can be extended to the case of two arbitrary functions A and B of the coordinates and of their conjugate momenta

$$[A, B]_{\text{PB}} \to \frac{1}{i\hbar}[\hat{A}, \hat{B}]$$

Since $A = A(q_1, \ldots, q_n, p_1, \ldots, p_n)$ and $B = B(q_1, \ldots, q_n, p_1, \ldots, p_n)$ are arbitrary functions of q and p, the replacement above holds if it holds for each term of the expansion

$$[p_{\rho_1} \cdots p_{\rho_{n_1}} q_{\sigma_1} \cdots q_{\sigma_{n_2}}, \; p_{\rho_1} \cdots p_{\rho_{m_1}} q_{\sigma_1} \cdots q_{\sigma_{m_2}}]_{\text{PB}}$$

$$\to \frac{1}{i\hbar}[\hat{p}_{\rho_1} \cdots \hat{p}_{\rho_{n_1}} \hat{q}_{\sigma_1} \cdots \hat{q}_{\sigma_{n_2}}, \; \hat{p}_{\rho_1} \cdots \hat{p}_{\rho_{m_1}} \hat{q}_{\sigma_1} \cdots \hat{q}_{\sigma_{m_2}}]$$

This property may be proven through repeated use of the identities

$$[AB, C]_{\text{PB}} = A[B, C]_{\text{PB}} + B[A, C]_{\text{PB}}$$
$$[\hat{A}\hat{B}, \hat{C}] = \hat{A}[\hat{B}, \hat{C}] + [\hat{A}, \hat{C}]\hat{B}$$

Notice that in the case of the Poisson brackets, $B[A, C]_{\text{PB}} = [A, C]_{\text{PB}}B$, since A, B, C are not operators. Therefore the replacement

$$[AB, C]_{\text{PB}} \to \frac{1}{i\hbar}[\hat{A}\hat{B}, \hat{C}]$$

requires that the appropriate *ordering* is taken in the Poisson bracket before promoting the functions to operators.

****

Chapter 6

Symmetries

Problem 6.1 Write out the commutator in the first line of Eqs. (6.19) in detail. Use the canonical commutation relations to reorder the fields to get two cancelling contributions. Show that one is left with the contribution in the second line of Eqs. (6.19).

Solution to Problem 6.1

We are asked to evaluate the commutator in the first of Eqs. (6.19)

$$[\hat{Q}_i, \hat{Q}_j] = \left(\frac{1}{i\hbar}\right)^2 \int_\Omega d^3x \int_\Omega d^3y \left[\hat{\pi}_a(\mathbf{x})[\underline{t}_i]_{ab}\hat{\phi}_b(\mathbf{x}), \ \hat{\pi}_r(\mathbf{y})[\underline{t}_j]_{rs}\hat{\phi}_s(\mathbf{y})\right]$$

with the aid of the canonical commutation relations in Eqs. (6.17)

$$[\hat{\phi}_k(\mathbf{x}), \hat{\pi}_l(\mathbf{x}')] = i\hbar\,\delta_{kl}\delta^{(3)}(\mathbf{x} - \mathbf{x}') \qquad ; \ (k,l) = 1,2,3$$

$$[\hat{\phi}_k(\mathbf{x}), \hat{\phi}_l(\mathbf{x}')] = [\hat{\pi}_k(\mathbf{x}), \hat{\pi}_l(\mathbf{x}')] = 0$$

Write out the commutator[1]

$$\mathcal{C}_{ij} \equiv \left[\hat{\pi}_a(\mathbf{x})[\underline{t}_i]_{ab}\hat{\phi}_b(\mathbf{x}), \ \hat{\pi}_r(\mathbf{y})[\underline{t}_j]_{rs}\hat{\phi}_s(\mathbf{y})\right]$$

$$= \hat{\pi}_a(\mathbf{x})[\underline{t}_i]_{ab}\hat{\phi}_b(\mathbf{x})\hat{\pi}_r(\mathbf{y})[\underline{t}_j]_{rs}\hat{\phi}_s(\mathbf{y}) - \hat{\pi}_r(\mathbf{y})[\underline{t}_j]_{rs}\hat{\phi}_s(\mathbf{y})\hat{\pi}_a(\mathbf{x})[\underline{t}_i]_{ab}\hat{\phi}_b(\mathbf{x})$$

Now use the C.C.R. to write

$$\mathcal{C}_{ij} = \hat{\pi}_a(\mathbf{x})[\underline{t}_i]_{ab} \left\{i\hbar\delta_{br}\delta^{(3)}(\mathbf{x} - \mathbf{y}) + \hat{\pi}_r(\mathbf{y})\hat{\phi}_b(\mathbf{x})\right\} [\underline{t}_j]_{rs}\hat{\phi}_s(\mathbf{y}) -$$

$$\hat{\pi}_r(\mathbf{y})[\underline{t}_j]_{rs} \left\{i\hbar\delta_{sa}\delta^{(3)}(\mathbf{x} - \mathbf{y}) + \hat{\pi}_a(\mathbf{x})\hat{\phi}_s(\mathbf{y})\right\} [\underline{t}_i]_{ab}\hat{\phi}_b(\mathbf{x})$$

---

[1]Recall that repeated Latin indices are summed from 1 to 3. For clarity, matrices are underlined in these arguments.

The second and fourth terms cancel, and so

$$\frac{C_{ij}}{i\hbar} = \delta^{(3)}(\mathbf{x}-\mathbf{y})\left\{\hat{\pi}_a(\mathbf{x})[\underline{t}_i]_{ab}\,\delta_{br}[\underline{t}_j]_{rs}\hat{\phi}_s(\mathbf{y}) - \hat{\pi}_r(\mathbf{y})[\underline{t}_j]_{rs}\delta_{sa}\,[\underline{t}_i]_{ab}\hat{\phi}_b(\mathbf{x})\right\}$$

This gives us the remainder of Eqs. (6.19)

$$[\hat{Q}_i, \hat{Q}_j] = \left(\frac{1}{i\hbar}\right)^2 \int_\Omega d^3x \int_\Omega d^3y \left[\hat{\pi}_a(\mathbf{x})[\underline{t}_i]_{ab}\hat{\phi}_b(\mathbf{x}), \; \hat{\pi}_r(\mathbf{y})[\underline{t}_j]_{rs}\hat{\phi}_s(\mathbf{y})\right]$$

$$= \frac{1}{i\hbar}\int_\Omega d^3x \int_\Omega d^3y \left\{\hat{\pi}_a(\mathbf{x})[\underline{t}_i]_{ab}\,\delta_{br}\delta^{(3)}(\mathbf{x}-\mathbf{y})\,[\underline{t}_j]_{rs}\hat{\phi}_s(\mathbf{y})-\right.$$

$$\left.\hat{\pi}_r(\mathbf{y})[\underline{t}_j]_{rs}\,\delta_{sa}\delta^{(3)}(\mathbf{x}-\mathbf{y})\,[\underline{t}_i]_{ab}\hat{\phi}_b(\mathbf{x})\right\}$$

$$= \frac{1}{i\hbar}\int_\Omega d^3x \,\hat{\underline{\pi}}(\mathbf{x})\,[\underline{t}_i, \underline{t}_j]\,\hat{\underline{\phi}}(\mathbf{x})$$

Problem 6.2 Show by direct matrix multiplication that the 3×3 matrices defined in Eq. (6.13), and presented explicitly in Eqs. (6.14), satisfy the commutation relations for the generators of SU(2) in Eq. (6.21).

Solution to Problem 6.2

The three matrices in Eqs. (6.14) are

$$\underline{t}_1 = \begin{bmatrix} 0 & 0 & 0 \\ 0 & 0 & -i \\ 0 & i & 0 \end{bmatrix} \quad ; \; \underline{t}_2 = \begin{bmatrix} 0 & 0 & i \\ 0 & 0 & 0 \\ -i & 0 & 0 \end{bmatrix} \quad ; \; \underline{t}_3 = \begin{bmatrix} 0 & -i & 0 \\ i & 0 & 0 \\ 0 & 0 & 0 \end{bmatrix}$$

Consider the commutator of the first two

$$[\underline{t}_1, \underline{t}_2] = \begin{bmatrix} 0 & 0 & 0 \\ 0 & 0 & -i \\ 0 & i & 0 \end{bmatrix}\begin{bmatrix} 0 & 0 & i \\ 0 & 0 & 0 \\ -i & 0 & 0 \end{bmatrix} - \begin{bmatrix} 0 & 0 & i \\ 0 & 0 & 0 \\ -i & 0 & 0 \end{bmatrix}\begin{bmatrix} 0 & 0 & 0 \\ 0 & 0 & -i \\ 0 & i & 0 \end{bmatrix}$$

$$= \begin{bmatrix} 0 & 1 & 0 \\ -1 & 0 & 0 \\ 0 & 0 & 0 \end{bmatrix} = i\underline{t}_3$$

In a similar fashion

$$[\underline{t}_2, \underline{t}_3] = \begin{bmatrix} 0 & 0 & i \\ 0 & 0 & 0 \\ -i & 0 & 0 \end{bmatrix} \begin{bmatrix} 0 & -i & 0 \\ i & 0 & 0 \\ 0 & 0 & 0 \end{bmatrix} - \begin{bmatrix} 0 & -i & 0 \\ i & 0 & 0 \\ 0 & 0 & 0 \end{bmatrix} \begin{bmatrix} 0 & 0 & i \\ 0 & 0 & 0 \\ -i & 0 & 0 \end{bmatrix}$$

$$= \begin{bmatrix} 0 & 0 & 0 \\ 0 & 0 & 1 \\ 0 & -1 & 0 \end{bmatrix} = i\underline{t}_1$$

Also

$$[\underline{t}_3, \underline{t}_1] = \begin{bmatrix} 0 & -i & 0 \\ i & 0 & 0 \\ 0 & 0 & 0 \end{bmatrix} \begin{bmatrix} 0 & 0 & 0 \\ 0 & 0 & -i \\ 0 & i & 0 \end{bmatrix} - \begin{bmatrix} 0 & 0 & 0 \\ 0 & 0 & -i \\ 0 & i & 0 \end{bmatrix} \begin{bmatrix} 0 & -i & 0 \\ i & 0 & 0 \\ 0 & 0 & 0 \end{bmatrix}$$

$$= \begin{bmatrix} 0 & 0 & -1 \\ 0 & 0 & 0 \\ 1 & 0 & 0 \end{bmatrix} = i\underline{t}_2$$

In this fashion, by direct matrix multiplication, we establish Eq. (6.21)

$$[\underline{t}_i, \underline{t}_j] = i\epsilon_{ijk}\underline{t}_k$$

Problem 6.3 Repeat Prob. 6.1 starting with the charges in Eq. (6.37). Use the canonical *anticommutation* relations in Eq. (6.36). Show that one again obtains the relation in Eq. (6.38).

Solution to Problem 6.3

We are asked to use the charges in Eq. (6.37)

$$\hat{Q}_j = \frac{1}{i\hbar} \int_\Omega d^3x\, \hat{\underline{\pi}}\, \frac{1}{2} \underline{\tau}_j \hat{\underline{\psi}}$$

to evaluate the commutator in the first of Eqs. (6.19)

$$[\hat{Q}_i, \hat{Q}_j] = \left(\frac{1}{i\hbar}\right)^2 \int_\Omega d^3x \int_\Omega d^3y \left[\hat{\pi}_a(\mathbf{x})[\tfrac{1}{2}\underline{\tau}_i]_{ab}\hat{\psi}_b(\mathbf{x}),\ \hat{\pi}_r(\mathbf{y})[\tfrac{1}{2}\underline{\tau}_j]_{rs}\hat{\psi}_s(\mathbf{y})\right]$$

Here the fields obey the canonical *anticommutation* relations in Eqs. (6.36)[2]

$$\{\hat{\psi}_a(\mathbf{x}), \hat{\pi}_b(\mathbf{x}')\} = i\hbar\, \delta_{ab}\delta^{(3)}(\mathbf{x} - \mathbf{x}') \qquad ; (a,b) = 1,2$$
$$\{\hat{\psi}_a(\mathbf{x}), \hat{\psi}_b(\mathbf{x}')\} = \{\hat{\pi}_a(\mathbf{x}), \hat{\pi}_b(\mathbf{x}')\} = 0$$

---

[2] We suppress the Dirac indices, and the canonical momentum is $\hat{\underline{\pi}}(\mathbf{x}) = i\hbar\, \hat{\underline{\psi}}^\dagger(\mathbf{x})$.

Write out the commutator[3]

$$\mathcal{C}_{ij} \equiv \left[\hat{\pi}_a(\mathbf{x})[\tfrac{1}{2}\mathcal{I}_i]_{ab}\hat{\psi}_b(\mathbf{x}), \ \hat{\pi}_r(\mathbf{y})[\tfrac{1}{2}\mathcal{I}_j]_{rs}\hat{\psi}_s(\mathbf{y})\right]$$

This gives

$$4\mathcal{C}_{ij} = \hat{\pi}_a(\mathbf{x})[\mathcal{I}_i]_{ab}\hat{\psi}_b(\mathbf{x})\hat{\pi}_r(\mathbf{y})[\mathcal{I}_j]_{rs}\hat{\psi}_s(\mathbf{y}) -$$
$$\hat{\pi}_r(\mathbf{y})[\mathcal{I}_j]_{rs}\hat{\psi}_s(\mathbf{y})\hat{\pi}_a(\mathbf{x})[\mathcal{I}_i]_{ab}\hat{\psi}_b(\mathbf{x})$$

Now use the C.A.R. to write

$$4\mathcal{C}_{ij} = \hat{\pi}_a(\mathbf{x})[\mathcal{I}_i]_{ab}\left\{i\hbar\delta_{br}\delta^{(3)}(\mathbf{x}-\mathbf{y}) - \hat{\pi}_r(\mathbf{y})\hat{\psi}_b(\mathbf{x})\right\}[\mathcal{I}_j]_{rs}\hat{\psi}_s(\mathbf{y}) -$$
$$\hat{\pi}_r(\mathbf{y})[\mathcal{I}_j]_{rs}\left\{i\hbar\delta_{sa}\delta^{(3)}(\mathbf{x}-\mathbf{y}) - \hat{\pi}_a(\mathbf{x})\hat{\psi}_s(\mathbf{y})\right\}[\mathcal{I}_i]_{ab}\hat{\psi}_b(\mathbf{x})$$

The second and fourth terms again cancel, and so

$$\frac{4\mathcal{C}_{ij}}{i\hbar} = \delta^{(3)}(\mathbf{x}-\mathbf{y})\left\{\hat{\pi}_a(\mathbf{x})[\mathcal{I}_i]_{ab}\,\delta_{br}[\mathcal{I}_j]_{rs}\hat{\psi}_s(\mathbf{y}) - \hat{\pi}_r(\mathbf{y})[\mathcal{I}_j]_{rs}\delta_{sa}[\mathcal{I}_i]_{ab}\hat{\psi}_b(\mathbf{x})\right\}$$

This gives us Eqs. (6.38)

$$[\hat{Q}_i, \hat{Q}_j] = \left(\frac{1}{i\hbar}\right)^2 \int_\Omega d^3x \int_\Omega d^3y \left[\hat{\pi}_a(\mathbf{x})[\tfrac{1}{2}\mathcal{I}_i]_{ab}\hat{\psi}_b(\mathbf{x}), \ \hat{\pi}_r(\mathbf{y})[\tfrac{1}{2}\mathcal{I}_j]_{rs}\hat{\psi}_s(\mathbf{y})\right]$$
$$= \frac{1}{4i\hbar}\int_\Omega d^3x \int_\Omega d^3y \left\{\hat{\pi}_a(\mathbf{x})[\mathcal{I}_i]_{ab}\,\delta_{br}\delta^{(3)}(\mathbf{x}-\mathbf{y})\,[\mathcal{I}_j]_{rs}\hat{\psi}_s(\mathbf{y}) -\right.$$
$$\left.\hat{\pi}_r(\mathbf{y})[\mathcal{I}_j]_{rs}\,\delta_{sa}\delta^{(3)}(\mathbf{x}-\mathbf{y})\,[\mathcal{I}_i]_{ab}\hat{\psi}_b(\mathbf{x})\right\}$$
$$= \frac{1}{i\hbar}\int_\Omega d^3x \ \hat{\underline{\pi}}(\mathbf{x})\,[\tfrac{1}{2}\mathcal{I}_i, \tfrac{1}{2}\mathcal{I}_j]\,\hat{\underline{\psi}}(\mathbf{x})$$

Problem 6.4 Show that Eq. (6.48) holds even when the fields obey canonical *anticommutation* relations.

Solution to Problem 6.4

We are asked to evaluate the expression

$$[\hat{\mathbf{T}}, \hat{\psi}_j(\mathbf{x})] = \frac{1}{i\hbar}\int_\Omega d^3y \left[\hat{\pi}_a(\mathbf{y})\tfrac{1}{2}[\mathcal{I}]_{ab}\hat{\psi}_b(\mathbf{y}), \ \hat{\psi}_j(\mathbf{x})\right]$$

where the fields obey the *anticommutation* relations of the previous problem.

---

[3]Here the repeated Latin indices are summed from 1 to 2.

Write out the commutator

$$\mathcal{C}_j \equiv \left[\hat{\pi}_a(\mathbf{y}) \frac{1}{2} [\underline{\tau}]_{ab} \hat{\psi}_b(\mathbf{y}), \ \hat{\psi}_j(\mathbf{x}) \right]$$

$$2\,\mathcal{C}_j = \hat{\pi}_a(\mathbf{y})[\underline{\tau}]_{ab} \hat{\psi}_b(\mathbf{y}) \hat{\psi}_j(\mathbf{x}) - \hat{\psi}_j(\mathbf{x}) \hat{\pi}_a(\mathbf{y})[\underline{\tau}]_{ab} \hat{\psi}_b(\mathbf{y})$$

Now use the C.A.R. to obtain

$$2\,\mathcal{C}_j = \hat{\pi}_a(\mathbf{y})[\underline{\tau}]_{ab} \hat{\psi}_b(\mathbf{y}) \hat{\psi}_j(\mathbf{x}) - \left\{ i\hbar\delta_{ja}\delta^{(3)}(\mathbf{x} - \mathbf{y}) - \hat{\pi}_a(\mathbf{y}) \hat{\psi}_j(\mathbf{x}) \right\} [\underline{\tau}]_{ab} \hat{\psi}_b(\mathbf{y})$$

The first and last terms cancel, and therefore

$$2\,\mathcal{C}_j = -i\hbar\delta_{ja}\,\delta^{(3)}(\mathbf{x} - \mathbf{y})[\underline{\tau}]_{ab} \hat{\psi}_b(\mathbf{y})$$

It follows that

$$[\hat{\mathbf{T}}, \ \hat{\psi}_j(\mathbf{x})] = -\frac{1}{2}[\underline{\tau}]_{jb} \hat{\psi}_b(\mathbf{x})$$

This is the component form of Eq. (6.48)

$$[\hat{\mathbf{T}}, \ \hat{\underline{\psi}}(\mathbf{x})] = -\frac{1}{2}\underline{\tau}\,\hat{\underline{\psi}}(\mathbf{x})$$

Problem 6.5 Suppose $\hat{\underline{\phi}}(\mathbf{x})$ is an n-component, column-vector, real scalar field; $\hat{\underline{\Pi}}(\mathbf{x})$ is the conjugate momentum density (a row matrix); and $\underline{M}^a$ with $a = (1, \cdots, p)$ is a set of $n \times n$ matrices. Define the charge $\hat{Q}^a$ by

$$\hat{Q}^a = \frac{1}{i\hbar} \int_\Omega d^3x \ \hat{\underline{\Pi}} \ \underline{M}^a \ \hat{\underline{\phi}}$$

Use the canonical commutation relations to show that

$$[\hat{Q}^a, \hat{Q}^b] = \frac{1}{i\hbar} \int_\Omega d^3x \ \hat{\underline{\Pi}} \ [\underline{M}^a, \underline{M}^b] \ \hat{\underline{\phi}} \qquad ; \ (a, b) = 1, \cdots, p$$

Solution to Problem 6.5

The proof follows as in Prob. 6.1. Consider the commutator

$$[\hat{Q}^a, \hat{Q}^b] = \left(\frac{1}{i\hbar}\right)^2 \int_\Omega d^3x \int_\Omega d^3y \left[\hat{\Pi}_i(\mathbf{x})[\underline{M}^a]_{ij}\hat{\phi}_j(\mathbf{x}), \ \hat{\Pi}_k(\mathbf{y})[\underline{M}^b]_{kl}\hat{\phi}_l(\mathbf{y}) \right]$$

where the fields obey the canonical commutation relations

$$[\hat{\phi}_i(\mathbf{x}), \hat{\Pi}_j(\mathbf{x}')] = i\hbar\,\delta_{ij}\delta^{(3)}(\mathbf{x}-\mathbf{x}') \qquad ; (i,j) = 1,\cdots,n$$
$$[\hat{\phi}_i(\mathbf{x}), \hat{\phi}_j(\mathbf{x}')] = [\hat{\Pi}_i(\mathbf{x}), \hat{\Pi}_j(\mathbf{x}')] = 0$$

Write out the commutator[4]

$$\mathcal{C}^{ab} \equiv \left[\hat{\Pi}_i(\mathbf{x})[\underline{M}^a]_{ij}\hat{\phi}_j(\mathbf{x}),\ \hat{\Pi}_k(\mathbf{y})[\underline{M}^b]_{kl}\hat{\phi}_l(\mathbf{y})\right]$$
$$= \hat{\Pi}_i(\mathbf{x})[\underline{M}^a]_{ij}\hat{\phi}_j(\mathbf{x})\hat{\Pi}_k(\mathbf{y})[\underline{M}^b]_{kl}\hat{\phi}_l(\mathbf{y}) -$$
$$\hat{\Pi}_k(\mathbf{y})[\underline{M}^b]_{kl}\hat{\phi}_l(\mathbf{y})\hat{\Pi}_i(\mathbf{x})[\underline{M}^a]_{ij}\hat{\phi}_j(\mathbf{x})$$

Now use the C.C.R. to write

$$\mathcal{C}^{ab} = \hat{\Pi}_i(\mathbf{x})[\underline{M}^a]_{ij}\left\{i\hbar\delta_{jk}\delta^{(3)}(\mathbf{x}-\mathbf{y}) + \hat{\Pi}_k(\mathbf{y})\hat{\phi}_j(\mathbf{x})\right\}[\underline{M}^b]_{kl}\hat{\phi}_l(\mathbf{y}) -$$
$$\hat{\Pi}_k(\mathbf{y})[\underline{M}^b]_{kl}\left\{i\hbar\delta_{li}\delta^{(3)}(\mathbf{x}-\mathbf{y}) + \hat{\Pi}_i(\mathbf{x})\hat{\phi}_l(\mathbf{y})\right\}[\underline{M}^a]_{ij}\hat{\phi}_j(\mathbf{x})$$

The second and fourth terms cancel, and so

$$\frac{\mathcal{C}^{ab}}{i\hbar} = \delta^{(3)}(\mathbf{x}-\mathbf{y})\left\{\hat{\Pi}_i(\mathbf{x})[\underline{M}^a]_{ij}\,\delta_{jk}[\underline{M}^b]_{kl}\hat{\phi}_l(\mathbf{y}) - \hat{\Pi}_k(\mathbf{y})[\underline{M}^b]_{kl}\delta_{li}\,[\underline{M}^a]_{ij}\hat{\phi}_j(\mathbf{x})\right\}$$

This gives the stated answer

$$[\hat{Q}^a, \hat{Q}^b] = \left(\frac{1}{i\hbar}\right)^2 \int_\Omega d^3x \int_\Omega d^3y \left[\hat{\Pi}_i(\mathbf{x})[\underline{M}^a]_{ij}\hat{\phi}_j(\mathbf{x}),\ \hat{\Pi}_k(\mathbf{y})[\underline{M}^b]_{kl}\hat{\phi}_l(\mathbf{y})\right]$$
$$= \frac{1}{i\hbar}\int_\Omega d^3x\,\underline{\hat{\Pi}}(\mathbf{x})\,[\underline{M}^a, \underline{M}^b]\,\underline{\hat{\phi}}(\mathbf{x}) \qquad ; (a,b) = 1,\cdots,p$$

Problem 6.6 (a) Extend the interaction-picture representation of the massive scalar fields in Eqs. (5.60) to the isovector case by adding a subscript for the internal space. Insert these field expansions in the definition of the isospin operator $\hat{\mathbf{T}} = (\hat{T}_1, \hat{T}_2, \hat{T}_3)$ in Eq. (6.16), and express this operator in terms of the creation and destruction operators;[5]

(b) Repeat for the Dirac fields in Eqs. (5.102) and (6.28) and the isospin operator in Eq. (6.37).

Solution to Problem 6.6

We add a subscript to the massive scalar fields in Eqs. (5.60) and (5.65) that takes three values $j = (1,2,3)$, denoting three distinct equal-mass

---

[4]Here repeated Latin indices are summed from 1 to n.
[5]Recall Eqs. (5.65) and (6.23).

species[6]

$$\hat{\phi}_j(\mathbf{x},t) = \sum_k \left(\frac{\hbar}{2\omega_k\Omega}\right)^{1/2}\left[c_{\mathbf{k}j}e^{i(\mathbf{k}\cdot\mathbf{x}-\omega_k t)} + c_{\mathbf{k}j}^\dagger e^{-i(\mathbf{k}\cdot\mathbf{x}-\omega_k t)}\right] \quad ; j = 1,2,3$$

$$\hat{\pi}_j(\mathbf{x},t) = \frac{1}{i}\sum_k \left(\frac{\hbar\omega_k}{2\Omega}\right)^{1/2}\left[c_{\mathbf{k}j}e^{i(\mathbf{k}\cdot\mathbf{x}-\omega_k t)} - c_{\mathbf{k}j}^\dagger e^{-i(\mathbf{k}\cdot\mathbf{x}-\omega_k t)}\right]$$

Here $\omega_k \equiv c\sqrt{\mathbf{k}^2 + m_s^2}$. The new creation and destruction operators satisfy the canonical commutation relations

$$[c_{\mathbf{k}i}, c_{\mathbf{k}j}^\dagger] = \delta_{ij}\delta_{\mathbf{kk}'}$$
$$[c_{\mathbf{k}i}^\dagger, c_{\mathbf{k}j}^\dagger] = [c_{\mathbf{k}i}, c_{\mathbf{k}j}] = 0$$

The isospin operator in Eq. (6.18) is given by

$$\hat{T}_i = \frac{1}{i\hbar}\int_\Omega d^3x\,\hat{\underline{\pi}}(\mathbf{x})\,\underline{t}_i\,\hat{\underline{\phi}}(\mathbf{x}) \quad ; i = 1,2,3$$
$$= \frac{1}{i\hbar}\int_\Omega d^3x\,\hat{\pi}_a(\mathbf{x})\,[\underline{t}_i]_{ab}\,\hat{\phi}_b(\mathbf{x})$$

where, in the second line, the matrix indices have been made explicit and repeated Latin indices are summed from 1 to 3. Insertion of the above expansions, and use of the orthonormality of the plane waves gives

$$\hat{T}_i = \frac{1}{2}\sum_k \left\{c_{\mathbf{k}a}^\dagger [\underline{t}_i]_{ab} c_{\mathbf{k}b} - c_{\mathbf{k}a} [\underline{t}_i]_{ab} c_{\mathbf{k}b}^\dagger \right.$$
$$\left. + c_{\mathbf{k}a}^\dagger [\underline{t}_i]_{ab} c_{-\mathbf{k},b}^\dagger e^{2i\omega_k t} - c_{\mathbf{k}a} [\underline{t}_i]_{ab} c_{-\mathbf{k},b} e^{-2i\omega_k t}\right\}$$

Now make use of the results from Eqs. (6.14)

$$[\underline{t}_i]_{ba} = -[\underline{t}_i]_{ab} \qquad ; [\underline{t}_i]_{aa} = 0$$

The last two terms in $\hat{T}_i$ are then odd under the change in dummy summation variables $(\mathbf{k},a,b) \to (-\mathbf{k},b,a)$, and hence they vanish. Thus

$$\hat{T}_i = \sum_k c_{\mathbf{k}a}^\dagger [\underline{t}_i]_{ab} c_{\mathbf{k}b} = \sum_k \underline{c}_{\mathbf{k}}^\dagger \underline{t}_i \underline{c}_{\mathbf{k}}$$

Here, in the last equality, we restore the matrix notation.

---

[6]We now use $(c^\dagger, c)$ for the creation and destruction operators.

(b) We can similarly add an index that takes two values $j = (1, 2)$ to the Dirac field in Eqs.(5.102) and (6.28)

$$\hat{\psi}_j(\mathbf{x}, t) = \frac{1}{\sqrt{\Omega}} \sum_{\mathbf{k}} \sum_{\lambda} \left[a_{\mathbf{k}\lambda j} u_\lambda(\mathbf{k}) e^{i(\mathbf{k}\cdot\mathbf{x} - \omega_k t)} + b^\dagger_{\mathbf{k}\lambda j} v_\lambda(-\mathbf{k}) e^{-i(\mathbf{k}\cdot\mathbf{x} - \omega_k t)} \right]$$
$$; \ j = 1, 2$$

where $\omega_k = c\sqrt{\mathbf{k}^2 + M^2}$, and $\lambda = (\uparrow, \downarrow)$ denotes the helicity.[7] The anticommutation relations for the particle and antiparticle creation and destruction operators for all modes are then given by

$$\{a_{\mathbf{k}\lambda j}, a^\dagger_{\mathbf{k}'\lambda' j'}\} = \{b_{\mathbf{k}\lambda j}, b^\dagger_{\mathbf{k}'\lambda' j'}\} = \delta_{\mathbf{k}\mathbf{k}'} \delta_{\lambda\lambda'} \delta_{jj'}$$
$$; \ \text{all other anticommutators vanish}$$

The isospin operator is given in Eq. (6.43)

$$\hat{T}_i = \int_\Omega d^3x \ \underline{\hat{\psi}}^\dagger(\mathbf{x}) \frac{1}{2} \underline{\tau}_i \ \underline{\hat{\psi}}(\mathbf{x})$$
$$= \int_\Omega d^3x \ \hat{\psi}^\dagger_a(\mathbf{x}) \frac{1}{2} [\underline{\tau}_i]_{ab} \ \hat{\psi}_b(\mathbf{x})$$

where, in the second line, the matrix indices have again been made explicit and repeated Latin indices are now summed from 1 to 2. Insertion of the above expansion, and use of the orthonormality of the plane waves and of the Dirac spinors gives

$$\hat{T}_i = \frac{1}{2} \sum_{\mathbf{k}} \sum_{\lambda} \left\{ a^\dagger_{\mathbf{k}\lambda p} [\underline{\tau}_i]_{pq} a_{\mathbf{k}\lambda q} + b_{\mathbf{k}\lambda p} [\underline{\tau}_i]_{pq} b^\dagger_{\mathbf{k}\lambda q} \right\}$$

It follows from the properties of the Pauli matrices that $\underline{\tau}_{pp} = 0$.[8] Thus

$$\hat{T}_i = \frac{1}{2} \sum_{\mathbf{k}} \sum_{\lambda} \left\{ a^\dagger_{\mathbf{k}\lambda p} [\underline{\tau}_i]_{pq} a_{\mathbf{k}\lambda q} - b^\dagger_{\mathbf{k}\lambda q} [\underline{\tau}^T_i]_{qp} b_{\mathbf{k}\lambda p} \right\}$$
$$= \sum_{\mathbf{k}} \sum_{\lambda} \left\{ \underline{a}^\dagger_{\mathbf{k}\lambda} \frac{1}{2} \underline{\tau}_i \underline{a}_{\mathbf{k}\lambda} + \underline{b}^\dagger_{\mathbf{k}\lambda} \left(-\frac{1}{2} \underline{\tau}^T_i \right) \underline{b}_{\mathbf{k}\lambda} \right\}$$

Here, in the last equality, we again restore the matrix notation.

---

[7]The Dirac indices are suppressed. Note
$$u^\dagger_\lambda(\mathbf{k}) u_{\lambda'}(\mathbf{k}) = v^\dagger_\lambda(\mathbf{k}) v_{\lambda'}(\mathbf{k}) = \delta_{\lambda\lambda'}$$
$$u^\dagger_\lambda(\mathbf{k}) v_{\lambda'}(\mathbf{k}) = v^\dagger_\lambda(\mathbf{k}) u_{\lambda'}(\mathbf{k}) = 0$$

[8]See, for example, Prob. 3.3.

Note that $-\underline{\tau}^T/2 = (-\underline{\tau}_1^T, -\underline{\tau}_2^T, -\underline{\tau}_3^T)/2$ also satisfies the SU(2) algebra in Eq. (6.39).

Problem 6.7 Make a transformation to a new set of real (hermitian) fields in Probs. 5.15–5.16, with $\phi^\star \equiv (\phi_1 + i\phi_2)/\sqrt{2}$ and $\phi \equiv (\phi_1 - i\phi_2)/\sqrt{2}$.

(a) Compute the new lagrangian density and new conserved currents;

(b) Find the new commutation relations and interaction-picture fields;

(c) Add a neutral scalar field ϕ_3 of the same mass, and recover the configuration in Prob. 6.6(a).

Solution to Problem 6.7

(a) The lagrangian density in Prob. 5.15 is

$$\mathcal{L}\left(\phi, \frac{\partial\phi}{\partial x_\mu}, \phi^\star, \frac{\partial\phi^\star}{\partial x_\mu}\right) = -c^2\left[\frac{\partial\phi^\star}{\partial x_\mu}\frac{\partial\phi}{\partial x_\mu} + m_s^2\phi^\star\phi\right]$$

where $(\phi^\star, \phi)$ are independent, charged scalar fields. Introduce the independent, real linear combinations (ϕ_1, ϕ_2) through

$$\phi^\star \equiv \frac{1}{\sqrt{2}}(\phi_1 + i\phi_2) \qquad ; \phi \equiv \frac{1}{\sqrt{2}}(\phi_1 - i\phi_2)$$

These relations are inverted to give

$$\phi_1 = \frac{1}{\sqrt{2}}(\phi + \phi^\star) \qquad ; \phi_2 = \frac{i}{\sqrt{2}}(\phi - \phi^\star)$$

In terms of these new coordinates, the lagrangian density is diagonalized

$$\mathcal{L}\left(\phi_1, \frac{\partial\phi_1}{\partial x_\mu}, \phi_2, \frac{\partial\phi_2}{\partial x_\mu}\right) = -\frac{c^2}{2}\sum_{i=1}^{2}\left[\left(\frac{\partial\phi_i}{\partial x_\mu}\right)^2 + m_s^2\phi_i^2\right]$$

Lagrange's equations now read

$$(\Box - m_s^2)\phi_i = 0 \qquad ; i = 1, 2$$

In terms of these new coordinates, the conserved electromagnetic current in Prob. 5.17 becomes

$$j_\mu(x) = \frac{c}{\hbar}\left[\phi_2\frac{\partial\phi_1}{\partial x_\mu} - \phi_1\frac{\partial\phi_2}{\partial x_\mu}\right]$$

(b) The new canonical momentum densities are

$$\pi_{\phi_i} \equiv \pi_i = \frac{\partial\phi_i}{\partial t}$$

The new canonical commutation relations follow as

$$[\hat{\pi}_i(\mathbf{x},t), \hat{\phi}_j(\mathbf{x}',t')]_{t=t'} = \frac{\hbar}{i}\delta_{ij}\delta^{(3)}(\mathbf{x}-\mathbf{x}') \qquad ; (i,j) = 1,2$$

A representation of these fields in the interaction picture is then given by (see Prob. 5.16)

$$\hat{\phi}_j(\mathbf{x},t) = \sum_{\mathbf{k}}\left(\frac{\hbar}{2\omega_k\Omega}\right)^{1/2}\left[c_{\mathbf{k}j}e^{i(\mathbf{k}\cdot\mathbf{x}-\omega_k t)} + c^\dagger_{\mathbf{k}j}e^{-i(\mathbf{k}\cdot\mathbf{x}-\omega_k t)}\right] \qquad ; j = 1,2$$

$$\hat{\pi}_j(\mathbf{x},t) = i\sum_{\mathbf{k}}\left(\frac{\hbar\omega_k}{2\Omega}\right)^{1/2}\left[c^\dagger_{\mathbf{k}j}e^{-i(\mathbf{k}\cdot\mathbf{x}-\omega_k t)} - c_{\mathbf{k}j}e^{i(\mathbf{k}\cdot\mathbf{x}-\omega_k t)}\right]$$

where the new creation and destruction operators satisfy

$$[c_{\mathbf{k}i}, c^\dagger_{\mathbf{k}'j}] = \delta_{ij}\delta_{\mathbf{k}\mathbf{k}'} \qquad ; (i,j) = 1,2$$

$$[c_{\mathbf{k}i}, c_{\mathbf{k}'j}] = [c^\dagger_{\mathbf{k}i}, c^\dagger_{\mathbf{k}'j}] = 0$$

These new operators are evidently related to the previous operators in Prob. 5.16 by

$$c_{\mathbf{k}1} = \frac{1}{\sqrt{2}}(a_{\mathbf{k}} + b_{\mathbf{k}}) \qquad ; c_{\mathbf{k}2} = \frac{i}{\sqrt{2}}(a_{\mathbf{k}} - b_{\mathbf{k}})$$

It follows that

$$c^\dagger_{\mathbf{k}1}c_{\mathbf{k}1} + c^\dagger_{\mathbf{k}2}c_{\mathbf{k}2} = a^\dagger_{\mathbf{k}}a_{\mathbf{k}} + b^\dagger_{\mathbf{k}}b_{\mathbf{k}}$$

Hence the hamiltonian and momentum in Prob. 5.16 become

$$\hat{H}_0 = \sum_{i=1}^{2}\sum_{\mathbf{k}}\hbar\omega_k\left(c^\dagger_{\mathbf{k}i}c_{\mathbf{k}i} + 1/2\right)$$

$$\hat{\mathbf{P}} = \sum_{i=1}^{2}\sum_{\mathbf{k}}\hbar\mathbf{k}\left(c^\dagger_{\mathbf{k}i}c_{\mathbf{k}i}\right)$$

(c) A neutral scalar field ϕ_3 of the same mass can be now be included by simply extending the sums from 1 to 3

$$\mathcal{L} = -\frac{c^2}{2} \sum_{i=1}^{3} \left[\left(\frac{\partial \phi_i}{\partial x_\mu} \right)^2 + m_s^2 \phi_i^2 \right]$$

$$\hat{\phi}_j(\mathbf{x}, t) = \sum_{\mathbf{k}} \left(\frac{\hbar}{2\omega_k \Omega} \right)^{1/2} \left[c_{\mathbf{k}j} e^{i(\mathbf{k} \cdot \mathbf{x} - \omega_k t)} + c_{\mathbf{k}j}^\dagger e^{-i(\mathbf{k} \cdot \mathbf{x} - \omega_k t)} \right] \qquad ; j = 1, 2, 3$$

$$\hat{\pi}_j(\mathbf{x}, t) = i \sum_{\mathbf{k}} \left(\frac{\hbar \omega_k}{2\Omega} \right)^{1/2} \left[c_{\mathbf{k}j}^\dagger e^{-i(\mathbf{k} \cdot \mathbf{x} - \omega_k t)} - c_{\mathbf{k}j} e^{i(\mathbf{k} \cdot \mathbf{x} - \omega_k t)} \right]$$

where, now

$$[c_{\mathbf{k}i}, c_{\mathbf{k}'j}^\dagger] = \delta_{ij} \delta_{\mathbf{k}\mathbf{k}'} \qquad\qquad ; (i, j) = 1, 2, 3$$

$$[c_{\mathbf{k}i}, c_{\mathbf{k}'j}] = [c_{\mathbf{k}i}^\dagger, c_{\mathbf{k}'j}^\dagger] = 0$$

Thus we recover the configuration in Prob. 6.6(a).

Note that the previous conserved current now becomes the third component of a cross product in this internal space with $\boldsymbol{\phi} = (\phi_1, \phi_2, \phi_3)$ [9]

$$j_\mu(x) = -\frac{c}{\hbar} \left[\boldsymbol{\phi} \times \frac{\partial \boldsymbol{\phi}}{\partial x_\mu} \right]_3$$

Problem 6.8 (a) Use the repeated commutator relation in Prob. 2.5 to establish the following matrix identity

$$\exp\left\{ -\frac{i}{2} \boldsymbol{\omega} \cdot \boldsymbol{\tau} \right\} \tau_i \exp\left\{ \frac{i}{2} \boldsymbol{\omega} \cdot \boldsymbol{\tau} \right\} = \exp\left\{ i\boldsymbol{\omega} \cdot \mathbf{t} \right\}_{ij} \tau_j$$

(b) Show $\exp\left\{ i\boldsymbol{\omega} \cdot \mathbf{t} \right\}_{ij} = a_{ij}(\boldsymbol{\omega})$ is a real, orthogonal, 3×3 rotation matrix [compare Eq.(3.22)];

(c) Hence prove that the combination $\bar{\psi} \boldsymbol{\tau} \psi$ transforms as an isovector under isospin rotations.

Solution to Problem 6.8

(a) From the relation in Prob. (2.5), we have

$$\exp\left\{ -\frac{i}{2} \boldsymbol{\omega} \cdot \boldsymbol{\tau} \right\} \tau_i \exp\left\{ \frac{i}{2} \boldsymbol{\omega} \cdot \boldsymbol{\tau} \right\} = \tau_i - \frac{i}{2} \omega_j [\tau_j, \tau_i] +$$

$$\left(-\frac{i}{2} \right)^2 \frac{1}{2!} \omega_j \omega_k [\tau_j, [\tau_k, \tau_i]] + \left(-\frac{i}{2} \right)^3 \frac{1}{3!} \omega_j \omega_k \omega_l [\tau_j, [\tau_k, [\tau_l, \tau_i]]] + \cdots$$

_____

[9] See Eqs. (6.10).

Use the commutation relation for the Pauli matrices

$$[\mathcal{I}_i, \mathcal{I}_j] = 2i\epsilon_{ijk}\mathcal{I}_k$$

This gives[10]

$$\exp\left\{-\frac{i}{2}\boldsymbol{\omega}\cdot\boldsymbol{\mathcal{I}}\right\}\mathcal{I}_i \exp\left\{\frac{i}{2}\boldsymbol{\omega}\cdot\boldsymbol{\mathcal{I}}\right\} = \mathcal{I}_i + \omega_j\epsilon_{jik}\mathcal{I}_k +$$
$$\frac{1}{2!}\omega_j\omega_k\epsilon_{kil}\epsilon_{jlm}\mathcal{I}_m + \frac{1}{3!}\omega_j\omega_k\omega_l\epsilon_{lim}\epsilon_{kmn}\epsilon_{jnp}\mathcal{I}_p + \cdots$$

Now introduce the rotation matrices from Eqs. (6.13)–(6.14)

$$[\underline{t}_i]_{jk} \equiv -i\epsilon_{ijk}$$

$$\underline{t}_1 = \begin{bmatrix} 0 & 0 & 0 \\ 0 & 0 & -i \\ 0 & i & 0 \end{bmatrix} \quad ; \underline{t}_2 = \begin{bmatrix} 0 & 0 & i \\ 0 & 0 & 0 \\ -i & 0 & 0 \end{bmatrix} \quad ; \underline{t}_3 = \begin{bmatrix} 0 & -i & 0 \\ i & 0 & 0 \\ 0 & 0 & 0 \end{bmatrix}$$

Then

$$\exp\left\{-\frac{i}{2}\boldsymbol{\omega}\cdot\boldsymbol{\mathcal{I}}\right\}\mathcal{I}_i \exp\left\{\frac{i}{2}\boldsymbol{\omega}\cdot\boldsymbol{\mathcal{I}}\right\} = \mathcal{I}_i + [i\boldsymbol{\omega}\cdot\underline{t}]_{ij}\mathcal{I}_j +$$
$$\frac{1}{2!}[i\boldsymbol{\omega}\cdot\underline{t}]_{ik}[i\boldsymbol{\omega}\cdot\underline{t}]_{kj}\mathcal{I}_j + \frac{1}{3!}[i\boldsymbol{\omega}\cdot\underline{t}]_{ik}[i\boldsymbol{\omega}\cdot\underline{t}]_{kl}[i\boldsymbol{\omega}\cdot\underline{t}]_{lj}\mathcal{I}_j + \cdots$$

Thus

$$\exp\left\{-\frac{i}{2}\boldsymbol{\omega}\cdot\boldsymbol{\mathcal{I}}\right\}\mathcal{I}_i \exp\left\{\frac{i}{2}\boldsymbol{\omega}\cdot\boldsymbol{\mathcal{I}}\right\} = \exp\left\{i\boldsymbol{\omega}\cdot\underline{t}\right\}_{ij}\mathcal{I}_j$$

(b) The matrices $\underline{t}_i$ are hermitian and imaginary; hence $\exp\left\{i\boldsymbol{\omega}\cdot\underline{t}\right\}$ is real. The latter matrix is also unitary

$$\exp\left\{i\boldsymbol{\omega}\cdot\underline{t}\right\}_{ik}^{\dagger}\exp\left\{i\boldsymbol{\omega}\cdot\underline{t}\right\}_{kj} = \exp\left\{i\boldsymbol{\omega}\cdot\underline{t}\right\}_{ki}^{*}\exp\left\{i\boldsymbol{\omega}\cdot\underline{t}\right\}_{kj} = \delta_{ij}$$

Define

$$\exp\left\{i\boldsymbol{\omega}\cdot\underline{t}\right\}_{ij} \equiv a_{ij}(\boldsymbol{\omega})$$

then $\underline{a}$ is a real rotation [compare Eq.(3.22)]

$$a_{ki}(\boldsymbol{\omega})a_{kj}(\boldsymbol{\omega}) = \delta_{ij}$$

---

[10]Recall that repeated Latin indices are here summed from 1 to 3; we freely relabel them. Note that the subscript i is always to be distinguished from the commonly occuring factor of $i = \sqrt{-1}$.

It follows that the $\underline{\tau}$ matrices transform according to

$$\exp\left\{-\frac{i}{2}\boldsymbol{\omega}\cdot\underline{\tau}\right\}\tau_i\exp\left\{\frac{i}{2}\boldsymbol{\omega}\cdot\underline{\tau}\right\} = a_{ij}(\boldsymbol{\omega})\,\tau_j$$

(c) Consider the isospin rotation in Eq. (6.31)

$$\underline{\psi} \to \exp\left\{\frac{i}{2}\boldsymbol{\omega}\cdot\underline{\tau}\right\}\underline{\psi}$$

Under this transformation, the quantity $\bar{\underline{\psi}}\,\underline{\tau}\,\underline{\psi}$ transforms as an isovector

$$\bar{\underline{\psi}}\,\tau_i\,\underline{\psi} \to \bar{\underline{\psi}}\,\exp\left\{-\frac{i}{2}\boldsymbol{\omega}\cdot\underline{\tau}\right\}\tau_i\exp\left\{\frac{i}{2}\boldsymbol{\omega}\cdot\underline{\tau}\right\}\underline{\psi}$$
$$= a_{ij}(\boldsymbol{\omega})\,\bar{\underline{\psi}}\,\tau_j\,\underline{\psi}$$

Problem 6.9 Consider a field theory composed of an isovector scalar field with the lagrangian density of Eq. (6.7), an isospinor Dirac field with the lagrangian density of Eq. (6.29), and an *interaction* of the form

$$\mathcal{L}_1 = g\,\bar{\underline{\psi}}\,\frac{1}{2}\underline{\tau}\,\underline{\psi}\cdot\boldsymbol{\phi}$$

(a) Use the result in Prob. 6.8 to show this interaction is invariant under isospin rotations;

(b) Construct the total isospin operator $\hat{\mathbf{T}}$ for the combined system. What are the eigenvalues of $\hat{\mathbf{T}}^2$ in the interacting system? What is the degeneracy of each isospin multiplet $|TM_T\rangle$?

(c) A scattering state starts as one free isovector scalar meson and one free isospinor Dirac particle. What values of T are accessed in the scattering process?

(d) Show the scattering operator $\hat{S}$ commutes with all components of $\hat{\mathbf{T}}$.[11] Use the Wigner-Eckart theorem to deduce the consequences for the matrix elements $\langle T'M_T'|\hat{S}|TM_T\rangle$.

Solution to Problem 6.9

(a) From Eqs. (6.18), (6.23), and (6.43), the isospin operators for the

---

[11] *Hint*: start from Eq. (4.17).

scalar and Dirac fields are

$$\hat{\mathbf{T}}^S = \frac{1}{i\hbar} \int_\Omega d^3x \; \hat{\underline{\pi}}(\mathbf{x}) \, \underline{\mathbf{t}} \, \hat{\underline{\phi}}(\mathbf{x})$$

$$\hat{\mathbf{T}}^D = \int_\Omega d^3x \; \hat{\underline{\psi}}^\dagger(\mathbf{x}) \frac{1}{2}\underline{\boldsymbol{\tau}} \, \hat{\underline{\psi}}(\mathbf{x})$$

Consider the isospin transformation[12]

$$\hat{R}(\boldsymbol{\omega}) = \exp\{-i\boldsymbol{\omega}\cdot\hat{\mathbf{T}}\} \qquad ; \; \hat{\mathbf{T}} \equiv \hat{\mathbf{T}}^S + \hat{\mathbf{T}}^D$$

Then from Eqs. (6.46), (6.49), and Prob. 6.8

$$e^{-i\boldsymbol{\omega}\cdot\hat{\mathbf{T}}} \, \hat{\underline{\phi}}(\mathbf{x}) \, e^{i\boldsymbol{\omega}\cdot\hat{\mathbf{T}}} = \exp\{i\boldsymbol{\omega}\cdot\underline{\mathbf{t}}\} \, \hat{\underline{\phi}}(\mathbf{x}) = \underline{a}(\boldsymbol{\omega}) \, \hat{\underline{\phi}}(\mathbf{x})$$

$$e^{-i\boldsymbol{\omega}\cdot\hat{\mathbf{T}}} \, \hat{\underline{\psi}}(\mathbf{x}) \, e^{i\boldsymbol{\omega}\cdot\hat{\mathbf{T}}} = \exp\left\{\frac{i}{2}\boldsymbol{\omega}\cdot\underline{\boldsymbol{\tau}}\right\} \hat{\underline{\psi}}(\mathbf{x})$$

It follows from Prob. 6.8 that with the given interaction lagrangian density $\hat{\mathcal{L}}_1 = (g/2)\hat{\underline{\bar{\psi}}} \, \underline{\boldsymbol{\tau}} \, \hat{\underline{\psi}} \cdot \hat{\underline{\phi}}$

$$\hat{R}(\boldsymbol{\omega})\hat{\mathcal{L}}_1\hat{R}^{-1}(\boldsymbol{\omega}) = \frac{g}{2} \left[a_{ij}(\boldsymbol{\omega})\hat{\underline{\bar{\psi}}}\,\underline{\tau}_j\,\hat{\underline{\psi}}\right]\left[a_{ik}(\boldsymbol{\omega})\,\hat{\underline{\phi}}_k\right]$$

$$= \frac{g}{2}\,\hat{\underline{\bar{\psi}}}\,\underline{\boldsymbol{\tau}}\,\hat{\underline{\psi}}\cdot\hat{\underline{\phi}}$$

Hence this interaction is left invariant under isospin rotations.

(b) From part (a), the total isospin operator is

$$\hat{\mathbf{T}} = \hat{\mathbf{T}}^S + \hat{\mathbf{T}}^D$$

This satisfies the commutation relations

$$[\hat{T}_i, \hat{T}_j] = i\epsilon_{ijk}\hat{T}_k \qquad ; \; (i,j,k) = 1,2,3$$
$$[\hat{\mathbf{T}}^2, \hat{T}_i] = 0$$

The theory of angular momentum then implies that these operators have the spectrum

$$\hat{\mathbf{T}}^2|TM_T\rangle = T(T+1)|TM_T\rangle$$
$$\hat{T}_3|TM_T\rangle = M_T|TM_T\rangle \qquad ; \; M_T = -T, -T+1, \cdots, T$$

There are $2T+1$ states in the subspace of a given T.

---

[12]Note that $[\hat{\mathbf{T}}^S, \hat{\mathbf{T}}^D] = 0$.

(c) The isovector scalar meson has $T^S = 1$ and the isospinor Dirac particle has $T^D = 1/2$. The values of T accessed in the scattering process are then

$$T = (1/2, 3/2) \qquad ; \text{ scattering}$$

(d) It follows from Eq. (4.17) that if the interaction hamiltonian density is invariant under isospin rotations, then the scattering operator $\hat{S}$ will be also. Hence

$$[\hat{\mathbf{T}}, \hat{S}] = 0$$

The scattering operator is therefore an irreducible tensor operator of rank zero with respect to isospin. The Wigner-Eckart theorem in Eq. (3.113) then implies

$$\langle \gamma' T' M_T' | \hat{S} | \gamma T M_T \rangle = (-1)^{T' - M_T'} \begin{pmatrix} T' & 0 & T \\ -M_T' & 0 & M_T \end{pmatrix} \langle \gamma' T' || S(0) || \gamma T \rangle$$

$$= \delta_{M_T' M_T} \, \delta_{T' T} \, S^T$$

The S-matrix elements are diagonal in T, diagonal in M_T, and independent of M_T.

Problem 6.10 Show that the Dirac mass term $\mathcal{L}_{\text{mass}} = -\hbar c \, \bar{\psi} M \psi$ is *not* invariant under the chiral transformation in Eq. (6.138).

Solution to Problem 6.10

The Dirac mass term under consideration is

$$\mathcal{L}_{\text{mass}} = -\hbar c M \, \bar{\psi} \, \psi = -\hbar c M \, \psi^\dagger \gamma_4 \, \psi$$

The infinitesimal form of the chiral transformation in Eq. (6.138) is

$$\psi \to \psi + \delta \psi = \psi + \frac{i}{2} \gamma_5 \, \boldsymbol{\tau} \cdot \boldsymbol{\varepsilon} \, \psi$$

Under this transformation the Dirac mass term gets changed to

$$\mathcal{L}_{\text{mass}} \to -\hbar c M \, \psi^\dagger \left(1 - \frac{i}{2} \gamma_5 \, \boldsymbol{\tau} \cdot \boldsymbol{\varepsilon} \right) \gamma_4 \left(1 + \frac{i}{2} \gamma_5 \, \boldsymbol{\tau} \cdot \boldsymbol{\varepsilon} \right) \psi$$

$$= -\hbar c M \, \bar{\psi} \left(1 + \frac{i}{2} \gamma_5 \, \boldsymbol{\tau} \cdot \boldsymbol{\varepsilon} \right) \left(1 + \frac{i}{2} \gamma_5 \, \boldsymbol{\tau} \cdot \boldsymbol{\varepsilon} \right) \psi$$

It follows that

$$\mathcal{L}_{\text{mass}} \to -\hbar c M \, \bar{\underline{\psi}} \left(1 + i\gamma_5 \, \underline{\boldsymbol{\tau}} \cdot \boldsymbol{\varepsilon}\right) \underline{\psi} + O(\varepsilon^2)$$

and the Dirac mass term $\mathcal{L}_{\text{mass}} = -\hbar c \, \bar{\underline{\psi}} M \underline{\psi}$ is evidently *not* invariant under the chiral transformation in Eq. (6.138).

Problem 6.11 This problem concerns the sigma-model. While each part is a significant problem on its own, it was decided to group them together because they are clearly related, and because the required algebra is, in fact, carried out in detail in [Walecka (2004)] (albeit with units where $\hbar = c = 1$).

(a) Set $\delta \mathcal{V}_{\text{csb}} = 0$ and use Noether's theorem to derive the conserved currents in Eqs. (6.157);

(b) Now include $\delta \mathcal{V}_{\text{csb}} = \epsilon \sigma$, and derive the field equations. Start from the currents in Eqs. (6.157) and derive the CVC and PCAC relations in Eqs. (6.167);

(d) Verify the form of the lagrangian density in Eq. (6.166) obtained by using the expansions about the value $\sigma = \sigma_0$ in Eqs. (6.164).

Solution to Problem 6.11

The σ-model lagrangian is given in Eq. (6.156) in the text

$$\mathcal{L} = -\hbar c \underline{\bar{\psi}} \left[\gamma_\mu \frac{\partial}{\partial x_\mu} - g(i\gamma_5 \, \underline{\boldsymbol{\tau}} \cdot \boldsymbol{\pi} + \sigma)\right] \underline{\psi} - \frac{c^2}{2} \left(\frac{\partial \boldsymbol{\pi}}{\partial x_\mu} \cdot \frac{\partial \boldsymbol{\pi}}{\partial x_\mu} + \frac{\partial \sigma}{\partial x_\mu} \frac{\partial \sigma}{\partial x_\mu}\right)$$
$$- V\left(\pi^2 + \sigma^2\right)$$

Under an infinitesimal rotation in the internal isospin space, the pion and nucleon fields transform as [see Eqs. (6.8) and (6.32)]

$$\boldsymbol{\pi} \to \boldsymbol{\pi} - \boldsymbol{\varepsilon} \times \boldsymbol{\pi}$$
$$\underline{\psi} \to \underline{\psi} + \frac{i}{2} \, \underline{\boldsymbol{\tau}} \cdot \boldsymbol{\varepsilon} \, \underline{\psi}$$

Since the lagrangian is invariant under this transformation, Noether's theorem implies that the isovector current

$$\tilde{\mathbf{J}}_\mu \cdot \boldsymbol{\varepsilon} = \frac{\partial \mathcal{L}}{\partial(\partial \boldsymbol{\pi}/\partial x_\mu)} \cdot \delta \boldsymbol{\pi} + \frac{\partial \mathcal{L}}{\partial(\partial \underline{\psi}/\partial x_\mu)} \delta \underline{\psi}$$

is conserved. With the use of the appropriate re-scaling factor $-1/\hbar c$, the conserved isovector current reads

$$\mathbf{J}_\mu = i\underline{\bar\psi}\gamma_\mu\frac{1}{2}\underline{\boldsymbol{\tau}}\,\underline{\psi} + \frac{c}{\hbar}\left(\frac{\partial\boldsymbol{\pi}}{\partial x_\mu}\times\boldsymbol{\pi}\right)$$

This reproduces the first of Eqs. (6.157).

Let us now consider an infinitesimal chiral transformation. In this case, the fields change according to Eqs. (6.138) and (6.153)

$$\underline{\psi}\to\underline{\psi}+\frac{i}{2}\gamma_5\,\underline{\boldsymbol{\tau}}\cdot\boldsymbol{\varepsilon}\,\underline{\psi}$$
$$\boldsymbol{\pi}\to\boldsymbol{\pi}-\boldsymbol{\varepsilon}\,\sigma$$
$$\sigma\to\sigma+\boldsymbol{\varepsilon}\cdot\boldsymbol{\pi}$$

As a consequence of the invariance of the σ-model lagrangian under this chiral transformation, there is a conserved axial-vector current, which is again obtained using Noether's theorem. The first of Eqs. (6.140) reports the nucleon contribution to this axial-vector current

$$\mathbf{J}_{5\mu}^{(N)} = i\underline{\bar\psi}\gamma_\mu\gamma_5\frac{1}{2}\underline{\boldsymbol{\tau}}\,\underline{\psi}$$

We proceed to calculate the contribution from the pion and the sigma

$$\tilde{\mathbf{J}}_{5\mu}^{(\pi/\sigma)}\cdot\boldsymbol{\varepsilon} = \frac{\partial\mathcal{L}}{\partial(\partial\boldsymbol{\pi}/\partial x_\mu)}\cdot\delta\boldsymbol{\pi} + \frac{\partial\mathcal{L}}{\partial(\partial\sigma/\partial x_\mu)}\delta\sigma$$
$$= c^2\left(\sigma\frac{\partial\boldsymbol{\pi}}{\partial x_\mu} - \boldsymbol{\pi}\frac{\partial\sigma}{\partial x_\mu}\right)\cdot\boldsymbol{\varepsilon}$$

With the proper re-scaling factor of $-1/\hbar c$, one obtains

$$\mathbf{J}_{5\mu}^{(\pi/\sigma)} = \frac{c}{\hbar}\left(\boldsymbol{\pi}\frac{\partial\sigma}{\partial x_\mu} - \sigma\frac{\partial\boldsymbol{\pi}}{\partial x_\mu}\right)$$

Consequently, the total axial-vector current takes the form of the second of Eqs. (6.157)

$$\mathbf{J}_{5\mu} = i\underline{\bar\psi}\gamma_\mu\gamma_5\frac{1}{2}\underline{\boldsymbol{\tau}}\,\underline{\psi} + \frac{c}{\hbar}\left(\boldsymbol{\pi}\frac{\partial\sigma}{\partial x_\mu} - \sigma\frac{\partial\boldsymbol{\pi}}{\partial x_\mu}\right)$$

(b) We add the intrinsic chiral-symmetry-breaking term of Eq. (6.160)

$$\mathcal{V}(\boldsymbol{\pi}^2+\sigma^2)\to\frac{\lambda}{4}\left[(\boldsymbol{\pi}^2+\sigma^2)-v^2\right]^2+\epsilon\sigma\qquad;\,\epsilon\to 0$$
$$\equiv\mathcal{V}_0(\boldsymbol{\pi}^2+\sigma^2)+\epsilon\sigma$$

We then derive Lagrange's equations for the fields. We have

$$\frac{\partial}{\partial x_\mu}\frac{\partial \mathcal{L}}{\partial(\partial \boldsymbol{\pi}/\partial x_\mu)} - \frac{\partial \mathcal{L}}{\partial \boldsymbol{\pi}} = -c^2 \Box\,\boldsymbol{\pi} - \hbar c\, ig\, \underline{\bar{\psi}}\gamma_5\,\boldsymbol{\tau}\,\underline{\psi} + \frac{\partial V_0}{\partial \boldsymbol{\pi}} = 0$$

$$\frac{\partial}{\partial x_\mu}\frac{\partial \mathcal{L}}{\partial(\partial \sigma/\partial x_\mu)} - \frac{\partial \mathcal{L}}{\partial \sigma} = -c^2 \Box\,\sigma - \hbar c\, g\, \underline{\bar{\psi}}\,\underline{\psi} + \frac{\partial V_0}{\partial \sigma} + \epsilon = 0$$

$$\frac{\partial}{\partial x_\mu}\frac{\partial \mathcal{L}}{\partial(\partial\,\underline{\psi}/\partial x_\mu)} - \frac{\partial \mathcal{L}}{\partial\,\underline{\psi}} = -\hbar c\, \underline{\bar{\psi}}\left[\gamma_\mu \frac{\overleftarrow{\partial}}{\partial x_\mu} + g(i\gamma_5\,\boldsymbol{\tau}\cdot\boldsymbol{\pi} + \sigma)\right] = 0$$

$$\frac{\partial}{\partial x_\mu}\frac{\partial \mathcal{L}}{\partial(\partial\,\underline{\bar{\psi}}/\partial x_\mu)} - \frac{\partial \mathcal{L}}{\partial\,\underline{\bar{\psi}}} = \hbar c\left[\gamma_\mu \frac{\partial}{\partial x_\mu} - g(i\gamma_5\,\boldsymbol{\tau}\cdot\boldsymbol{\pi} + \sigma)\right]\underline{\psi} = 0$$

To derive the CVC and PCAC relations of Eqs. (6.167), we need to use the expressions for the vector and axial-vector currents obtained in (a) and calculate the four divergence, simplifying the resulting expressions using the field equations given above.[13] However, the extra term in the lagrangian $\delta V_{\rm csb} = \epsilon\sigma$ is invariant under isospin transformations, and therefore, from Noether's theorem the previous isovector current remains conserved

$$\frac{\partial \mathbf{J}_\mu}{\partial x_\mu} = 0$$

We may therefore concentrate on the four-divergence of the axial-vector current

$$\frac{\partial \mathbf{J}_{5\mu}}{\partial x_\mu} = \frac{\partial\bar{\psi}}{\partial x_\mu}i\gamma_\mu\gamma_5\frac{1}{2}\boldsymbol{\tau}\,\underline{\psi} + \bar{\psi}\,i\gamma_\mu\gamma_5\frac{1}{2}\boldsymbol{\tau}\,\frac{\partial\psi}{\partial x_\mu} + \frac{c}{\hbar}\left(\boldsymbol{\pi}\Box\sigma - \sigma\Box\boldsymbol{\pi}\right)$$

Since Noether's theorem implies that the current is conserved when the symmetry breaking term $\delta V_{\rm csb} = \epsilon\sigma$ vanishes, the terms in the expression above which are independent of ϵ must cancel out,[14] and the only non-vanishing contribution must originate from $\Box\sigma$, which indeed depends on ϵ, as one sees from the field equation for σ. The relevant contribution of $\Box\sigma$ is ϵ/c^2 and the four-divergence reads

$$\frac{\partial \mathbf{J}_{5\mu}}{\partial x_\mu} \to \frac{c}{\hbar}\left(\boldsymbol{\pi}\frac{\epsilon}{c^2}\right) = \frac{\epsilon\boldsymbol{\pi}}{\hbar c}$$

which is precisely the PCAC relation in Eq. (6.167).

(c) To expand the potential about the new minimum (see Figs. 6.2–6.4 in the text), we first use the expansion $\sigma = \sigma_0 + \varphi$ in Eq. (6.164) inside the

---

[13]The presence of an additional term $\delta V_{\rm csb} = \epsilon\sigma$ in the lagrangian density does not modify the currents obtained in Noether's theorem.

[14]In case of doubt, substitute the field equations in the four-divergence and verify it.

potential $\mathcal{V} = \mathcal{V}_0 + \epsilon\sigma$ in Eq. (6.160)

$$\mathcal{V} = \frac{\lambda}{4}\left[\pi^2 + (\sigma_0 + \varphi)^2 - v^2\right]^2 + \epsilon(\sigma_0 + \varphi)$$

$$= \left[\frac{\lambda}{4}\left(\sigma_0^2 - v^2\right)^2 + \epsilon\sigma_0\right] + \left[\epsilon + \sigma_0\lambda\left(\sigma_0^2 - v^2\right)\right]\varphi +$$

$$\frac{\lambda}{2}(3\sigma_0^2 - v^2)\varphi^2 + \frac{\lambda}{2}(\sigma_0^2 - v^2)\pi^2 + \lambda\sigma_0\varphi(\pi^2 + \varphi^2) + \frac{\lambda}{4}(\pi^2 + \varphi^2)^2$$

Here v^2 is determined by the minimum condition in Eq. (6.163)

$$\left[\frac{d\mathcal{V}(\sigma, 0)}{d\sigma}\right]_{\sigma_0} = 0$$

$$\Rightarrow \qquad \lambda\sigma_0(\sigma_0^2 - v^2) = -\epsilon \qquad\qquad ; \text{ minimum}$$

The following definitions of (M, m_π^2, m_s^2) are introduced in Eqs. (6.165)

$$\sigma_0 \equiv -\frac{M}{g} \qquad ; \epsilon \equiv \frac{M}{g}m_\pi^2 c^2 \qquad ; \lambda \equiv \frac{c^2}{2}\left(\frac{g}{M}\right)^2 (m_s^2 - m_\pi^2)$$

The coefficients of the meson fields in $\mathcal{V}$ are then expressed as

$$\left[\epsilon + \sigma_0\lambda\left(\sigma_0^2 - v^2\right)\right] = 0$$

$$\frac{\lambda}{2}(\sigma_0^2 - v^2) = \frac{m_\pi^2 c^2}{2}$$

$$\frac{\lambda}{2}(3\sigma_0^2 - v^2) = \frac{m_s^2 c^2}{2}$$

$$\lambda\sigma_0 = -\frac{c^2}{2}\frac{g}{M}(m_s^2 - m_\pi^2)$$

$$\left[\frac{\lambda}{4}\left(\sigma_0^2 - v^2\right)^2 + \epsilon\sigma_0\right] = \text{constant}$$

When expanded about the new minimum, the potential thus takes the form

$$\mathcal{V}(\pi, \varphi) = \frac{m_s^2 c^2}{2}\varphi^2 + \frac{m_\pi^2 c^2}{2}\pi^2 - \frac{c^2}{2}\frac{g}{M}(m_s^2 - m_\pi^2)\varphi(\pi^2 + \varphi^2)$$

$$+ \frac{c^2}{8}\left(\frac{g}{M}\right)^2 (m_s^2 - m_\pi^2)(\pi^2 + \varphi^2)^2 + \text{constant}$$

which agrees with the corresponding expression in Eq. (6.166).[15]

Finally, the remaining terms in $\mathcal{L}(\pi, \varphi)$ in Eq. (6.166) are obtained trivially simply using the definitions in Eqs. (6.164).

---

[15]Remember $\mathcal{L} = \mathcal{T} - \mathcal{V}$.

Problem 6.12 This problem explicitly exhibits the $SU(2)_L \otimes SU(2)_R$ symmetry of the σ-model. Define left- and right-handed Dirac fields, and a meson matrix, by [recall Eqs. (6.145)]

$$\hat{\psi}_L \equiv P_L \hat{\psi} \qquad\qquad ; \hat{\psi}_R \equiv P_R \hat{\psi}$$

$$\underline{\chi} \equiv \frac{1}{\sqrt{2}} \left(\underline{1}\sigma + i\underline{\tau} \cdot \underline{\pi} \right)$$

(a) Show the σ-model lagrangian can be written as

$$\mathcal{L} = -\hbar c \left[\hat{\bar{\psi}}_L \gamma_\mu \frac{\partial}{\partial x_\mu} \hat{\psi}_L + \hat{\bar{\psi}}_R \gamma_\mu \frac{\partial}{\partial x_\mu} \hat{\psi}_R - \sqrt{2}\, g \left(\hat{\bar{\psi}}_R \underline{\chi}\, \hat{\psi}_L + \hat{\bar{\psi}}_L \underline{\chi}^\dagger\, \hat{\psi}_R \right) \right]$$
$$- \frac{c^2}{2} \mathrm{tr} \left[\left(\frac{\partial \underline{\chi}}{\partial x_\mu} \right)^* \left(\frac{\partial \underline{\chi}}{\partial x_\mu} \right) \right] - \mathcal{V} \left[\mathrm{tr}\, (\underline{\chi}^\dagger \underline{\chi}) \right]$$

Here "tr" is the trace of the isospin matrix, and $\underline{v}_\mu^* \equiv (\underline{v}^\dagger, +i\underline{v}_0^\dagger)$;[16]

(b) Let $\underline{R} = \exp\{(i/2)\underline{\omega} \cdot \underline{\tau}\}$ be a global SU(2) matrix. Show that $\mathcal{L}$ is invariant under the SU(2) transformation

$$\hat{\psi}_R \to \underline{R}\, \hat{\psi}_R \qquad ; \underline{\chi} \to \underline{R}\,\underline{\chi} \qquad ; \hat{\psi}_L \to \hat{\psi}_L$$

(c) Show that $\mathcal{L}$ is also invariant under the *independent* global SU(2) transformation

$$\hat{\psi}_L \to \underline{L}\, \hat{\psi}_L \qquad ; \underline{\chi} \to \underline{\chi}\, \underline{L}^\dagger \qquad ; \hat{\psi}_R \to \hat{\psi}_R$$

Solution to Problem 6.12

Recall the definitions in Eqs. (6.145)

$$P_L \equiv \frac{1}{2}(1 + \gamma_5) \qquad\qquad ; P_R \equiv \frac{1}{2}(1 - \gamma_5)$$
$$P_L^2 = P_L \qquad\qquad ; P_R^2 = P_R$$
$$P_L P_R = P_R P_L = 0$$

(a) The lagrangian density of the σ-model is defined in Eq. (6.156)

$$\mathcal{L} = -\hbar c \hat{\bar{\psi}} \left[\gamma_\mu \frac{\partial}{\partial x_\mu} - g\, (i\gamma_5 \underline{\tau} \cdot \underline{\pi} + \sigma) \right] \hat{\psi} - \frac{c^2}{2} \left(\frac{\partial \underline{\pi}}{\partial x_\mu} \cdot \frac{\partial \underline{\pi}}{\partial x_\mu} + \frac{\partial \sigma}{\partial x_\mu} \frac{\partial \sigma}{\partial x_\mu} \right)$$
$$- \mathcal{V}(\underline{\pi}^2 + \sigma^2)$$

---

[16] We set $\delta \mathcal{V}_{\mathrm{csb}} \equiv 0$, and the metric is not complex conjugated in $\underline{v}_\mu^*$. Note that the trace is invariant under cyclic permutations of its arguments, $\mathrm{tr}\, \underline{\tau} = 0$, and $\mathrm{tr}\, \underline{1} = 2$.

Introduce the following quantities[17]

$$\psi_L \equiv P_L \psi \qquad\qquad ; \ \psi_R \equiv P_R \psi$$

$$\underline{\chi} \equiv \frac{1}{\sqrt{2}} \left(\underline{1} \sigma + i \underline{\tau} \cdot \boldsymbol{\pi} \right)$$

If "tr" is the trace of the isospin matrix, with $\text{tr} \, \underline{1} = 2$ and $\text{tr} \, \underline{\tau} = 0$, then

$$\text{tr} \, \underline{\chi}^\dagger \underline{\chi} = \frac{1}{2} \text{tr} \, \left(\underline{1} \sigma - i \underline{\tau} \cdot \boldsymbol{\pi} \right) \left(\underline{1} \sigma + i \underline{\tau} \cdot \boldsymbol{\pi} \right)$$

$$= \frac{1}{2} \text{tr} \, \underline{1} \left(\sigma^2 + \boldsymbol{\pi}^2 \right) = \sigma^2 + \boldsymbol{\pi}^2$$

Also, if $\underline{v}^\star_\mu \equiv (\underline{v}^\dagger, + i \underline{v}_0^\dagger)$, then

$$\text{tr} \left[\left(\frac{\partial \underline{\chi}}{\partial x_\mu} \right)^\star \left(\frac{\partial \underline{\chi}}{\partial x_\mu} \right) \right] = \frac{1}{2} \text{tr} \, \left(\underline{1} \frac{\partial \sigma}{\partial x_\mu} - i \underline{\tau} \cdot \frac{\partial \boldsymbol{\pi}}{\partial x_\mu} \right) \left(\underline{1} \frac{\partial \sigma}{\partial x_\mu} + i \underline{\tau} \cdot \frac{\partial \boldsymbol{\pi}}{\partial x_\mu} \right)$$

$$= \frac{1}{2} \text{tr} \, \underline{1} \left(\frac{\partial \boldsymbol{\pi}}{\partial x_\mu} \cdot \frac{\partial \boldsymbol{\pi}}{\partial x_\mu} + \frac{\partial \sigma}{\partial x_\mu} \frac{\partial \sigma}{\partial x_\mu} \right)$$

$$= \frac{\partial \boldsymbol{\pi}}{\partial x_\mu} \cdot \frac{\partial \boldsymbol{\pi}}{\partial x_\mu} + \frac{\partial \sigma}{\partial x_\mu} \frac{\partial \sigma}{\partial x_\mu}$$

Next, evaluate[18]

$$\bar{\underline{\psi}}_L \gamma_\mu \frac{\partial}{\partial x_\mu} \underline{\psi}_L = \bar{\underline{\psi}} \frac{1}{2} (1 - \gamma_5) \gamma_\mu \frac{1}{2} (1 + \gamma_5) \frac{\partial}{\partial x_\mu} \underline{\psi}$$

$$= \bar{\underline{\psi}} \gamma_\mu \frac{1}{2} (1 + \gamma_5) \frac{\partial}{\partial x_\mu} \underline{\psi}$$

In a similar fashion

$$\bar{\underline{\psi}}_R \gamma_\mu \frac{\partial}{\partial x_\mu} \underline{\psi}_R = \bar{\underline{\psi}} \frac{1}{2} (1 + \gamma_5) \gamma_\mu \frac{1}{2} (1 - \gamma_5) \frac{\partial}{\partial x_\mu} \underline{\psi}$$

$$= \bar{\underline{\psi}} \gamma_\mu \frac{1}{2} (1 - \gamma_5) \frac{\partial}{\partial x_\mu} \underline{\psi}$$

Therefore

$$\bar{\underline{\psi}}_L \gamma_\mu \frac{\partial}{\partial x_\mu} \underline{\psi}_L + \bar{\underline{\psi}}_R \gamma_\mu \frac{\partial}{\partial x_\mu} \underline{\psi}_R = \bar{\underline{\psi}} \gamma_\mu \frac{\partial}{\partial x_\mu} \underline{\psi}$$

---

[17] For simplicity, we suppress the hats on the Dirac fields in this solution.
[18] Recall the additional γ_4 in $\bar{\underline{\psi}} = \underline{\psi}^\dagger \gamma_4$.

Finally, consider

$$\sqrt{2}\left(\bar{\underline{\psi}}_R\,\underline{\chi}\,\underline{\psi}_L + \bar{\underline{\psi}}_L\,\underline{\chi}^\dagger\,\underline{\psi}_R\right) = \bar{\underline{\psi}}\,(\underline{1}\sigma + i\underline{\tau}\cdot\boldsymbol{\pi})\,\frac{1}{2}(1+\gamma_5)\underline{\psi} +$$
$$\bar{\underline{\psi}}\,(\underline{1}\sigma - i\underline{\tau}\cdot\boldsymbol{\pi})\,\frac{1}{2}(1-\gamma_5)\underline{\psi}$$
$$= \bar{\underline{\psi}}\,(\underline{1}\sigma + i\gamma_5\underline{\tau}\cdot\boldsymbol{\pi})\,\underline{\psi}$$

A combination of these results allows us to express the σ-model lagrangian density in the stated form

$$\mathcal{L} = -\hbar c\left[\bar{\underline{\psi}}_L\gamma_\mu\frac{\partial}{\partial x_\mu}\underline{\psi}_L + \bar{\underline{\psi}}_R\gamma_\mu\frac{\partial}{\partial x_\mu}\underline{\psi}_R - \sqrt{2}\,g\left(\bar{\underline{\psi}}_R\,\underline{\chi}\,\underline{\psi}_L + \bar{\underline{\psi}}_L\,\underline{\chi}^\dagger\,\underline{\psi}_R\right)\right]$$
$$-\frac{c^2}{2}\mathrm{tr}\left[\left(\frac{\partial\underline{\chi}}{\partial x_\mu}\right)^*\left(\frac{\partial\underline{\chi}}{\partial x_\mu}\right)\right] - \mathcal{V}\left[\mathrm{tr}\left(\underline{\chi}^\dagger\underline{\chi}\right)\right]$$

(b) Let $\underline{R} = \exp\{(i/2)\boldsymbol{\omega}\cdot\underline{\tau}\}$ be a global SU(2) matrix, and consider the SU(2) transformation

$$\underline{\psi}_R \to \underline{R}\,\underline{\psi}_R \qquad ;\ \underline{\chi} \to \underline{R}\,\underline{\chi} \qquad ;\ \underline{\psi}_L \to \underline{\psi}_L$$

It follows that for this transformation

$$\underline{\chi}^\dagger \to \underline{\chi}^\dagger\,\underline{R}^\dagger$$

Since $\boldsymbol{\omega}$ is a constant, the transformation can be moved through any derivatives, and

$$\underline{R}^\dagger\underline{R} = 1$$

It then follows by inspection that $\mathcal{L}$ is left invariant under this transformation.

(c) Consider the *independent* global SU(2) transformation

$$\underline{\psi}_L \to \underline{L}\,\underline{\psi}_L \qquad ;\ \underline{\chi} \to \underline{\chi}\,\underline{L}^\dagger \qquad ;\ \underline{\psi}_R \to \underline{\psi}_R$$

For this transformation

$$\underline{\chi}^\dagger \to \underline{L}\,\underline{\chi}^\dagger$$

Since $\boldsymbol{\omega}$ is a constant, the transformation can again be moved through any derivatives, and

$$\underline{L}^\dagger\underline{L} = 1$$

Furthermore the trace "tr" is invariant under cyclic permutations, It then again follows by inspection that $\mathcal{L}$ is left invariant under this transformation.

Problem 6.13 The generator operators $\hat{G}^a$ for the SU(3) symmetry of the Sakata model are obtained from the fundamental matrices in Eqs. (6.74) through Eq. (6.79).

(a) The *rank* of a group is the number of mutually commuting generators. Show from the matrices involved that SU(3) is a rank 2 group;

(b) The baryon number $\hat{B}$ in the Sakata model arises from the unit 3×3 matrix $\underline{1}$. Conclude that the strangeness operator $\hat{S}$ is obtained from $(\sqrt{3}\underline{\lambda}^8 - \underline{1})/3$, and the hypercharge operator $\hat{Y} = (\hat{B} + \hat{S})$ from $(2\underline{1} + \sqrt{3}\underline{\lambda}^8)/3$;

(c) The third component of isospin $\hat{T}_3$ arises from $\underline{\lambda}^3/2$. Hence prove from the matrices involved that in the Sakata model the charge operator $\hat{Q}$ is given by

$$\hat{Q} = \hat{T}_3 + \frac{1}{2}\hat{Y}$$

This is the Gell-Mann–Nishijima relation.

Solution to Problem 6.13

(a) The Gell-Mann matrices $\underline{\lambda}^a$ for SU(3) are given in Eqs. (6.74). They satisfy the following Lie algebra

$$[\frac{1}{2}\underline{\lambda}^a, \frac{1}{2}\underline{\lambda}^b] = if^{abc}\frac{1}{2}\underline{\lambda}^c \qquad ; (a,b,c) = 1, 2, \cdots, 8$$

where the f^{abc} are the structure constants; they are real, antisymmetric in the indices (abc), and evaluated in [Amore and Walecka (2013)]

$$f^{123} = 1$$
$$f^{147} = f^{165} = f^{246} = f^{572} = f^{345} = f^{376} = \frac{1}{2}$$
$$f^{458} = f^{678} = \frac{\sqrt{3}}{2}$$

It is evident that while the two diagonal matrices $(\underline{\lambda}^3, \underline{\lambda}^8)$ commute with each other, there is no additional matrix that commutes with both of them.

The generators in the Sakata model are given in Eq. (6.79)

$$\hat{G}^a = \int_\Omega d^3x \, \hat{\underline{\psi}}^\dagger(\mathbf{x}) \frac{1}{2}\underline{\lambda}^a \, \hat{\underline{\psi}}(\mathbf{x})$$

Their commutator is given by [compare Eq. (6.38) and Prob. 6.3]

$$[\hat{G}^a, \hat{G}^a] = \int_\Omega d^3x \, \hat{\underline{\psi}}^\dagger(\mathbf{x}) [\frac{1}{2}\underline{\lambda}^a, \frac{1}{2}\underline{\lambda}^b] \, \hat{\underline{\psi}}(\mathbf{x})$$

Thus the largest set of mutually commuting generators consists of two members $(\hat{G}^3, \hat{G}^8)$, and the rank of the SU(3) symmetry group of the Sakata model is 2.[19]

(b) All three baryons in the Sakata model carry baryon number $B = 1$. The baryon operator $\hat{B}$ is then obtained from the unit 3×3 matrix

$$\hat{B} = \int_\Omega d^3x \, \hat{\underline{\psi}}^\dagger(\mathbf{x}) \underline{1} \, \hat{\underline{\psi}}(\mathbf{x})$$

The matrices $(\underline{\lambda}^3, \underline{\lambda}^8)$ are given by

$$\underline{\lambda}^3 = \begin{pmatrix} 1 & & \\ & -1 & \\ & & 0 \end{pmatrix} \qquad ; \underline{\lambda}^8 = \begin{pmatrix} 1/\sqrt{3} & & \\ & 1/\sqrt{3} & \\ & & -2/\sqrt{3} \end{pmatrix}$$

The third baryon in the Sakata model has strangeness $S = -1$, and thus the strangeness operator $\hat{S}$ is obtained from the matrix

$$\begin{pmatrix} 0 & & \\ & 0 & \\ & & -1 \end{pmatrix} = \frac{1}{\sqrt{3}} \begin{pmatrix} 1/\sqrt{3} & & \\ & 1/\sqrt{3} & \\ & & -2/\sqrt{3} \end{pmatrix} - \frac{1}{3} \begin{pmatrix} 1 & & \\ & 1 & \\ & & 1 \end{pmatrix}$$

The strangeness operator is therefore

$$\hat{S} = \int_\Omega d^3x \, \hat{\underline{\psi}}^\dagger(\mathbf{x}) \frac{1}{3} \left(\sqrt{3}\underline{\lambda}^8 - \underline{1} \right) \hat{\underline{\psi}}(\mathbf{x})$$

The *hypercharge* operator $\hat{Y} \equiv \hat{B} + \hat{S}$ is then obtained from the matrix

$$\begin{pmatrix} 1 & & \\ & 1 & \\ & & 0 \end{pmatrix} = \frac{2}{3} \begin{pmatrix} 1 & & \\ & 1 & \\ & & 1 \end{pmatrix} + \frac{1}{\sqrt{3}} \begin{pmatrix} 1/\sqrt{3} & & \\ & 1/\sqrt{3} & \\ & & -2/\sqrt{3} \end{pmatrix}$$

---

[19]It is only mutually commuting operators that can be simultaneously diagonalized (see [Walecka (2013)]).

It is given by

$$\hat{Y} = \hat{B} + \hat{S} = \int_\Omega d^3x \, \underline{\hat{\psi}}^\dagger(\mathbf{x}) \frac{1}{3} \left(2\underline{1} + \sqrt{3}\underline{\lambda}^8 \right) \underline{\hat{\psi}}(\mathbf{x})$$

(c) The first two baryons in the Sakata model form an isodoublet, and hence the third component of isospin is obtained from the matrix

$$\begin{pmatrix} 1/2 & & \\ & -1/2 & \\ & & 0 \end{pmatrix} = \frac{1}{2}\underline{\lambda}^3$$

The corresponding operator is

$$\hat{T}_3 = \int_\Omega d^3x \, \underline{\hat{\psi}}^\dagger(\mathbf{x}) \frac{1}{2}\underline{\lambda}^3 \, \underline{\hat{\psi}}(\mathbf{x})$$

The first baryon carries charge $Q = +1$. The appropriate charge matrix is thus

$$\begin{pmatrix} 1 & & \\ & 0 & \\ & & 0 \end{pmatrix} = \begin{pmatrix} 1/2 & & \\ & -1/2 & \\ & & 0 \end{pmatrix} + \frac{1}{2}\begin{pmatrix} 1 & & \\ & 1 & \\ & & 0 \end{pmatrix}$$

Hence the corresponding operators satisfy the relation

$$\hat{Q} = \hat{T}_3 + \frac{1}{2}\hat{Y}$$

This is the Gell-Mann–Nishijima relation.

Problem 6.14 Consider the proof of the fundamental SU(3) matrix relations in Eqs. (6.76), which is here carried out to first order in ε^a where $\omega^a \equiv \varepsilon^a \to 0$. Write

$$\underline{r}(\varepsilon) = 1 + \frac{i}{2}\varepsilon^a \underline{\lambda}^a$$

(a) Show to $O(\varepsilon)$ that

$$\underline{r}(\varepsilon)^\dagger = \underline{r}(\varepsilon)^{-1} \qquad ; \underline{r}(\varepsilon)^{-1} = \underline{r}(-\varepsilon) \qquad ; \underline{r}(0) = 1$$
$$\underline{r}(\varepsilon)\underline{r}(\varepsilon') = \underline{r}(\varepsilon'') \qquad ; \varepsilon'' = \varepsilon' + \varepsilon$$

(b) Use the definition of the determinant in Eq. (D.16) to show to $O(\varepsilon)$

$$\det \underline{r}(\varepsilon) = 1 + \frac{i}{2}\varepsilon^a \, \lambda^a_{i_p i_p} = 1 + \frac{i}{2}\varepsilon^a \, \mathrm{tr}\,\underline{\lambda}^a = 1$$

Solution to Problem 6.14

Here $\underline{\lambda}^a$ are the eight hermitian, traceless, 3×3 matrices in Eq. (6.74), ε^a are real infinitesimals, and the repeated index a is summed from 1 to 8.[20] We work through $O(\varepsilon)$.

(a) The first set of properties follows directly:

- Evidently

$$\underline{r}(0) = 1$$

- Take the adjoint of $\underline{r}(\varepsilon)$

$$\underline{r}^\dagger(\varepsilon) = 1 - \frac{i}{2}\varepsilon^a \underline{\lambda}^{a\dagger} = 1 - \frac{i}{2}\varepsilon^a \underline{\lambda}^a$$
$$\underline{r}^\dagger(\varepsilon)\underline{r}(\varepsilon) = 1 + O(\varepsilon^2)$$

Thus

$$\underline{r}(\varepsilon)^\dagger = \underline{r}(\varepsilon)^{-1}$$

- The same demonstration works for $\underline{r}(-\varepsilon)$, and hence

$$\underline{r}(-\varepsilon) = \underline{r}(\varepsilon)^{-1}$$

- Consider the product

$$\underline{r}(\varepsilon)\underline{r}(\varepsilon') = 1 + \frac{i}{2}(\varepsilon' + \varepsilon)^a \underline{\lambda}^a + O(\varepsilon^2)$$

Therefore

$$\underline{r}(\varepsilon)\underline{r}(\varepsilon') = \underline{r}(\varepsilon'') \qquad ; \; \varepsilon'' = \varepsilon' + \varepsilon$$

(b) Through $O(\varepsilon)$, the determinant of $\underline{r}(\varepsilon)$ arises from the product of diagonal elements

$$\det \underline{r}(\varepsilon) = \prod_{i_p=1}^{3} \left(1 + \frac{i}{2}\varepsilon^a \underline{\lambda}^a\right)_{i_p i_p} + O(\varepsilon^2)$$

$$= 1 + \frac{i}{2}\varepsilon^a \sum_{i_p=1}^{3} \lambda^a_{i_p i_p}$$

$$= 1 + \frac{i}{2}\varepsilon^a \operatorname{tr}\underline{\lambda}^a = 1$$

This establishes the relations in Eqs. (6.76) through $O(\varepsilon)$.

---

[20] The 1 in $\underline{r}(\varepsilon)$ is the unit matrix.

Chapter 7

Feynman Rules

Problem 7.1 Write out the fields in Eqs. (7.2)–(7.3), and verify Eqs. (7.10).

Solution to Problem 7.1

The destruction (positive frequency) and creation (negative frequency) parts of the fields in Eqs. (7.2)–(7.3) are

$$\hat{\phi}^{(+)}(\mathbf{x},t) = \sum_{\mathbf{k}} \left(\frac{\hbar}{2\omega_k^s \Omega}\right)^{1/2} \left[c_{\mathbf{k}} e^{i(\mathbf{k}\cdot\mathbf{x}-\omega_k^s t)}\right] \qquad ; \ \omega_k^s = c\sqrt{\mathbf{k}^2 + m_s^2}$$

$$\hat{\phi}^{(-)}(\mathbf{x},t) = \sum_{\mathbf{k}} \left(\frac{\hbar}{2\omega_k^s \Omega}\right)^{1/2} \left[c_{\mathbf{k}}^\dagger e^{-i(\mathbf{k}\cdot\mathbf{x}-\omega_k^s t)}\right]$$

$$\hat{\psi}^{(+)}(\mathbf{x},t) = \frac{1}{\sqrt{\Omega}} \sum_{\mathbf{k}} \sum_{\lambda} \left[a_{\mathbf{k}\lambda} u(\mathbf{k}\lambda) e^{i(\mathbf{k}\cdot\mathbf{x}-\omega_k t)}\right] \qquad ; \ \omega_k = c\sqrt{\mathbf{k}^2 + M^2}$$

$$\hat{\psi}^{(-)}(\mathbf{x},t) = \frac{1}{\sqrt{\Omega}} \sum_{\mathbf{k}} \sum_{\lambda} \left[b_{\mathbf{k}\lambda}^\dagger v(-\mathbf{k}\lambda) e^{-i(\mathbf{k}\cdot\mathbf{x}-\omega_k t)}\right]$$

$$\hat{\bar{\psi}}^{(+)}(\mathbf{x},t) = \frac{1}{\sqrt{\Omega}} \sum_{\mathbf{k}} \sum_{\lambda} \left[b_{\mathbf{k}\lambda} \bar{v}(-\mathbf{k}\lambda) e^{i(\mathbf{k}\cdot\mathbf{x}-\omega_k t)}\right]$$

$$\hat{\bar{\psi}}^{(-)}(\mathbf{x},t) = \frac{1}{\sqrt{\Omega}} \sum_{\mathbf{k}} \sum_{\lambda} \left[a_{\mathbf{k}\lambda}^\dagger \bar{u}(\mathbf{k}\lambda) e^{-i(\mathbf{k}\cdot\mathbf{x}-\omega_k t)}\right]$$

The contraction is given in Eq. (7.8) as the vacuum expectation value of the time-ordered P-product

$$\hat{U} \cdot \hat{V} \cdot = \langle 0 | P(\hat{U}\hat{V}) | 0 \rangle$$

Since one needs both a creation and destruction operator of the same type to obtain a non-zero vacuum expectation value, Eqs. (7.10) follow immediately

$$\hat{\phi}^{(+)} \cdot \hat{\phi}^{(+)\bullet} = \hat{\phi}^{(-)} \cdot \hat{\phi}^{(-)\bullet} = 0$$

$$\hat{\psi}^{(+)} \cdot \hat{\psi}^{(+)\bullet} = \hat{\psi}^{(-)} \cdot \hat{\psi}^{(-)\bullet} = \hat{\bar{\psi}}^{(+)} \cdot \hat{\psi}^{(+)\bullet} = \hat{\bar{\psi}}^{(-)} \cdot \hat{\psi}^{(-)\bullet} =$$

$$\hat{\psi}^{(+)} \cdot \hat{\psi}^{(-)\bullet} = \hat{\bar{\psi}}^{(+)} \cdot \hat{\psi}^{(-)\bullet} = 0$$

Problem 7.2 (a) Evaluate the matrix elements of the scalar field in Eqs. (7.12), and show that one obtains the appropriate sums over modes for the Feynman propagator as given in Eqs. (F.43) and (F.39);

(b) Use the distributive law to then verify Eq. (7.13);

(c) Repeat parts (a) and (b) for the Dirac field in Eqs. (7.15)–(7.18), where the appropriate sums over modes are given in Eqs. (F.47)–(F.48).

Solution to Problem 7.2

(a) We are to evaluate the matrix elements of the fields in Eqs. (7.12), making use of the decompositions in Prob. 7.1. One has

$$\langle 0|\hat{\phi}^{(+)}(x)\hat{\phi}^{(-)}(y)|0\rangle = \sum_k \frac{\hbar}{2\omega_k^s \Omega} e^{ik \cdot (x-y)} = \frac{\hbar}{ic}\Delta_+(x-y)$$

Here $k \cdot x = \mathbf{k} \cdot \mathbf{x} - \omega_k^s t$ and we have identified Δ_+ as in Eq. (F.43). In a similar fashion

$$\langle 0|\hat{\phi}^{(+)}(y)\hat{\phi}^{(-)}(x)|0\rangle = \sum_k \frac{\hbar}{2\omega_k^s \Omega} e^{-ik \cdot (x-y)} = -\frac{\hbar}{ic}\Delta_-(x-y)$$

(b) With the use of the distributive law we have

$$\langle 0|T[\hat{\phi}(x)\hat{\phi}(y)]|0\rangle = \langle 0|\hat{\phi}^{(+)}(x)\hat{\phi}^{(-)}(y)|0\rangle = \frac{\hbar}{ic}\Delta_+(x-y) \qquad ; t_x > t_y$$

$$= \langle 0|\hat{\phi}^{(+)}(y)\hat{\phi}^{(-)}(x)|0\rangle = -\frac{\hbar}{ic}\Delta_-(x-y) \qquad ; t_y > t_x$$

It is evident from Eq. (F.39) that this is just the Feynman propagator, thus verifying Eq. (7.13)

$$\hat{\phi}(x) \cdot \hat{\phi}(y)^\bullet = \langle 0|T[\hat{\phi}(x)\,\hat{\phi}(y)]|0\rangle = \frac{\hbar}{ic}\Delta_F(x-y; m^2)$$

(c) For the Dirac field, an analogous argument gives

$$\langle 0|\hat{\psi}_\alpha^{(+)}(x)\hat{\bar{\psi}}_\beta^{(-)}(y)|0\rangle = \frac{1}{\Omega}\sum_{\mathbf{k}}\sum_\lambda u_\alpha(\mathbf{k}\lambda)\bar{u}_\beta(\mathbf{k}\lambda)\,e^{ik\cdot(x-y)}$$

$$= \frac{1}{\Omega}\sum_{\mathbf{k}}\left[\frac{M-i\gamma_\mu k_\mu}{2\omega_k/c}\right]_{\alpha\beta}e^{ik\cdot(x-y)}$$

$$= i\left[\gamma_\mu\frac{\partial}{\partial x_\mu}-M\right]_{\alpha\beta}\Delta_+(x-y)$$

where use has been made of the positive-energy projection operator in Eqs. (F.22), and the subsequent analysis in Eqs. (F.47). In a similar manner

$$-\langle 0|\hat{\bar{\psi}}_\beta^{(+)}(y)\hat{\psi}_\alpha^{(-)}(x)|0\rangle = -\frac{1}{\Omega}\sum_{\mathbf{k}}\sum_\lambda v_\alpha(-\mathbf{k}\lambda)\bar{v}_\beta(-\mathbf{k}\lambda)\,e^{-ik\cdot(x-y)}$$

$$= -\frac{1}{\Omega}\sum_{\mathbf{k}}\left[\frac{-M-i\gamma_\mu k_\mu}{2\omega_k/c}\right]_{\alpha\beta}e^{-ik\cdot(x-y)}$$

$$= -i\left[\gamma_\mu\frac{\partial}{\partial x_\mu}-M\right]_{\alpha\beta}\Delta_-(x-y)$$

The distributive law then reproduces the Feynman propagator for the Dirac particle in Eqs. (F.49) and (7.17)

$$\langle 0|P[\hat{\psi}_\alpha(x),\,\hat{\bar{\psi}}_\beta(y)]|0\rangle = i\left[\gamma_\mu\frac{\partial}{\partial x_\mu}-M\right]_{\alpha\beta}\Delta_F(x-y;\,M^2)$$

$$\equiv iS_{\alpha\beta}^F(x-y)$$

Problem 7.3 Show that if the massive particles (N,ϕ) are stable, then the first-order scattering operator $\hat{S}^{(1)}$ vanishes in the Dirac-scalar theory.

Solution to Problem 7.3

The first-order scattering operator in the Dirac-scalar theory is

$$\hat{S}^{(1)} = \frac{ig}{\hbar c}\int d^4x :\hat{\bar{\psi}}(x)\hat{\psi}(x): \hat{\phi}(x)$$

Insert the field expansions. The integral over all space-time then gives

$$\int d^4x\,e^{i(k_1+k_2+l)\cdot x} = (2\pi)^4\delta^{(4)}(k_1+k_2+l)$$

where the particles' four-momenta (k_1,k_2,l) may have either sign, depending on the process described by this vertex. If the massive particles (N,ϕ)

are stable, there is no real process involving a single $\bar{N}N\phi$ vertex which conserves four-momentum, and hence this delta-function vanishes.

Problem 7.4 (a) Find the matrix elements S_{fi} to second order in g in the Dirac-scalar theory for the processes $N+\bar{N} \to \phi+\phi$ and $\phi+\phi \to N+\bar{N}$;

(b) Use these results to extend the Feynman rules in momentum space to take into account incoming and outgoing antifermion lines in this theory:

- Include a factor of $(1/\Omega)^{1/2}\bar{v}(-\mathbf{k}\lambda)$ for each incoming $\bar{N}$-line with incoming four-momentum k (opposite to the line's direction);
- Include a factor of $(1/\Omega)^{1/2}v(-\mathbf{k}\lambda)$ for each outgoing $\bar{N}$-line with outgoing four-momentum k (opposite to the line's direction). See Fig. 8.7 in the text.

Solution to Problem 7.4

(a) We follow the arguments in section 7.2.1. For the process $N+\bar{N} \to \phi+\phi$, the initial and final states are

$$|i\rangle = a^\dagger_{\mathbf{k}_1\lambda} b^\dagger_{\mathbf{k}_3\lambda'}|0\rangle \qquad ; \ |f\rangle = c^\dagger_{\mathbf{k}_2} c^\dagger_{\mathbf{k}_4}|0\rangle$$

Then, just as in Eq. (7.37),

$$S_{fi} = \left(\frac{ig}{\hbar c}\right)^2 \frac{1}{\Omega^2} \frac{\hbar}{\sqrt{4\omega_2^s \omega_4^s}} \int d^4x_1 \int d^4x_2$$
$$\left[\bar{v}_\alpha(-\mathbf{k}_3\lambda')iS^F_{\alpha\beta}(x_1 - x_2)u_\beta(\mathbf{k}_1\lambda)e^{i(k_1-k_2)\cdot x_2}e^{i(k_3-k_4)\cdot x_1} + \right.$$
$$\left. \bar{v}_\alpha(-\mathbf{k}_3\lambda')iS^F_{\alpha\beta}(x_1 - x_2)u_\beta(\mathbf{k}_1\lambda)e^{i(k_1-k_4)\cdot x_2}e^{i(k_3-k_2)\cdot x_1}\right]$$

For the process $\phi+\phi \to N+\bar{N}$, the initial and final states are

$$|i\rangle = c^\dagger_{\mathbf{k}_1} c^\dagger_{\mathbf{k}_3}|0\rangle \qquad ; \ |f\rangle = a^\dagger_{\mathbf{k}_2\lambda} b^\dagger_{\mathbf{k}_4\lambda'}|0\rangle$$

Then, again just as in Eq. (7.37),

$$S_{fi} = \left(\frac{ig}{\hbar c}\right)^2 \frac{1}{\Omega^2} \frac{\hbar}{\sqrt{4\omega_1^s \omega_3^s}} \int d^4x_1 \int d^4x_2$$
$$\left[\bar{u}_\alpha(\mathbf{k}_2\lambda)iS^F_{\alpha\beta}(x_1 - x_2)v_\beta(-\mathbf{k}_4\lambda')e^{i(k_3-k_4)\cdot x_2}e^{i(k_1-k_2)\cdot x_1} + \right.$$
$$\left. \bar{u}_\alpha(\mathbf{k}_2\lambda)iS^F_{\alpha\beta}(x_1 - x_2)v_\beta(-\mathbf{k}_4\lambda')e^{i(k_1-k_4)\cdot x_2}e^{i(k_3-k_2)\cdot x_1}\right]$$

(b) It is evident from these results that the Feynman rules in momentum space can be extended to take into account incoming and outgoing antifermion lines in this theory:

- Include a factor of $(1/\Omega)^{1/2}\bar{v}(-\mathbf{k}\lambda)$ for each incoming $\bar{N}$-line with incoming four-momentum k (opposite to the line's direction);
- Include a factor of $(1/\Omega)^{1/2}v(-\mathbf{k}\lambda)$ for each outgoing $\bar{N}$-line with outgoing four-momentum k (opposite to the line's direction). See Fig. 8.7 in the text.

Problem 7.5 (a) Draw all connected Feynman diagrams through order g^4 in the Dirac-scalar theory for the processes $\phi + N \to \phi + N$, $N + N \to N + N$, and the self-energies of (N, ϕ);

(b) Use the Feynman rules in momentum space to identify all the elements in the diagrams for $N(k_1) + N(k_2) \to N(k_3) + N(k_4)$.

Solution to Problem 7.5

(a) The Feynman diagrams for the process $\phi + N \to \phi + N$ through order g^4 are illustrated in Figs. 7.1–7.6.[1] The solid lines are used for the fermion, while the dashed lines for the scalar. The diagrams are to be summed to produce the scattering amplitude. In particular, Fig. 7.1 displays the contributions of order g^2, corresponding to Figs. 7.1–7.2 of [Walecka (2010)].

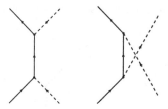

Fig. 7.1 Feynman diagrams for the process $\phi + N \to \phi + N$ of order g^2.

The remaining figures correspond to the different contributions of order g^4, organized in different groups: in Fig. 7.2 are represented the diagrams containing self-energy corrections to the fermion lines;[2] in Fig. 7.3 are rep-

---

[1]The Feynman diagrams displayed in this problem have been drawn using the program jaxodraw [Binosi and Theussi (2004); Binosi *et al.* (2009)], which is a java implementation of the latex-style axodraw [Vermaseren (1994)].

[2]For simplicity, we suppress all graphs containing the corresponding mass counter-term insertions.

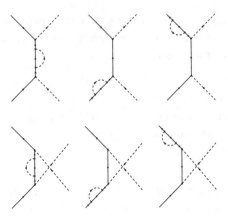

Fig. 7.2   Feynman diagrams for the process $\phi + N \to \phi + N$ of order g^4 containing fermion self-energy contributions.

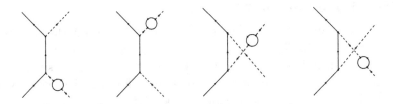

Fig. 7.3   Feynman diagrams for the process $\phi + N \to \phi + N$ of order g^4 containing scalar self-energy contributions.

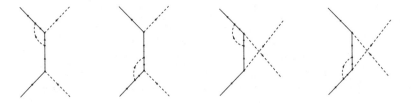

Fig. 7.4   Feynman diagrams for the process $\phi + N \to \phi + N$ of order g^4 containing vertex corrections.

resented the diagrams containing self-energy corrections to the scalar lines; while in Fig. 7.4 are represented all the diagrams containing corrections to the $NN\phi$ vertex. In the diagrams of Figs. 7.5 the scalar particle is absorbed and emitted on an internal fermionic leg, while in the diagrams of Fig. 7.6

the process $\phi + N \to \phi + N$ takes place through a effective three-scalar vertex.

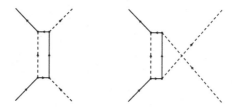

Fig. 7.5   Additional Feynman diagrams for the process $\phi + N \to \phi + N$ of order g^4.

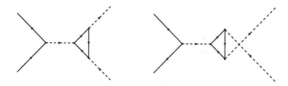

Fig. 7.6   Additional Feynman diagrams for the process $\phi + N \to \phi + N$ of order g^4.

The Feynman diagrams for the process $N + N \to N + N$ through order g^4 are illustrated in in Figs. 7.7–7.11. In particular, Fig. 7.7 displays the diagrams of order g^2. It is assumed here that the two nucleons are identical; this results in the appearance of the exchange amplitude, which enters with opposite sign.

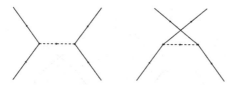

Fig. 7.7   Feynman diagrams for the process $N + N \to N + N$ of order g^2.

Figures 7.8–7.10 display the contributions of order g^4, respectively containing fermion self-energy corrections, scalar self-energy corrections, and vertex corrections.

Figure 7.11 shows the additional Feynman diagrams for the process $N + N \to N + N$ of order g^4. These are the ladder and crossed-ladder diagrams for $N + N \rightarrow N + N$ of order g^4.

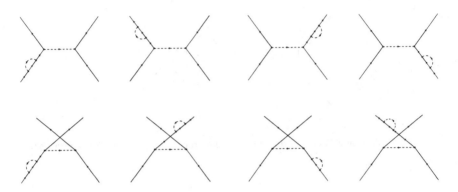

Fig. 7.8 Feynman diagrams for the process $N + N \rightarrow N + N$ of order g^4 containing fermion self-energy contributions.

Fig. 7.9 Feynman diagrams for the process $N + N \rightarrow N + N$ of order g^4 containing scalar self-energy contributions.

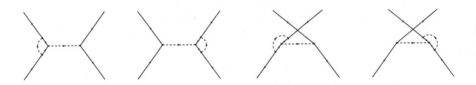

Fig. 7.10 Feynman diagrams for the process $N + N \rightarrow N + N$ of order g^4 containing vertex corrections.

In Figs. 7.12 and 7.13 we show the Feynman diagrams for the fermion self-energy of orders g^2 and g^4 respectively. Note the presence of the *improper* self-energy in the fourth-order contribution.

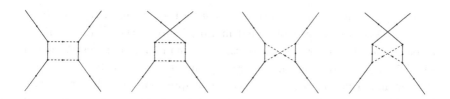

Fig. 7.11   Additional Feynman diagrams for the process $N + N \to N + N$ of order g^4.

In Figs. 7.14 and 7.15 we show the Feynman diagrams for the scalar self-energy of orders g^2 and g^4 respectively. Again, note the presence of the *improper* self-energy in the fourth-order contribution.

Fig. 7.12   Fermion self-energy correction of order g^2.

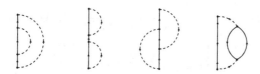

Fig. 7.13   Fermion self-energy corrections of order g^4.

Fig. 7.14   Scalar self-energy correction of order g^2.

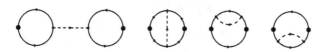

Fig. 7.15   Scalar self-energy corrections of order g^4.

(b) The second-order S-matrix element for the process $N(k_1) + N(k_2) \rightarrow N(k_3) + N(k_4)$ has been obtained in Eq. (7.46) of the text, and it corresponds to the contributions of the Feynman diagrams represented in Fig. 7.7. This expression can be obtained directly from the diagrams, using the Feynman rules of section 7.4, as one can appreciate from Figures 7.3 and 7.4 of the text, which describe the process through order g^2 in coordinate and momentum space respectively.

Through order g^4 we have many more diagrams contributing to the process, so we will just focus on a pair of diagrams: we consider the diagrams of Fig. 7.9. With the application of the Feynman rules for the Dirac-scalar theory we may write down explicitly the contributions to the S-matrix element stemming from the diagrams in Fig. 7.9 as follows[3]

$$
\delta S_{fi}^{(4)} = (2\pi)^4 \delta^{(4)}(k_1 + k_2 - k_3 - k_4) \frac{1}{\Omega^2} \left(\frac{ig}{\hbar c} \right)^4 \left(\frac{\hbar}{ic} \right)^2 \left(\frac{1}{i} \right)^2 (-1) \times
$$

$$
\left\{ \bar{u}(\mathbf{k_3}\lambda_3) u(\mathbf{k_1}\lambda_1) \frac{1}{[(k_3 - k_1)^2 + m_s^2]^2} \, \bar{u}(\mathbf{k_4}\lambda_4) u(\mathbf{k_2}\lambda_2) \times \right.
$$

$$
\int \frac{d^4 p}{(2\pi)^4} \mathrm{Tr} \left[\frac{1}{i\gamma_\mu p_\mu + M} \frac{1}{i\gamma_\mu(p + k_3 - k_1)_\mu + M} \right] -
$$

$$
\bar{u}(\mathbf{k_4}\lambda_4) u(\mathbf{k_1}\lambda_1) \frac{1}{[(k_4 - k_1)^2 + m_s^2]^2} \, \bar{u}(\mathbf{k_3}\lambda_3) u(\mathbf{k_2}\lambda_2) \times
$$

$$
\left. \int \frac{d^4 p}{(2\pi)^4} \mathrm{Tr} \left[\frac{1}{i\gamma_\mu p_\mu + M} \frac{1}{i\gamma_\mu(p + k_4 - k_1)_\mu + M} \right] \right\}
$$

Notice that from the expression above we may read off the scalar self-energy corrections of order g^2, which appear in Fig. 7.14.

This process can be repeated for all the remaining diagrams of order g^4, thus obtaining the complete expression for the S-matrix element of order g^4.

Problem 7.6 (a) Consider the Dirac-(scalar)^2 theory containing two independent scalar fields (ϕ_1, ϕ_2), with coupling constants and masses (g_1, m_1) and (g_2, m_2) respectively. Assume that $m_1 \neq m_2$ and that all the particles are stable. The interaction hamiltonian density is now

$$
\mathcal{H}_1 = \mathcal{H}_1^{(1)} + \mathcal{H}_1^{(2)}
$$

$$
\mathcal{H}_1^{(i)} = -g_i \bar{\psi}\psi\phi_i \qquad ; \, i = 1, 2
$$

Show that the scattering operator takes the form

---

[3] As in the text, "Tr'" denotes the trace of the Dirac matrix product.

$$\hat{S} = \sum_{l=0}^{\infty} \sum_{m=0}^{\infty} \left(\frac{ig_1}{\hbar c}\right)^l \left(\frac{ig_2}{\hbar c}\right)^m \frac{1}{l!} \frac{1}{m!} \int d^4x_1 \cdots \int d^4x_l \int d^4y_1 \cdots \int d^4y_m \times$$

$$P[:\hat{\bar{\psi}}(x_1)\hat{\psi}(x_1): \cdots :\hat{\bar{\psi}}(x_l)\hat{\psi}(x_l)::\hat{\bar{\psi}}(y_1)\hat{\psi}(y_1): \cdots :\hat{\bar{\psi}}(y_m)\hat{\psi}(y_m):] \times$$

$$P[\hat{\phi}_1(x_1)\cdots\hat{\phi}_1(x_l)] P[\hat{\phi}_2(y_1)\cdots\hat{\phi}_2(y_m)]$$

The convention here is that the term with $l = 0$ is the scattering operator with $\mathcal{H}_1^{(1)} = 0$, and the term with $m = 0$ is the scattering operator with $\mathcal{H}_1^{(2)} = 0$.[4]

(b) Apply Wick's theorem up through second order in the coupling constants (that is, up through order $l + m = 2$).

Solution to Problem 7.6

The starting point of our discussion is Eq. (7.1) of the text

$$\hat{S} = \sum_{n=0}^{\infty} \left(-\frac{i}{\hbar}\right)^n \frac{1}{n!} \int_{-\infty}^{\infty} dt_1 \cdots \int_{-\infty}^{\infty} dt_n \, T\left[\hat{H}_I(t_1)\hat{H}_I(t_2)\cdots\hat{H}_I(t_n)\right]$$

where, in the present case,

$$\hat{H}_I(t) = \hat{H}_I^{(1)}(t) + \hat{H}_I^{(2)}(t)$$

$$= -g_1 \int d^3x :\hat{\bar{\psi}}(\mathbf{x},t)\hat{\psi}(\mathbf{x},t): \hat{\phi}_1(\mathbf{x},t) - g_2 \int d^3x :\hat{\bar{\psi}}(\mathbf{x},t)\hat{\psi}(\mathbf{x},t): \hat{\phi}_2(\mathbf{x},t)$$

If we substitute the expression for $\hat{H}(t)$ inside Eq. (7.1), we see that the scattering operator will contain terms of the form

$$T\left\{[\hat{H}_I^{(1)}(t_1) + \hat{H}_I^{(2)}(t_1)] \cdots [\hat{H}_I^{(1)}(t_n) + \hat{H}_I^{(2)}(t_n)]\right\}$$

The factor inside the T-product can be expanded, and its components may be organized in different groups:

- Terms which contain only the operator $\hat{H}_I^{(1)}$;
- Terms which contain a product of $n - 1$ operators $\hat{H}_I^{(1)}$ and one operator $\hat{H}_I^{(2)}$ (all at different times);
- $\cdots$
- Terms which contain only the operator $\hat{H}_I^{(2)}$.

---

[4]*Hint*: Show all the terms with a given partition $(\hat{\phi}_1)^l(\hat{\phi}_2)^m$ are identical by a change of dummy variables, and then appeal to the binomial theorem.

It is easy to convince oneself that a group with l factors $\hat{H}_I^{(1)}$ and m factors $\hat{H}_I^{(2)}$ will contain $(l+m)!/l!\,m!$ terms (clearly in our case $l+m=n$).

Since we are dealing with operators and not c-numbers we have to use extra care in handling these expressions; however, given that these terms are inside a T-product and that an integration over the times has to be performed, we are allowed to change the ordering inside the T-product and then make a change of dummy variables. In this way we may organize the expression as

$$T\left\{[\hat{H}_I^{(1)}(t_1)+\hat{H}_I^{(2)}(t_1)]\cdots[\hat{H}_I^{(1)}(t_n)+\hat{H}_I^{(2)}(t_n)]\right\} \longrightarrow$$
$$T\left[\hat{H}_I^{(1)}(t_1)\ldots\hat{H}_I^{(1)}(t_n)\right]$$
$$+\frac{n!}{1!(n-1)!}T\left[\hat{H}_I^{(1)}(t_1)\cdots\hat{H}_I^{(1)}(t_{n-1})\hat{H}_I^{(2)}(t_n)\right]$$
$$+\frac{n!}{2!(n-2)!}T\left[\hat{H}_I^{(1)}(t_1)\cdots\hat{H}_I^{(1)}(t_{n-2})\hat{H}_I^{(2)}(t_{n-1})\hat{H}_I^{(2)}(t_n)\right]$$
$$+\cdots+T\left[\hat{H}_I^{(2)}(t_1)\cdots\hat{H}_I^{(2)}(t_n)\right]$$

Therefore Eq. (7.1) takes the form

$$\hat{S}=\sum_{n=0}^{\infty}\left(-\frac{i}{\hbar c}\right)^n\frac{1}{n!}\sum_{l=0}^{n}\frac{n!}{l!(n-l)!}(-g_1)^l(-g_2)^{n-l}\int d^4x_1\ldots\int d^4x_l \times$$
$$\int d^4y_1\ldots\int d^4y_{n-l}\,T\left[\mathcal{H}_1^{(1)}(x_1)\ldots\mathcal{H}_1^{(1)}(x_l)\mathcal{H}_1^{(2)}(y_1)\ldots\mathcal{H}_1^{(2)}(y_{n-l})\right]$$

To obtain the final form we may eliminate the sum over n in favor of the sum over m, with $l+m\equiv n$ and then observe that the T-product can be substituted with the P-product since the scalar density is bilinear in the fermion fields. Thus we obtain the final formula

$$\hat{S}=\sum_{l=0}^{\infty}\sum_{m=0}^{\infty}\left(\frac{ig_1}{\hbar c}\right)^l\left(\frac{ig_2}{\hbar c}\right)^m\frac{1}{l!}\frac{1}{m!}\int d^4x_1\cdots\int d^4x_l\int d^4y_1\cdots\int d^4y_m \times$$
$$P[:\hat{\bar{\psi}}(x_1)\hat{\psi}(x_1):\;\cdots\;:\hat{\bar{\psi}}(x_l)\hat{\psi}(x_l)::\hat{\bar{\psi}}(y_1)\hat{\psi}(y_1):\;\cdots\;:\hat{\bar{\psi}}(y_m)\hat{\psi}(y_m):]\times$$
$$P[\hat{\phi}_1(x_1)\cdots\hat{\phi}_1(x_l)]\,P[\hat{\phi}_2(y_1)\cdots\hat{\phi}_2(y_m)]$$

(b) Since all particles are stable, the first-order $\hat{S}$-operator vanishes

$$\hat{S}^{(1)}=0$$

The second-order $\hat{S}$-operator follows from the expression obtained in (a)

$$\hat{S}^{(2)} = \frac{1}{2!} \left(\frac{ig_1}{\hbar c}\right)^2 \int d^4x_1 \int d^4x_2 \times$$

$$P[:\hat{\bar{\psi}}(x_1)\hat{\psi}(x_1)::\hat{\bar{\psi}}(x_2)\hat{\psi}(x_2):]P[\hat{\phi}_1(x_1)\hat{\phi}_1(x_2)] +$$

$$\frac{1}{2!} \left(\frac{ig_2}{\hbar c}\right)^2 \int d^4y_1 \int d^4y_2 \times$$

$$P[:\hat{\bar{\psi}}(y_1)\hat{\psi}(y_1)::\hat{\bar{\psi}}(y_2)\hat{\psi}(y_2):]P[\hat{\phi}_2(y_1)\hat{\phi}_2(y_2)] +$$

$$\frac{2!}{2!} \left(\frac{ig_1}{\hbar c}\right)\left(\frac{ig_2}{\hbar c}\right) \int d^4x_1 \int d^4y_1 \times$$

$$P[:\hat{\bar{\psi}}(x_1)\hat{\psi}(x_1)::\hat{\bar{\psi}}(y_1)\hat{\psi}(y_1):]\,\hat{\phi}_1(x_1)\hat{\phi}_2(y_1)$$

With the use of Wick's theorem, as done in Eqs. (7.33) and (7.34) of the text, one obtains the final expression

$$\hat{S}^{(2)} = \frac{1}{2!} \int d^4x_1 \int d^4x_2 \left[:\hat{\bar{\psi}}_\alpha(x_1)\hat{\psi}_\alpha(x_1)\hat{\bar{\psi}}_\beta(x_2)\hat{\psi}_\beta(x_2): +\right.$$

$$:\hat{\bar{\psi}}_\alpha(x_1)iS_{\alpha\beta}^F(x_1-x_2)\hat{\psi}_\beta(x_2): + :\hat{\bar{\psi}}_\beta(x_2)iS_{\beta\alpha}^F(x_2-x_1)\hat{\psi}_\alpha(x_1): -$$

$$\left. iS_{\alpha\beta}^F(x_1-x_2)iS_{\beta\alpha}^F(x_2-x_1)\right] \times$$

$$\left\{\left(\frac{ig_1}{\hbar c}\right)^2 \left[:\hat{\phi}_1(x_1)\hat{\phi}_1(x_2): + \frac{\hbar}{ic}\Delta_F^{(1)}(x_1-x_2)\right] +\right.$$

$$\left(\frac{ig_2}{\hbar c}\right)^2 \left[:\hat{\phi}_2(x_1)\hat{\phi}_2(x_2): + \frac{\hbar}{ic}\Delta_F^{(2)}(x_1-x_2)\right] +$$

$$\left. 2!\left(\frac{ig_1}{\hbar c}\right)\left(\frac{ig_2}{\hbar c}\right)\hat{\phi}_1(x_1)\hat{\phi}_2(x_2)\right\}$$

Problem 7.7 (a) Use the results in Prob. 7.6(b) to calculate the amplitudes for the processes $\phi_1+N \to \phi_1+N$, $\phi_1+N \to \phi_2+N$, $N+N \to N+N$, and the self-energies of the (ϕ_1, ϕ_2, N);

(b) As in the text, use these results to deduce the Feynman rules for this theory.

Solution to Problem 7.7

(a) The calculation of the amplitudes for the processes $\phi_1+N \to \phi_1+N$, $\phi_2 + N \to \phi_2 + N$ and $N + N \to N + N$ can be done straightforwardly using the results of the Dirac-scalar theory discussed in the text, which contains a single scalar, with the appropriate substitutions. In coordinate space, the second-order matrix elements of the scattering operator for the process $\phi_l + N \to \phi_l + N$, with $l = 1, 2$, read [compare Eq. (7.37)]

$$S_{fi}^{(2)} = \frac{1}{\Omega^2} \left(\frac{ig_l}{\hbar c} \right)^2 \frac{\hbar}{\sqrt{4\omega_{k_1}^{(l)} \omega_{k_3}^{(l)}}} \int d^4x_1 \int d^4x_2 \times$$

$$\bar{u}_\alpha(\mathbf{k}_4\lambda') i S_{\alpha\beta}^F(x_1 - x_2) u_\beta(\mathbf{k}_2\lambda) \times$$

$$\left[e^{-i(k_4 - k_1) \cdot x_1} e^{-i(k_3 - k_2)x_2} + e^{-i(k_4 + k_3) \cdot x_1} e^{i(k_1 + k_2)x_2} \right] \qquad ; l = 1, 2$$

and for the process $N + N \to N + N$ [compare Eq. (7.45)][5]

$$S_{fi}^{(2)} = \frac{1}{\Omega^2} \int d^4x_1 \int d^4x_2 \times$$

$$\left\{ [\bar{u}(\mathbf{k}_4\lambda_4)u(\mathbf{k}_2\lambda_2)][\bar{u}(\mathbf{k}_3\lambda_3)u(\mathbf{k}_1\lambda_1)] e^{i(k_2 - k_4) \cdot x_1} e^{i(k_1 - k_3) \cdot x_2} - \right.$$

$$\left. [\bar{u}(\mathbf{k}_4\lambda_4)u(\mathbf{k}_1\lambda_1)][\bar{u}(\mathbf{k}_3\lambda_3)u(\mathbf{k}_2\lambda_2)] e^{i(k_1 - k_4) \cdot x_2} e^{i(k_2 - k_3) \cdot x_1} \right\} \times$$

$$\left\{ \left(\frac{ig_1}{\hbar c} \right)^2 \frac{\hbar}{ic} \Delta_F^{(1)}(x_1 - x_2) + \left(\frac{ig_2}{\hbar c} \right)^2 \frac{\hbar}{ic} \Delta_F^{(2)}(x_1 - x_2) \right\}$$

The process $\phi_1 + N \to \phi_2 + N$ has no equivalent in the model discussed in the text. In this case, the initial and final states for this process are

$$|i\rangle = c_{\mathbf{k}_1}^{(1)\dagger} a_{\mathbf{k}_2\lambda}^\dagger |0\rangle \qquad ; |f\rangle = c_{\mathbf{k}_3}^{(2)\dagger} a_{\mathbf{k}_4\lambda'}^\dagger |0\rangle$$

and one needs to evaluate the appropriate matrix elements. These matrix elements, however, are the same that have been used in the evaluation carried out for the processes $\phi_l + N \to \phi_l + N$, and therefore we just need to substitute their expression in the parts of the S-matrix which describe the process under consideration.

A simple substitution, followed by the usual changes $x_1 \leftrightarrow x_2$ and $\alpha \leftrightarrow \beta$, allows us to obtain the final expression

$$S_{fi}^{(2)} = \frac{1}{\Omega^2} \frac{\hbar}{\sqrt{4\omega_{k_1}^{(1)} \omega_{k_3}^{(2)}}} \left(\frac{ig_1}{\hbar c} \right) \left(\frac{ig_2}{\hbar c} \right) \int d^4x_1 \int d^4x_2 \times$$

$$\left[\bar{u}_\alpha(\mathbf{k}_4, \lambda') i S_{\alpha\beta}^F(x_1 - x_2) u_\beta(\mathbf{k}_2, \lambda) \, e^{-i(k_4 - k_1) \cdot x_1 + i(k_2 - k_3) \cdot x_2} + \right.$$

$$\left. \bar{u}_\alpha(\mathbf{k}_4, \lambda') i S_{\alpha\beta}^F(x_1 - x_2) u_\beta(\mathbf{k}_2, \lambda) \, e^{-i(k_3 + k_4) \cdot x_1} e^{i(k_1 + k_2) \cdot x_2} \right]$$

---

[5]It is assumed that the nucleons are identical.

With the use of the four-dimensional Fourier transform of the Feynman propagator given in Eq. (7.39), and performance of the integrations over x_1 and x_2, we obtain

$$S_{fi}^{(2)} = (2\pi)^4 \delta^{(4)}(k_1 + k_2 - k_3 - k_4)\frac{1}{\Omega^2}\frac{\hbar}{\sqrt{4\omega_{k_1}^{(1)}\omega_{k_3}^{(2)}}}\left(\frac{ig_1}{\hbar c}\right)\left(\frac{ig_2}{\hbar c}\right)\left(\frac{1}{i}\right) \times$$

$$\bar{u}(\mathbf{k}_4\lambda')\left[\frac{1}{i(k_2 - k_3)_\mu\gamma_\mu + M} + \frac{1}{i(k_1 + k_2)_\mu\gamma_\mu + M}\right]u(\mathbf{k}_2\lambda)$$

As before, the calculation of the self-energies of ϕ_1, ϕ_2 and N can be done using the results of the Dirac-scalar model discussed in the text, with the appropriate adjustments. For the scalar self-energies, we have the matrix element with $l = 1, 2$ [compare Eq. (7.53)]

$$S_{fi}^{(2)} = \left(\frac{ig_l}{\hbar c}\right)^2 \frac{1}{\Omega}\frac{\hbar}{\sqrt{4\omega_{k_1}^{(l)}\omega_{k_2}^{(l)}}}\int d^4x_1 \int d^4x_2 \times$$

$$(-1)\text{Tr}\left[iS^F(x_1 - x_2)iS^F(x_2 - x_1)\right]e^{ik_1\cdot x_1}e^{-ik_2 x_2} \qquad ; l = 1, 2$$

After inserting the four-dimensional Fourier transforms of the fermion propagators, one obtains [see Eq. (7.54) and Fig. 7.5(b) in the text]

$$S_{fi}^{(2)} = (2\pi)^4 \delta^{(4)}(k_1 - k_2)\frac{ic}{\Omega}\frac{1}{\sqrt{4\omega_{k_1}^{(l)}\omega_{k_2}^{(l)}}}\Pi^{(l)}(k_1)$$

$$\Pi^{(l)}(k_1) = \frac{ig_l^2}{\hbar c^3}\int\frac{d^4t}{(2\pi)^4}\text{Tr}\left[\frac{1}{i\gamma_\mu(t + k_1)_\mu + M}\frac{1}{i\gamma_\mu t_\mu + M}\right] \qquad ; l = 1, 2$$

Although there is a mixing $\phi_1 \rightleftharpoons \phi_2$ through the fermion loop, which we do not pursue here (compare appendix J of Vol. I), the corresponding amplitude $S_{fi}^{(2)}$ vanishes by energy-momentum conservation since $m_1 \neq m_2$.

The self-energy of the fermion is obtained by generalizing Eq. (7.48) to the present model, and it reads

$$S_{fi}^{(2)} = \frac{1}{\Omega}\frac{\hbar}{ic}\int d^4x_1 \int d^4x_2\,\bar{u}_\beta(\mathbf{k}_2\lambda_2)iS_{\beta\alpha}^F(x_2 - x_1)u_\alpha(\mathbf{k}_1\lambda_1)e^{-ik_2\cdot x_2 + ik_1\cdot x_1} \times$$

$$\left\{\left(\frac{ig_1}{\hbar c}\right)^2 \Delta_F^{(1)}(x_1 - x_2) + \left(\frac{ig_2}{\hbar c}\right)^2 \Delta_F^{(2)}(x_1 - x_2)\right\}$$

In momentum space, the expression is

$$S_{fi}^{(2)} = -(2\pi)^4 \delta^{(4)}(k_1 - k_2)\frac{i}{\Omega}\bar{u}(\mathbf{k}_2\lambda_2)\Sigma(k_1)u(\mathbf{k}_1\lambda_1)$$

where [compare Fig. 7.5(a) in the text]

$$\Sigma(k_1) \equiv \frac{ig_1^2}{\hbar c^3} \int \frac{d^4q}{(2\pi)^4} \frac{1}{i\gamma_\mu(k_1+q)_\mu + M} \frac{1}{q^2 + m_1^2}$$
$$+ \frac{ig_2^2}{\hbar c^3} \int \frac{d^4q}{(2\pi)^4} \frac{1}{i\gamma_\mu(k_1+q)_\mu + M} \frac{1}{q^2 + m_2^2}$$

This generalizes Eq. (7.51) in the text to the case of two scalars.

(b) With the use of the results obtained in part (a), we may formulate the Feynman rules for the model that we are studying:

(1) Place down the coordinates $(x_1, x_2, \ldots, x_n)$. Draw all topologically distinct connected diagrams with one incoming N-line, one outgoing N-line and one scalar line at each vertex. The scalar line may be of two kinds, since there are two different scalars;
(2) Include a factor $(ig_l/\hbar c)$ for each vertex involving a scalar ϕ_l $(l = 1, 2)$;
(3) Include a factor $(\hbar/ic)\Delta_F^{(l)}(x_j - x_k)$ for each internal ϕ_l line $(l = 1, 2)$;
(4) Include a factor $iS_F(x_j - x_k)$ for each internal directed N-line;
(5) Include a factor $[\hbar/2\omega^{(l)}\Omega]^{1/2}e^{-ik\cdot x}$ for each outgoing $\phi^{(l)}$-line $(l = 1, 2)$;
(6) Include a factor $[\hbar/2\omega^{(l)}\Omega]^{1/2}e^{ik\cdot x}$ for each incoming $\phi^{(l)}$-line $(l = 1, 2)$;
(7) Include a factor $(1/\Omega)^{1/2}\bar{u}(k\lambda)e^{-ik\cdot x}$ for each outgoing N-line;
(8) Include a factor $(1/\Omega)^{1/2}u(k\lambda)e^{ik\cdot x}$ for each incoming N-line;
(9) Take a Dirac matrix product along each fermion line;
(10) Include a factor of (-1) for each closed fermion loop;
(11) Integrate over $d^4x_1\, d^4x_2 \ldots d^4x_n$.

In an analogous way, the Feynman diagrams describing this process in momentum space can also be drawn, and the corresponding factors to be associated to each element of the diagram are easily found (this is left to the reader). In this way the appropriate Feynman rules in momentum space are obtained.

Problem 7.8 An effective field theory that has proven useful in nuclear physics is composed of a Dirac nucleon described by the field in Eq. (6.28), interacting both with the neutral scalar field of Eqs. (5.117)–(5.118), and the neutral vector meson field of Probs. C.6–C.7.[6] The lagrangian density for this composite system is taken to be

$$\mathcal{L} = -\hbar c \bar{\underline{\psi}} \left[\gamma_\mu \left(\frac{\partial}{\partial x_\mu} - \frac{ig_v}{\hbar c} V_\mu \right) + \left(M - \frac{g_s}{\hbar c} \phi \right) \right] \underline{\psi} + \mathcal{L}_0^S + \mathcal{L}_0^V$$

---

[6]See Probl. 9.5.

(a) Show the interaction hamiltonian density is

$$\mathcal{H}_1 = -g_s \bar{\underline{\psi}} \underline{\psi} \phi - i g_v \bar{\underline{\psi}} \gamma_\mu \underline{\psi} V_\mu$$

(b) Show the Feynman propagator for the nucleon field is

$$\langle 0|P[\hat{\underline{\psi}}_i(x)\hat{\bar{\underline{\psi}}}_j(y)]|0\rangle = i\delta_{ij} S_F(x-y) \qquad ; (i,j) = 1,2$$

where these indices refer to isospin;

(c) Write out the scattering operator in the interaction picture to second order in the coupling constants (g_s, g_v);

(d) Apply Wick's theorem to the result in (c).

Solution to Problem 7.8

(a) Write the lagrangian density as

$$\mathcal{L} = \mathcal{L}_0 + \mathcal{L}_1$$
$$\mathcal{L}_0 = -\hbar c \bar{\underline{\psi}} \left(\gamma_\mu \frac{\partial}{\partial x_\mu} + M \right) \underline{\psi} + \mathcal{L}_0^S + \mathcal{L}_0^V$$
$$\mathcal{L}_1 = \bar{\underline{\psi}} \left(i g_v V_\mu + g_s \phi \right) \underline{\psi}$$

Since there are no time derivatives in the interaction, the canonical procedure gives

$$\mathcal{H} = \mathcal{H}_0 + \mathcal{H}_1$$
$$\mathcal{H}_0 = \mathcal{H}_0^D + \mathcal{H}_0^S + \mathcal{H}_0^V$$
$$\mathcal{H}_1 = -\mathcal{L}_1 = -g_s \bar{\underline{\psi}} \underline{\psi} \phi - i g_v \bar{\underline{\psi}} \gamma_\mu \underline{\psi} V_\mu$$

(b) From Eq. (6.28), the Dirac field has two isospin indices (we suppress the Dirac indices)

$$\underline{\psi} = \psi_p \underline{\eta}_p + \psi_n \underline{\eta}_n = \begin{pmatrix} \psi_p \\ \psi_n \end{pmatrix}$$

The Feynman propagator is then given in Eq. (F.49)

$$\langle 0|P[\hat{\underline{\psi}}_j(x_\mu), \hat{\bar{\underline{\psi}}}_k(x'_\mu)]|0\rangle = i S_{jk}^F(x_\mu - x'_\mu)$$

where we now exhibit only the isospin indices. Clearly, to get a non-zero vacuum expectation value, one must have $j = k$. Hence

$$S_{jk}^F(x_\mu - x'_\mu) = \delta_{jk} S^F(x_\mu - x'_\mu)$$

(c) To simplify matters, we shall normal-order the interaction as in Eq. (5.116)[7]

$$\hat{\mathcal{H}}_I = -g_s : \hat{\bar{\psi}}\,\hat{\psi} : \hat{\phi} - ig_v : \hat{\bar{\psi}}\gamma_\mu\hat{\psi} : \hat{V}_\mu$$

The appropriate extension in this case of the scattering operator $\hat{S}^{(2)}$ in Eq. (7.32) is then

$$\hat{S}^{(2)} = \int d^4x_1 \int d^4x_2 \times$$

$$\left\{ \left(\frac{ig_s}{\hbar c}\right)^2 \frac{1}{2!} P[:\hat{\bar{\psi}}(x_1)\hat{\psi}(x_1)::\hat{\bar{\psi}}(x_2)\hat{\psi}(x_2):]P[\hat{\phi}(x_1)\hat{\phi}(x_2)]+ \right.$$

$$\left(\frac{-g_v}{\hbar c}\right)^2 \frac{1}{2!} P[:\hat{\bar{\psi}}(x_1)\gamma_\mu\hat{\psi}(x_1)::\hat{\bar{\psi}}(x_2)\gamma_\nu\hat{\psi}(x_2):]P[\hat{V}_\mu(x_1)\hat{V}_\nu(x_2)]+$$

$$\left. -\frac{ig_s g_v}{(\hbar c)^2} P[:\hat{\bar{\psi}}(x_1)\hat{\psi}(x_1)::\hat{\bar{\psi}}(x_2)\gamma_\mu\hat{\psi}(x_2):]\,\hat{\phi}(x_1)\hat{V}_\mu(x_2) \right\}$$

Here, in the last line, we have identified two identical cross terms in the scattering operator.

(d) Application of Wick's theorem gives the appropriate generalization of the result in Eq. (7.35). We write this as

$$\hat{S}^{(2)} = \hat{S}^{(2)}_{SS} + \hat{S}^{(2)}_{VV} + \hat{S}^{(2)}_{SV}$$

The first term is the direct analogy of Eq. (7.35)

$$\hat{S}^{(2)}_{SS} = \left(\frac{ig_s}{\hbar c}\right)^2 \frac{1}{2!} \int d^4x_1 \int d^4x_2 \left[:\hat{\bar{\psi}}(x_1)\hat{\psi}(x_1)\,\hat{\bar{\psi}}(x_2)\hat{\psi}(x_2):\,+ \right.$$

$$:\hat{\bar{\psi}}(x_1)\,i\underline{S}_F(x_1-x_2)\,\hat{\psi}(x_2):\,+:\hat{\bar{\psi}}(x_2)\,i\underline{S}^F(x_2-x_1)\,\hat{\psi}(x_1):\,-$$

$$\left.\mathrm{Tr}\left\{i\underline{S}_F(x_1-x_2)\,i\underline{S}_F(x_2-x_1)\right\}\right]\left[:\hat{\phi}(x_1)\hat{\phi}(x_2):\,+\frac{\hbar}{ic}\Delta_F(x_1-x_2)\right]$$

The second term is the analog of Eq. (8.42) in QED

$$\hat{S}^{(2)}_{VV} = \left(\frac{-g_v}{\hbar c}\right)^2 \frac{1}{2!} \int d^4x_1 \int d^4x_2 \left[:\hat{\bar{\psi}}(x_1)\gamma_\mu\hat{\psi}(x_1)\,\hat{\bar{\psi}}(x_2)\gamma_\nu\hat{\psi}(x_2):\,+ \right.$$

$$:\hat{\bar{\psi}}(x_1)\gamma_\mu\,i\underline{S}_F(x_1-x_2)\,\gamma_\nu\hat{\psi}(x_2):\,+:\hat{\bar{\psi}}(x_2)\gamma_\nu\,i\underline{S}^F(x_2-x_1)\,\gamma_\mu\hat{\psi}(x_1):\,-$$

$$\left.\mathrm{Tr}\left\{i\underline{S}_F(x_1-x_2)\gamma_\nu\,i\underline{S}_F(x_2-x_1)\gamma_\mu\right\}\right]\left[:\hat{V}_\mu(x_1)\hat{V}_\nu(x_2):\,+\frac{\hbar c}{i}\Delta^F_{\mu\nu}(x_1-x_2)\right]$$

---

[7]See, however, Sec. G.7.2 and Prob. 7.11.

where we have made use of Eq. (F.51). The interference term is

$$\hat{S}_{SV}^{(2)} = -\frac{ig_s g_v}{(\hbar c)^2} \int d^4 x_1 \int d^4 x_2 \Big[:\hat{\bar{\psi}}(x_1)\hat{\psi}(x_1)\,\hat{\bar{\psi}}(x_2)\gamma_\mu\hat{\psi}(x_2): +$$

$$:\hat{\bar{\psi}}(x_1)\,i\underline{S}_F(x_1 - x_2)\,\gamma_\mu\hat{\psi}(x_2): + :\hat{\bar{\psi}}(x_2)\gamma_\mu\,i\underline{S}^F(x_2 - x_1)\,\hat{\psi}(x_1): -$$

$$\mathrm{Tr}\left\{ i\underline{S}_F(x_1 - x_2)\gamma_\mu\,i\underline{S}_F(x_2 - x_1) \right\} \Big]\,\hat{\phi}(x_1)\hat{V}_\mu(x_2)$$

Problem 7.9 Make use of the results in Prob. 7.8, the vector fields in Probs. C.6–C.7, and the vector propagator in Prob. F.5(b).

(a) Take a matrix element of the second-order scattering operator, and extend the construction of the coordinate-space amplitude for the process $N + N \to N + N$;

(b) Construct the coordinate-space amplitudes for the processes $V + N \to V + N$, and $\phi + N \to V + N$;

(c) Insert Fourier transforms, and construct the corresponding amplitudes in momentum space.

Solution to Problem 7.9

(a) With the use of the results in Prob. 7.8, the second-order scattering operator for the process $N(k_1) + N(k_2) \to N(k_3) + N(k_4)$ can be written as[8]

$$S_{fi} = S_{fi}^S + S_{fi}^V$$

Here the contribution from scalar exchange S_{fi}^S is the expression given in Eq. (7.45)

$$S_{fi}^S = \left(\frac{ig_s}{\hbar c}\right)^2 \frac{1}{\Omega^2} \int d^4 x_1 \int d^4 x_2 \frac{\hbar}{ic}\Delta_F(x_1 - x_2) \times$$

$$\left\{ [\bar{u}(\mathbf{k}_3\lambda_3)u(\mathbf{k}_1\lambda_1)]\,[\bar{u}(\mathbf{k}_4\lambda_4)u(\mathbf{k}_2\lambda_2)]\,e^{i(k_1-k_3)\cdot x_2}e^{i(k_2-k_4)\cdot x_1} - \right.$$

$$\left. [\bar{u}(\mathbf{k}_4\lambda_4)u(\mathbf{k}_1\lambda_1)]\,[\bar{u}(\mathbf{k}_3\lambda_3)u(\mathbf{k}_2\lambda_2)]\,e^{i(k_1-k_4)\cdot x_2}e^{i(k_2-k_3)\cdot x_1} \right\}$$

---

[8]We assume here that the nucleons are identical.

The additional vector contribution follows in analogy as

$$S_{fi}^V = \left(\frac{-g_v}{\hbar c}\right)^2 \frac{1}{\Omega^2} \int d^4 x_1 \int d^4 x_2 \frac{\hbar c}{i} \Delta_{\mu\nu}^F(x_1 - x_2) \times$$

$$\left\{ [\bar{u}(\mathbf{k}_3\lambda_3)\gamma_\nu u(\mathbf{k}_1\lambda_1)] [\bar{u}(\mathbf{k}_4\lambda_4)\gamma_\mu u(\mathbf{k}_2\lambda_2)] e^{i(k_1-k_3)\cdot x_2} e^{i(k_2-k_4)\cdot x_1} - \right.$$

$$\left. [\bar{u}(\mathbf{k}_4\lambda_4)\gamma_\nu u(\mathbf{k}_1\lambda_1)] [\bar{u}(\mathbf{k}_3\lambda_3)\gamma_\mu u(\mathbf{k}_2\lambda_2)] e^{i(k_1-k_4)\cdot x_2} e^{i(k_2-k_3)\cdot x_1} \right\}$$

(b) The second-order amplitude for the process $\phi(k_1) + N(k_2) \to \phi(k_3) + N(k_4)$ is given in Eq. (7.37)

$$S_{fi}^{SS} = \left(\frac{ig_s}{\hbar c}\right)^2 \frac{1}{\Omega^2} \frac{\hbar}{\sqrt{4\omega_1^s \omega_3^s}} \int d^4 x_1 \int d^4 x_2$$

$$\left[\bar{u}_\alpha(\mathbf{k}_4\lambda') i S_{\alpha\beta}^F(x_1 - x_2) u_\beta(\mathbf{k}_2\lambda) e^{i(k_2+k_1)\cdot x_2} e^{-i(k_4+k_3)\cdot x_1} + \right.$$

$$\left. \bar{u}_\alpha(\mathbf{k}_4\lambda') i S_{\alpha\beta}^F(x_1 - x_2) u_\beta(\mathbf{k}_2\lambda) e^{i(k_2-k_3)\cdot x_2} e^{-i(k_4-k_1)\cdot x_1} \right]$$

where the diagonal nucleon isospin matrix elements are unity, and the Dirac indices are restored. With the aid of the vector fields in Prob. C.7, the amplitude for $V_T + N \to V_T + N$ is written in an analogous fashion as

$$S_{fi}^{VV} = \left(\frac{-g_v}{\hbar c}\right)^2 \frac{1}{\Omega^2} \frac{\hbar c^2}{\sqrt{4\omega_1^v \omega_3^v}} \int d^4 x_1 \int d^4 x_2$$

$$\left[\bar{u}_\alpha(\mathbf{k}_4\lambda') \not{\epsilon}_f\, i S_{\alpha\beta}^F(x_1 - x_2) \not{\epsilon}_i u_\beta(\mathbf{k}_2\lambda) e^{i(k_2+k_1)\cdot x_2} e^{-i(k_4+k_3)\cdot x_1} + \right.$$

$$\left. \bar{u}_\alpha(\mathbf{k}_4\lambda') \not{\epsilon}_i\, i S_{\alpha\beta}^F(x_1 - x_2) \not{\epsilon}_f u_\beta(\mathbf{k}_2\lambda) e^{i(k_2-k_3)\cdot x_2} e^{-i(k_4-k_1)\cdot x_1} \right]$$

As is evident from the fields in Prob. C.6–C.7, the amplitudes involving V_L differ only in the factors for the external particles. To go from $V_T \to V_L$, one makes the substitution $(\hbar c^2/2\omega_k^v)^{1/2} \to (\hbar\omega_k^v/2m_v^2)^{1/2}$ and employs the longitudinal polarization four-vector $e_\mu(\mathbf{k}) = (\mathbf{e}_{\mathbf{k}3}, ikc/\omega_k)$ with $k_\mu e_\mu = 0$.
The amplitude for $\phi(k_1) + N(k_2) \to V_T(k_3) + N(k_4)$ is

$$S_{fi}^{SV} = -\frac{ig_s g_v}{(\hbar c)^2} \frac{1}{\Omega^2} \frac{\hbar}{\sqrt{4\omega_1^s \omega_3^v}} \int d^4 x_1 \int d^4 x_2$$

$$\left[\bar{u}_\alpha(\mathbf{k}_4\lambda') \not{\epsilon}\, i S_{\alpha\beta}^F(x_2 - x_1) u_\beta(\mathbf{k}_2\lambda) e^{i(k_2+k_1)\cdot x_1} e^{-i(k_4+k_3)\cdot x_2} + \right.$$

$$\left. \bar{u}_\alpha(\mathbf{k}_4\lambda') i S_{\alpha\beta}^F(x_1 - x_2) \not{\epsilon}\, u_\beta(\mathbf{k}_2\lambda) e^{i(k_2-k_3)\cdot x_2} e^{-i(k_4-k_1)\cdot x_1} \right]$$

The amplitude to produce V_L is obtained with the previously indicated substitutions.

(c) With the insertion of Fourier transforms, and performance of the spatial integrations, one obtains the second-order S-matrix element for the Dirac-scalar scattering process $\phi(k_1) + N(k_2) \to \phi(k_3) + N(k_4)$ given in Eq. (7.42)

$$S_{fi}^{SS} = (2\pi)^4 \delta^{(4)}(k_1 + k_2 - k_3 - k_4)\frac{1}{\Omega^2}\frac{\hbar}{\sqrt{4\omega_1^s \omega_3^s}}\left(\frac{ig_s}{\hbar c}\right)^2\left(\frac{1}{i}\right) \times$$

$$\bar{u}(\mathbf{k_4}\lambda')\left[\frac{1}{i\gamma_\mu(k_2 + k_1)_\mu + M} + \frac{1}{i\gamma_\mu(k_2 - k_3)_\mu + M}\right]u(\mathbf{k_2}\lambda)$$

Here we revert to a matrix notation and again suppress the Dirac and isospin indices. It is evident from the results in part (b) that the corresponding result for transverse-vector scattering $V_T(k_1) + N(k_2) \to V_T(k_3) + N(k_4)$ is

$$S_{fi}^{VV} = (2\pi)^4 \delta^{(4)}(k_1 + k_2 - k_3 - k_4)\frac{1}{\Omega^2}\frac{\hbar c^2}{\sqrt{4\omega_1^v \omega_3^v}}\left(\frac{-g_v}{\hbar c}\right)^2\left(\frac{1}{i}\right) \times$$

$$\bar{u}(\mathbf{k_4}\lambda')\left[\not{e}_f\frac{1}{i\gamma_\mu(k_2 + k_1)_\mu + M}\not{e}_i + \not{e}_i\frac{1}{i\gamma_\mu(k_2 - k_3)_\mu + M}\not{e}_f\right]u(\mathbf{k_2}\lambda)$$

For transverse-vector production in $\phi(k_1) + N(k_2) \to V_T(k_3) + N(k_4)$, the result is

$$S_{fi}^{SV} = (2\pi)^4 \delta^{(4)}(k_1 + k_2 - k_3 - k_4)\frac{1}{\Omega^2}\frac{\hbar c}{\sqrt{4\omega_1^s \omega_3^v}}\frac{-ig_s g_v}{(\hbar c)^2}\left(\frac{1}{i}\right) \times$$

$$\bar{u}(\mathbf{k_4}\lambda')\left[\not{e}\frac{1}{i\gamma_\mu(k_2 + k_1)_\mu + M} + \frac{1}{i\gamma_\mu(k_2 - k_3)_\mu + M}\not{e}\right]u(\mathbf{k_2}\lambda)$$

The amplitudes involving V_L are obtained with the previously indicated substitutions.

The second-order momentum-space S-matrix element for the nucleon-nucleon scattering process $N(k_1) + N(k_2) \to N(k_3) + N(k_4)$ taking place through scalar exchange is derived in Eq. (7.46)

$$S_{fi}^S = (2\pi)^4 \delta^{(4)}(k_1 + k_2 - k_3 - k_4)\frac{1}{\Omega^2}\left(\frac{ig_s}{\hbar c}\right)^2\left(\frac{\hbar}{ic}\right) \times$$

$$\left[\bar{u}(\mathbf{k_3}\lambda_3)u(\mathbf{k_1}\lambda_1)\frac{1}{(k_1 - k_3)^2 + m_s^2}\bar{u}(\mathbf{k_4}\lambda_4)u(\mathbf{k_2}\lambda_2) - \right.$$

$$\left. \bar{u}(\mathbf{k_4}\lambda_4)u(\mathbf{k_1}\lambda_1)\frac{1}{(k_1 - k_4)^2 + m_s^2}\bar{u}(\mathbf{k_3}\lambda_3)u(\mathbf{k_2}\lambda_2)\right]$$

The corresponding result for vector exchange then follows from part (b) as

$$
S^V_{fi} = (2\pi)^4 \delta^{(4)}(k_1 + k_2 - k_3 - k_4) \frac{1}{\Omega^2} \left(\frac{-g_v}{\hbar c}\right)^2 \left(\frac{\hbar c}{i}\right) \times
$$

$$
\left[\bar{u}(\mathbf{k}_3\lambda_3)\gamma_\mu u(\mathbf{k}_1\lambda_1) \frac{1}{(k_1 - k_3)^2 + m_v^2} \bar{u}(\mathbf{k}_4\lambda_4)\gamma_\mu u(\mathbf{k}_2\lambda_2) - \right.
$$

$$
\left. \bar{u}(\mathbf{k}_4\lambda_4)\gamma_\mu u(\mathbf{k}_1\lambda_1) \frac{1}{(k_1 - k_4)^2 + m_v^2} \bar{u}(\mathbf{k}_3\lambda_3)\gamma_\mu u(\mathbf{k}_2\lambda_2) \right]
$$

Here we have used the vector propagator from Eq. (F.51), and eliminated the term $q_\mu q_\nu / m_v^2$ through current conservation on the Dirac legs.[9]

Problem 7.10 (a) Work in analogy to the analysis in the text, and interpret the results in in Prob. 7.9 in terms of Feynman diagrams;

(b) Use these results to deduce a set of Feynman rules for the Dirac-scalar-vector theory.

Solution to Problem 7.10

(a) In part (a) of Prob. 7.9, the matrix element of the second-order scattering operator for the process $N(k_1) + N(k_2) \to N(k_3) + N(k_4)$ is calculated for a model containing a nucleon field, as well as neutral scalar and vector meson fields. To second order, the process can occur either with the exchange of a virtual scalar meson or with the exchange of a virtual vector meson. These two processes are described by the expressions for S^S_{fi} and S^V_{fi} obtained in Prob. 7.9. In particular, the expression for S^S_{fi} is just the one obtained in Eq. (7.45), with the obvious identification $g_s = g$,[10] and therefore the interpretation of this process in terms of Feynman diagrams is just the one displayed in Figs. 7.3–7.4 of the text.

The Feynman diagrams which correspond to the expression for S^V_{fi} are similar to the ones of Figs. 7.3–7.4 in the text, with the difference that now a virtual vector meson is exchanged between nucleons.

The matrix element of the scattering operator for the process $\phi + N \to \phi + N$ to second order coincides with the one calculated in Eq. (7.37), describing the absorption of a scalar by the nucleon at one vertex and the emission of a scalar at a different vertex. The Feynman diagrams for this process are shown in Figs. 7.1–7.2 in the text.

---

[9] Use $\bar{u}(\mathbf{k}_3\lambda_3)(i\slashed{k}_3 - i\slashed{k}_1)u(\mathbf{k}_1\lambda_1) = \bar{u}(\mathbf{k}_4\lambda_4)(i\slashed{k}_4 - i\slashed{k}_2)u(\mathbf{k}_2\lambda_2) = 0$.

[10] As a matter of fact, when $g_v = 0$ the model reduces to the model discussed in the text.

The processes $V_T + N \to V_T + N$ and $V_L + N \to V_L + N$ are described by similar Feynman diagrams, where however the meson lines now represent the neutral vector meson. Finally, in the case of the process $\phi(k_1) + N(k_2) \to V_L(k_3) + N(k_4)$ the incoming meson line with momentum k_1 corresponds to a scalar meson, while the outgoing meson line with momentum k_3 corresponds to the vector meson V_L.

(b) With the use of the results in Prob. 7.9, and the analysis of part (a), we are now able to obtain the Feynman rules for the Dirac-scalar-vector theory. Since this theory contains the Dirac-scalar theory discussed in the text, we only need to supplement the Feynman rules derived there with the appropriate rules for the new elements.

The extended Feynman rules in coordinate space read:

(1) Place down the coordinates $(x_1, x_2, \ldots, x_n)$. Draw all topologically distinct connected diagrams with one incoming N-line, one outgoing N-line and either one scalar line or one vector line at each vertex.

(2) Include a factor $(ig_s/\hbar c)$ for each vertex involving a scalar ϕ;

(3) Include a factor $(-g_v\gamma_\mu/\hbar c)$ for each vertex involving a vector V_μ;

(4) Include a factor $(\hbar/ic)\Delta_F(x_j - x_k)$ for each internal scalar line connecting x_k and x_j;

(5) Include a factor $(\hbar c/i)\Delta^F_{\mu\nu}(x_j - x_k)$ for each internal vector line connecting x_k and x_j;

(6) Include a factor $iS_F(x_j - x_k)$ for each internal directed N-line;

(7) Include a factor $(\hbar/2\omega^s_k\Omega)^{1/2}e^{-ik\cdot x}$ for each outgoing scalar line;

(8) Include a factor $(\hbar/2\omega^s_k\Omega)^{1/2}e^{ik\cdot x}$ for each incoming scalar line;

(9) Include a factor $(1/\Omega)^{1/2}\bar{u}(k\lambda)e^{-ik\cdot x}$ for each outgoing N-line;

(10) Include a factor $(1/\Omega)^{1/2}u(k\lambda)e^{ik\cdot x}$ for each incoming N-line;

(11) Include a factor $(\hbar c^2/2\omega^v_k\Omega)^{1/2}e_\mu(\mathbf{k}, s)e^{-ik\cdot x}$ for each outgoing transverse vector line, with $e_\mu(\mathbf{k}, s) = (\mathbf{e}_{\mathbf{k}s}, 0)$ and $k_\mu e_\mu = 0$;

(12) Include a factor $(\hbar c^2/2\omega^v_k\Omega)^{1/2}e_\mu(\mathbf{k}, s)e^{ik\cdot x}$ for each incoming transverse vector line;

(13) Include a factor $(\hbar\omega^v_k/2m^2_v\Omega)^{1/2}e_\mu(\mathbf{k})e^{-ik\cdot x}$ for each outgoing longitudinal vector line, with $e_\mu(\mathbf{k}) = (\mathbf{e}_{\mathbf{k}3}, ikc/\omega^v_k)$ and $k_\mu e_\mu = 0$;

(14) Include a factor $(\hbar\omega^v_k/2m^2_v\Omega)^{1/2}e_\mu(k)e^{ik\cdot x}$ for each incoming longitudinal vector line;

(15) Include a factor δ_{jk} along a fermion line for isospin;

(16) Take a Dirac matrix product along each fermion line;

(17) Include a factor of (-1) for each closed fermion loop;

(18) Integrate over $d^4x_1\, d^4x_2\ldots d^4x_n$.

Problem 7.11 (a) Relax the assumption that the scalar density is normal-ordered in the Dirac-scalar example in Eq. (7.32). Show that the contraction *within* the scalar density is to be interpreted as follows

$$\hat{\bar{\psi}}(x)^{\bullet} \hat{\psi}(x)^{\bullet} = (-1)i \operatorname{Tr} S_F(0^-)$$
$$= \langle 0|\hat{\bar{\psi}}(x)\hat{\psi}(x)|0\rangle \equiv \rho_S$$

Here $S_F(0^-)$ is $S_F(x - x')$ with $\mathbf{x}' \to \mathbf{x}$ and $t' \to t^+$. The scalar density ρ_S is a constant independent of x;

(b) Apply Wick's theorem, and show the result is to add the "tadpole" diagrams shown in Fig. 7.16. Assign all factors associated with the tadpole contribution to the self-energy in Fig. 7.16(a);

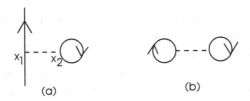

(a)   (b)

Fig. 7.16   Feynman diagrams for the tadpole contribution to: (a) the self-energy of the N; (b) the vacuum-vacuum amplitude.

(c) Insert Fourier transforms, and show the only consequence of the tadpole diagram in Fig. 7.16(a) is to add the following constant to the self-mass of the N

$$\delta M_{\text{tad}} = -\frac{g^2}{\hbar c^3} \frac{\rho_S}{m_s^2}$$

(d) Show the additional term in part (c) will be completely cancelled by the counter-term used in mass renormalization;

(e) Argue that the additional vacuum-vacuum contribution also disappears from physical results.

Solution to Problem 7.11

(a) Suppose the scalar density is *not* normal-ordered. Write

$$\hat{\bar{\psi}}(x)\hat{\psi}(x) = :\hat{\bar{\psi}}(x)\hat{\psi}(x): + \hat{\bar{\psi}}(x)^{\bullet}\hat{\psi}(x)^{\bullet}$$

The result is to append the additional contraction $\hat{\bar{\psi}}(x)^{\bullet}\hat{\psi}(x)^{\bullet}$ when Wick's theorem is applied to the expression in Eq. (7.32). Since the vacuum expectation value of the normal-ordered product vanishes, one has

$$\hat{\bar{\psi}}(x)^{\bullet}\,\hat{\psi}(x)^{\bullet} = \langle 0|\hat{\bar{\psi}}(x)\hat{\psi}(x)|0\rangle$$

Therefore

$$\hat{\bar{\psi}}(x)^{\bullet}\,\hat{\psi}(x)^{\bullet} = -\text{Lim}_{t'\to t^+}\,\langle 0|P[\hat{\psi}(\mathbf{x},t)\hat{\bar{\psi}}(\mathbf{x},t')]|0\rangle$$
$$= (-1)i\,\text{Tr}\,S_F(0^-)$$
$$\equiv \rho_S$$

where the last line defines ρ_S. From Eq. (F.50), this is[11]

$$\rho_S = 2i\,\text{Lim}_{\epsilon\to 0}\int\frac{d^3k}{(2\pi)^3}\int\frac{dk_0}{2\pi}\,e^{i\epsilon k_0}\,\text{Tr}\,\frac{M - ik_\mu\gamma_\mu}{k^2 - k_0^2 + M^2 - i\eta}$$
$$= -4\int\frac{d^3k}{(2\pi)^3}\frac{M}{\sqrt{\mathbf{k}^2 + M^2}}$$

The last line arises from closing the k_0-integral in the upper-half plane, and using the residue theorem. This is just the scalar density summed over the negative-energy states

$$\rho_S = \langle 0|\hat{\bar{\psi}}(x)\hat{\psi}(x)|0\rangle = \frac{2}{\Omega}\sum_{\mathbf{k},\lambda}\bar{v}(-\mathbf{k}\lambda)v(-\mathbf{k}\lambda)$$

(b) The additional contraction in Wick's theorem in part (a) results in the following additional contribution to the self-energy S-matrix element in Eq. (7.48)

$$S_{fi}^{\text{tad}} = \left(\frac{ig_s}{\hbar c}\right)^2\frac{1}{\Omega}\int d^4x_1\int d^4x_2\left(\frac{\hbar}{ic}\right)\Delta_F(x_1 - x_2)\,(-1)i\,\text{Tr}\,S_F(0^-)\times$$
$$\bar{u}(\mathbf{k}_2\lambda_2)e^{-ik_2\cdot x_1}u(\mathbf{k}_1\lambda_1)e^{ik_1\cdot x_1}$$

This is exactly the expression one would write down for the tadpole contribution in Fig. 7.16(a), using the coordinate-space Feynman rules in section 7.4.

(c) A repetition of the arguments in Eqs. (7.49)–(7.51) then gives for the momentum-space matrix element

$$S_{fi}^{\text{tad}} = -(2\pi)^4\delta^{(4)}(k_1 - k_2)\frac{i}{\Omega}\bar{u}(\mathbf{k}_2\lambda_2)\,\delta M_{\text{tad}}\,u(\mathbf{k}_1\lambda_1)$$
$$\delta M_{\text{tad}} = -\frac{g_s^2}{\hbar c^3}\frac{\rho_S}{m_s^2}$$

---

[11]There is a factor of 2 from isospin.

(d) The additional constant mass shift δM_{tad} is now removed with mass renormalization, exactly as in Eqs. (7.70)–(7.71).

(e) The additional vacuum-vacuum contribution in Fig. 7.16(b) is a disconnected diagram, and thus as shown in the text, it also disappears from physical results.

Problem 7.12 Discuss the mass renormalization of the ϕ in the Dirac-scalar theory.

Solution to Problem 7.12

The solution here closely parallels that in section 7.6. Go back to the starting scalar hamiltonian density with no interactions. Re-label the mass in that expression by m_0^2, and the mass term $\tilde{\mathcal{H}}_0^{\text{mass}}$. The starting hamiltonian density then takes the form [see Eq. (5.63)][12]

$$\tilde{\mathcal{H}}_0^{\text{mass}} = \frac{1}{2}m_0^2 c^2 \phi^2$$

Now we would like the mass here, in terms of which everything is subsequently expressed, to be the *experimental* mass m_{exp}^2 and not the "bare mass" m_0^2, which is not an observable. This can be achieved by adding and subtracting a term in δm^2 where[13]

$$m_{\text{exp}}^2 \equiv m_0^2 + \delta m^2$$

Correspondingly, we identify a *new* mass term $\mathcal{H}_0^{\text{mass}}$ in the starting hamiltonian density through

$$\tilde{\mathcal{H}}_0^{\text{mass}} = \mathcal{H}_0^{\text{mass}} + \mathcal{H}_1^{\text{mass}}$$
$$\mathcal{H}_0^{\text{mass}} = \frac{1}{2}m^2 c^2 \phi^2$$
$$\mathcal{H}_1^{\text{mass}} = -\frac{1}{2}\delta m^2 c^2 \phi^2$$

Now $\mathcal{H}_0^{\text{mass}}$, expressed in terms of $m^2 \equiv m_{\text{exp}}^2$, can be included in the starting hamiltonian, just as we have been doing. There is, however, a price to be paid. There is now an additional *interaction density* $\mathcal{H}_1^{\text{mass}}$ that must be included, and the interaction hamiltonian in Eq. (7.2) gets

---

[12] For simplicity, we suppress the label "s" on the scalar mass in this solution.
[13] Note that δm^2 can have either sign.

extended to read

$$\hat{H}_I(t) = -g \int_\Omega d^3x \, \hat{\bar{\psi}}(\mathbf{x},t)\hat{\psi}(\mathbf{x},t)\hat{\phi}(\mathbf{x},t) - \frac{1}{2}\delta m^2 c^2 \int_\Omega d^3x \, \hat{\phi}(\mathbf{x},t)\hat{\phi}(\mathbf{x},t)$$

The scattering operator $\hat{S}$ is constructed just as before. One simply has to deal with the effects arising from the additional term in the interaction. We assume that the "mass counter-term" δm^2 has a power-series expansion in the coupling constant of the form

$$\delta m^2 = [\delta m^2]^{(2)} + [\delta m^2]^{(4)} + \cdots$$
$$= \tilde{c}_2 \, g^2 + \tilde{c}_4 \, g^4 + \cdots$$

In this case, there will be a contribution of order g^2 arising from the term $\hat{S}^{(1)}$ in Eq. (7.1), where the notation here indicates the order in $\hat{H}_I(t)$.[14] The contribution of the mass counter-term will then be

$$\hat{S}^{(1)}_{\text{mass}} = \frac{i}{\hbar c}\frac{1}{2}\delta m^2 c^2 \int_\Omega d^4x : \hat{\phi}(\mathbf{x},t)\hat{\phi}(\mathbf{x},t):$$

where we again assume, just as in Eq. (7.32), that the scalar density is normal-ordered. The matrix element of this operator between the states in Eq. (7.52) leads to an additional contribution in Eq. (7.54), so that it now takes the form[15]

$$S_{fi} = (2\pi)^4\delta^{(4)}(k_1 - k_2)\frac{ic}{\Omega}\frac{1}{\sqrt{4\omega_1\omega_2}}\left[\Pi(k_1) + \delta m^2\right]$$

$$\Pi(k_1) \equiv \frac{ig^2}{\hbar c^3}\int \frac{d^4t}{(2\pi)^4}\text{Tr}\left[\frac{1}{i\gamma_\mu(t + k_1)_\mu + M}\frac{1}{i\gamma_\mu t_\mu + M}\right]$$

For a particle "on its mass shell", where $k_1^2 = -m^2$, the quantity $\Pi(k_1)$ is just a number. The mass counter-term will then *exactly cancel* the term in $\Pi(k_1)$ for a real particle provided it is chosen to satisfy

$$\left[\Pi(k_1) + \delta m^2\right]_{k_1^2 = -m^2} = 0$$

As a result, there is no mass shift of the ϕ as the interaction is turned on (at least to this order), and the mass m^2 in the starting hamiltonian remains the mass in the interacting theory. We again note that mass renormalization

---

[14]Of course, this mass term must also be included in the higher-order terms in $\hat{S}$. A mass counter-term insertion is usually indicated by a cross on a line in a Feynman diagram.

[15]Note the factor of 2 coming from the matrix element of $: \hat{\phi}(\mathbf{x},t)\hat{\phi}(\mathbf{x},t):$.

is not a luxury, but is an essential aspect of any relativistic quantum field theory.[16]

Problem 7.13 Suppose the masses in the Dirac-scalar case are such that the ϕ is unstable, and the decay $\phi \to N + \bar{N}$ can actually take place.

(a) Show the amplitude S_{fi} for $\phi(p) \to N(k_1) + \bar{N}(k_2)$ is given to lowest order by

$$S_{fi} = (2\pi)^4 \delta^{(4)}(p - k_1 - k_2) \frac{1}{\Omega^{3/2}} \left(\frac{\hbar}{2\omega_p} \right)^{1/2} \left(\frac{ig_s}{\hbar c} \right) \bar{u}(\mathbf{k}_1\lambda_1) v(-\mathbf{k}_2\lambda_2)$$

(b) Show that before going to the continuum limit, in a big box of volume Ω with p.b.c., this expression is actually[17]

$$S_{fi} = 2\pi \delta(W_f - W_i)[\Omega \delta_{\mathbf{K}_f,\mathbf{K}_i}] \frac{1}{\Omega^{3/2}} \left(\frac{\hbar}{2\omega_p} \right)^{1/2} ig_s \bar{u}(\mathbf{k}_1\lambda_1) v(-\mathbf{k}_2\lambda_2)$$

where (W_i, W_f) are the total initial and final energies, and $\hbar(\mathbf{K}_i, \mathbf{K}_f)$ are the total initial and final momenta.

Solution to Problem 7.13

(a) Here the initial and final states are

$$|i\rangle = c_{\mathbf{p}}^\dagger |0\rangle \qquad\qquad ; \ |f\rangle = a_{\mathbf{k}_1\lambda_1}^\dagger b_{\mathbf{k}_2\lambda_2}^\dagger |0\rangle$$

The first-order scattering operator is

$$\hat{S}^{(1)} = \frac{ig_s}{\hbar c} \int d^4x : \hat{\bar{\psi}}(x) \, \hat{\psi}(x) : \hat{\phi}(x)$$

With the use of the fields in Eqs. (7.2), the S-matrix element is then computed to be

$$S_{fi}^{(2)} = \frac{(2\pi)^4}{\Omega^{3/2}} \delta^{(4)}(p - k_1 - k_2) \left(\frac{\hbar}{2\omega_p} \right)^{1/2} \frac{ig_s}{\hbar c} \bar{u}(\mathbf{k}_1\lambda_1) v(-\mathbf{k}_2\lambda_2)$$

(b) In obtaining this result, we have used the continuum limit for the space-time integral

$$\int d^4x \, e^{i(p-k_1-k_2)\cdot x} = (2\pi)^4 \delta^{(4)}(p - k_1 - k_2)$$

---

[16]The renormalization of the Dirac-scalar theory is discussed more generally in [Serot and Walecka (1986)].

[17]Compare Probl. 7.2.

In a big box with p.b.c., this is really

$$\int_{-\infty}^{\infty} d(ct) \int_{\Omega} d^3x \, e^{i(\mathbf{K}_i-\mathbf{K}_f)\cdot\mathbf{x}} e^{-ict(W_i-W_f)/\hbar} = 2\pi\hbar c\, \delta(W_i - W_f)[\Omega\delta_{\mathbf{K}_f,\mathbf{K}_i}]$$

Hence, before the continuum limit,

$$S_{fi} = 2\pi\delta(W_f - W_i)[\Omega\delta_{\mathbf{K}_f,\mathbf{K}_i}]\frac{1}{\Omega^{3/2}}\left(\frac{\hbar}{2\omega_p}\right)^{1/2} ig_s\bar{u}(\mathbf{k}_1\lambda_1)v(-\mathbf{k}_2\lambda_2)$$

Problem 7.14 (a) Use the arguments in chapter 7 of Vol. I to show that the differential decay rate for $\phi \to N + \bar{N}$ that follows from Prob. 7.13 is

$$d\omega_{fi} = \frac{2\pi}{\hbar}\delta(W_f - W_i)\frac{1}{\Omega}|T_{fi}|^2 \frac{\Omega d^3k_1}{(2\pi)^3} \qquad ; \mathbf{p} = \mathbf{k}_1 + \mathbf{k}_2$$

Here we factor out $\Omega^{-(n_f+n_i)/2}$ in the definition of T_{fi} [see Eq. (7.38)].
(b) Show that for a ϕ at rest this gives

$$d\omega_{fi} = \frac{1}{64\pi^2}\left(\frac{g_s^2}{\hbar c^3}\right)\left(\frac{m_\phi c^2}{\hbar}\right)\left(1 - \frac{4m_N^2}{m_\phi^2}\right)^{1/2}|\bar{u}(\mathbf{k}_1\lambda_1)v(-\mathbf{k}_2\lambda_2)|^2 d\Omega_{k_1}$$

Solution to Problem 7.14

(a) Define the T-matrix for this process by

$$S_{fi} \equiv -2\pi i\delta(W_f - W_i)[\Omega\delta_{\mathbf{K}_f,\mathbf{K}_i}]\frac{1}{\Omega^{3/2}}T_{fi}$$

$$\to -\frac{i}{\hbar c}\frac{(2\pi)^4}{\Omega^{3/2}}\delta^{(4)}(p - k_1 - k_2)T_{fi} \qquad ; \Omega \to \infty$$

Thus

$$T_{fi} = -\left(\frac{\hbar}{2\omega_p}\right)^{1/2} g_s\bar{u}(\mathbf{k}_1\lambda_1)v(-\mathbf{k}_2\lambda_2)$$

Now from Eq. (4.91), the transition rate is

$$\omega_{fi} = \frac{2\pi}{\hbar}\delta(W_i - W_f)[\Omega\delta_{\mathbf{K}_f,\mathbf{K}_i}]^2\frac{1}{\Omega^3}|T_{fi}|^2$$

This is the transition rate into one final state. In a big box with p.b.c., the sum over final momentum states gives

$$\sum_{k_1}\sum_{k_2}[\delta_{\mathbf{K}_f,\mathbf{K}_i}]^2 = \sum_{k_1}\sum_{k_2}\delta_{\mathbf{K}_f,\mathbf{K}_i}$$

$$= \sum_{k_1} \longrightarrow \frac{\Omega d^3 k_1}{(2\pi)^3} \qquad\qquad ; \mathbf{p} = \mathbf{k}_1 + \mathbf{k}_2$$

Hence the differential decay rate for $\phi \to N(k_1) + \bar{N}(k_2)$ is

$$d\omega_{fi} = \frac{2\pi}{\hbar}\delta(W_f - W_i)\frac{1}{\Omega}|T_{fi}|^2\frac{\Omega d^3 k_1}{(2\pi)^3} \qquad\qquad ; \mathbf{p} = \mathbf{k}_1 + \mathbf{k}_2$$

(b) The total initial and final energies in the rest frame of the ϕ, which is the C-M system for the final $(N\bar{N})$, are

$$W_i = m_\phi c^2$$

$$W_f = 2\hbar c\sqrt{k_1^2 + M^2}$$

Here m_ϕ is the ϕ rest mass. It follows that

$$\frac{\partial W_f}{\partial k_1} = 2\hbar c\frac{k_1}{\sqrt{k_1^2 + M^2}} = 4(\hbar c)^2\frac{k_1}{m_\phi c^2} \qquad\qquad ; W_f = W_i$$

We can also solve for k_1, where $k_1 = \sqrt{\mathbf{k}_1^2}$,

$$2\hbar c\sqrt{k_1^2 + M^2} = m_\phi c^2$$

$$k_1 = \frac{m_\phi c^2}{2\hbar c}\left(1 - \frac{4m_N^2}{m_\phi^2}\right)^{1/2}$$

The differential decay rate is then

$$d\omega_{fi} = \frac{2\pi}{\hbar}\delta(W_f - W_i)dW_f\left(\frac{\partial W_f}{\partial k_1}\right)^{-1}|T_{fi}|^2\frac{k_1^2 d\Omega_{k_1}}{(2\pi)^3}$$

$$= \frac{1}{(2\pi)^2}\frac{m_\phi c^2}{4\hbar^3 c^2}\frac{m_\phi c^2}{2\hbar c}\left(1 - \frac{4m_N^2}{m_\phi^2}\right)^{1/2}\frac{\hbar^2 g_s^2}{2m_\phi c^2}|\bar{u}(\mathbf{k}_1\lambda_1)v(-\mathbf{k}_2\lambda_2)|^2 d\Omega_{k_1}$$

In the second line, we have done the integral over the energy-conserving delta-function, and used $\hbar\omega_p = m_\phi c^2$ in the ϕ rest frame. A combination of factors gives the stated answer

$$dw_{fi} = \frac{1}{16\pi} \left(\frac{g_s^2}{4\pi\hbar c^3}\right) \left(\frac{m_\phi c^2}{\hbar}\right) \left(1 - \frac{4m_N^2}{m_\phi^2}\right)^{1/2} |\bar{u}(\mathbf{k}_1\lambda_1)v(-\mathbf{k}_2\lambda_2)|^2 d\Omega_{k_1}$$

****

(*Aside*) We might as well finish the calculation. The sum over final fermion spins follows as in Eqs. (8.61)–(8.69) and (8.101)

$$\sum_{\lambda_1}\sum_{\lambda_2} |\bar{u}(\mathbf{k}_1\lambda_1)v(-\mathbf{k}_2\lambda_2)|^2 = \frac{1}{4\omega_1\omega_2/c^2} \mathrm{Tr}\,(M - i\slashed{k}_1)(-M - i\slashed{k}_2)$$

$$= \frac{1}{\omega_1\omega_2/c^2}(-M^2 - k_1 \cdot k_2)$$

where the four-vectors are given in the ϕ rest frame by

$$k_1 = (\mathbf{k}_1, i\omega_1/c) \quad ; \quad k_2 = (-\mathbf{k}_1, i\omega_2/c) \quad ; \quad \omega_1/c = \omega_2/c = \sqrt{\mathbf{k}_1^2 + M^2}$$

Thus

$$\sum_{\lambda_1}\sum_{\lambda_2} |\bar{u}(\mathbf{k}_1\lambda_1)v(-\mathbf{k}_2\lambda_2)|^2 = 2\frac{\mathbf{k}_1^2}{\mathbf{k}_1^2 + M^2}$$

With the use of the previous kinematics, this gives

$$\sum_{\lambda_1}\sum_{\lambda_2} |\bar{u}(\mathbf{k}_1\lambda_1)v(-\mathbf{k}_2\lambda_2)|^2 = 2\left(1 - \frac{4m_N^2}{m_\phi^2}\right)$$

The integration over solid angle gives

$$\int d\Omega_{k_1} = 4\pi$$

Hence the total decay rate is

$$\omega_{fi} = \frac{1}{2}\left(\frac{g_s^2}{4\pi\hbar c^3}\right)\left(\frac{m_\phi c^2}{\hbar}\right)\left(1 - \frac{4m_N^2}{m_\phi^2}\right)^{3/2}$$

****

Problem 7.15 (a) Verify the $N + N$ scattering amplitude in Eq. (7.44);
(b) Verify the expression for the ϕ self-energy in Eq. (7.53).

Solution to Problem 7.15

(a) The initial and final states are[18]

$$|i\rangle = a^\dagger_{\mathbf{k}_1\lambda_1}a^\dagger_{\mathbf{k}_2\lambda_2}|0\rangle \qquad ; |f\rangle = a^\dagger_{\mathbf{k}_3\lambda_3}a^\dagger_{\mathbf{k}_4\lambda_4}|0\rangle$$

Since these states do not contain scalar operators, the matrix elements select the scalar contribution to $\hat{S}^{(2)}$ that contains the scalar propagator, since

$$\langle 0| :\hat{\phi}(x_1)\hat{\phi}(x_2): +\frac{\hbar}{ic}\Delta_F(x_1-x_2)|0\rangle = \frac{\hbar}{ic}\Delta_F(x_1-x_2)$$

The only non-vanishing matrix element involving fermions is

$$\langle 0|a_{\mathbf{k}_4\lambda_4}a_{\mathbf{k}_3\lambda_3} :\hat{\bar\psi}_\alpha(x_1)\hat\psi_\alpha(x_1)\hat{\bar\psi}_\beta(x_2)\hat\psi_\beta(x_2): a^\dagger_{\mathbf{k}_1\lambda_1}a^\dagger_{\mathbf{k}_2\lambda_2}|0\rangle$$

This matrix element can be evaluated in terms of the contractions

$$\langle 0|a_{\mathbf{k}\lambda}\bar\psi_\alpha(x)|0\rangle = \frac{1}{\sqrt\Omega}\bar u_\alpha(\mathbf{k},\lambda)e^{-ik\cdot x}$$

$$\langle 0|\psi_\alpha(x)a^\dagger_{\mathbf{k}\lambda}|0\rangle = \frac{1}{\sqrt\Omega}u_\alpha(\mathbf{k},\lambda)e^{ik\cdot x}$$

In direct analogy with Wick's theorem, it is then only the fully contracted term that contributes to the fermion matrix element[19]

$$\langle 0|a_{\mathbf{k}_4\lambda_4}a_{\mathbf{k}_3\lambda_3} :\hat{\bar\psi}_\alpha(x_1)\hat\psi_\alpha(x_1)\hat{\bar\psi}_\beta(x_2)\hat\psi_\beta(x_2): a^\dagger_{\mathbf{k}_1\lambda_1}a^\dagger_{\mathbf{k}_2\lambda_2}|0\rangle = (-1)\times$$

$$\langle 0|a_{\mathbf{k}_4\lambda_4}a_{\mathbf{k}_3\lambda_3} :\hat{\bar\psi}_\alpha(x_1)\hat{\bar\psi}_\beta(x_2)\hat\psi_\alpha(x_1)\hat\psi_\beta(x_2): a^\dagger_{\mathbf{k}_1\lambda_1}a^\dagger_{\mathbf{k}_2\lambda_2}|0\rangle = (-1)\times$$

$$\left[\langle 0|a_{\mathbf{k}_3\lambda_3}\bar\psi_\alpha(x_1)|0\rangle\langle 0|a_{\mathbf{k}_4\lambda_4}\bar\psi_\beta(x_2)|0\rangle - \langle 0|a_{\mathbf{k}_3\lambda_3}\bar\psi_\beta(x_2)|0\rangle\langle 0|a_{\mathbf{k}_4\lambda_4}\bar\psi_\alpha(x_1)|0\rangle\right]\times$$

$$\left[\langle 0|\psi_\beta(x_2)a^\dagger_{\mathbf{k}_1\lambda_1}|0\rangle\langle 0|\psi_\alpha(x_1)a^\dagger_{\mathbf{k}_2,\lambda_2}|0\rangle - \langle 0|\psi_\beta(x_2)a^\dagger_{\mathbf{k}_2\lambda_2}|0\rangle\langle 0|\psi_\alpha(x_1)a^\dagger_{\mathbf{k}_1\lambda_1}|0\rangle\right]$$

It follows that the fermion matrix element is given by

$$\langle 0|a_{\mathbf{k}_4\lambda_4}a_{\mathbf{k}_3\lambda_3} :\hat{\bar\psi}_\alpha(x_1)\hat\psi_\alpha(x_1)\hat{\bar\psi}_\beta(x_2)\hat\psi_\beta(x_2): a^\dagger_{\mathbf{k}_1\lambda_1}a^\dagger_{\mathbf{k}_2\lambda_2}|0\rangle =$$

$$\frac{1}{\Omega^2}\left\{[\bar u(\mathbf{k}_3,\lambda_3)u(\mathbf{k}_1,\lambda_1)]\,[\bar u(\mathbf{k}_4,\lambda_4)u(\mathbf{k}_2,\lambda_2)]\,e^{i(k_1-k_3)\cdot x_1}e^{i(k_2-k_4)\cdot x_2} -\right.$$

$$[\bar u(\mathbf{k}_4,\lambda_4)u(\mathbf{k}_1,\lambda_1)]\,[\bar u(\mathbf{k}_3,\lambda_3)u(\mathbf{k}_2,\lambda_2)]\,e^{i(k_2-k_3)\cdot x_1}e^{i(k_1-k_4)\cdot x_2} +$$

$$[\bar u(\mathbf{k}_4,\lambda_4)u(\mathbf{k}_2,\lambda_2)]\,[\bar u(\mathbf{k}_3,\lambda_3)u(\mathbf{k}_1,\lambda_1)]\,e^{i(k_2-k_4)\cdot x_1}e^{i(k_1-k_3)\cdot x_2} -$$

$$\left.[\bar u(\mathbf{k}_4,\lambda_4)u(\mathbf{k}_1,\lambda_1)]\,[\bar u(\mathbf{k}_3,\lambda_3)u(\mathbf{k}_2,\lambda_2)]\,e^{i(k_1-k_4)\cdot x_1}e^{i(k_2-k_3)\cdot x_2}\right\}$$

---

[18]We assume the nucleons are identical and all the $\mathbf{k}_i$ are distinct in the scattering process.

[19]See the discussion in Sec. 11.2.2.3.

We may now put the pieces together and write the S-matrix element

$$S_{fi} = \left(\frac{ig}{\hbar c}\right)^2 \frac{1}{\Omega^2} \frac{1}{2!} \int d^4 x_1 \int d^4 x_2 \frac{\hbar}{ic} \Delta_F(x_1 - x_2) \times$$

$$\left\{ [\bar{u}(\mathbf{k}_3, \lambda_3)u(\mathbf{k}_1, \lambda_1)]\, [\bar{u}(\mathbf{k}_4, \lambda_4)u(\mathbf{k}_2, \lambda_2)]\, e^{i(k_1 - k_3)\cdot x_1} e^{i(k_2 - k_4)\cdot x_2} - \right.$$

$$[\bar{u}(\mathbf{k}_4, \lambda_4)u(\mathbf{k}_1, \lambda_1)]\, [\bar{u}(\mathbf{k}_3, \lambda_3)u(\mathbf{k}_2, \lambda_2)]\, e^{i(k_2 - k_3)\cdot x_1} e^{i(k_1 - k_4)\cdot x_2} +$$

$$[\bar{u}(\mathbf{k}_4, \lambda_4)u(\mathbf{k}_2, \lambda_2)]\, [\bar{u}(\mathbf{k}_3, \lambda_3)u(\mathbf{k}_1, \lambda_1)]\, e^{i(k_2 - k_4)\cdot x_1} e^{i(k_1 - k_3)\cdot x_2} -$$

$$\left. [\bar{u}(\mathbf{k}_4, \lambda_4)u(\mathbf{k}_1, \lambda_1)]\, [\bar{u}(\mathbf{k}_3, \lambda_3)u(\mathbf{k}_2, \lambda_2)]\, e^{i(k_1 - k_4)\cdot x_1} e^{i(k_2 - k_3)\cdot x_2} \right\}$$

which coincides with Eq. (7.44) in the text.

(b) We now come to the calculation of the self-energy of the scalar ϕ. In this case, the initial and final states are

$$|i\rangle = c^\dagger_{\mathbf{k}_1}|0\rangle \qquad ; |f\rangle = c^\dagger_{\mathbf{k}_2}|0\rangle$$

Since these states do not contain any fermion creation operators, the matrix element of the scattering operator $\hat{S}^{(2)}$ will select only the contribution which contains the fermion propagators

$$\langle 0| \left[:\hat{\bar{\psi}}_\alpha(x_1)\hat{\psi}_\alpha(x_1)\hat{\bar{\psi}}_\beta(x_2)\hat{\psi}_\beta(x_2): + :\hat{\bar{\psi}}_\alpha(x_1)iS^F_{\alpha\beta}(x_1 - x_2)\hat{\psi}_\beta(x_2): + \right.$$

$$\left. :\hat{\bar{\psi}}_\beta(x_2)iS^F_{\beta\alpha}(x_2 - x_1)\hat{\psi}_\alpha(x_1): - iS^F_{\alpha\beta}(x_1 - x_2)iS^F_{\beta\alpha}(x_2 - x_1) \right] |0\rangle$$

$$= -iS^F_{\alpha\beta}(x_1 - x_2)iS^F_{\beta\alpha}(x_2 - x_1)$$

The scalar contribution to the matrix element reads

$$\langle 0|c_{\mathbf{k}_2}:\hat{\phi}(x_1)\hat{\phi}(x_2): c^\dagger_{\mathbf{k}_1}|0\rangle = \frac{1}{\Omega}\left(\frac{\hbar}{\sqrt{4\omega_1\omega_2}}\right)\left[e^{ik_1\cdot x_1}e^{-ik_2\cdot x_2} + e^{-ik_2\cdot x_1}e^{ik_1\cdot x_2}\right]$$

Assembling the pieces, we have

$$S^{(2)}_{fi} = \frac{1}{2!}\left(\frac{ig}{\hbar c}\right)^2 \frac{1}{\Omega}\frac{\hbar}{\sqrt{4\omega_1\omega_2}} \int d^4 x_1 \int d^4 x_2 \times$$

$$(-1)iS^F_{\alpha\beta}(x_1 - x_2)iS^F_{\beta\alpha}(x_2 - x_1)\left[e^{ik_1\cdot x_1}e^{-ik_2\cdot x_2} + e^{-ik_2\cdot x_1}e^{ik_1\cdot x_2}\right]$$

$$= \left(\frac{ig}{\hbar c}\right)^2 \frac{1}{\Omega}\frac{\hbar}{\sqrt{4\omega_1\omega_2}} \int d^4 x_1 \int d^4 x_2 \times$$

$$(-1)\text{Tr}\left[iS^F(x_1 - x_2)iS^F(x_2 - x_1)\right] e^{ik_1\cdot x_1}e^{-ik_2\cdot x_2}$$

where the last line has been obtained with a change of variable $x_1 \leftrightarrows x_2$. This expression coincides with Eq. (7.53) in the text.

Problem 7.16 Use the notation for the creation and destruction operators in Eqs. (8.10). Evaluate the following matrix elements:

(a) $\langle 0|a_{k_4\lambda_4}c_{k_3s_3}a^\dagger_{k_8\lambda_8}a_{k_7\lambda_7}c^\dagger_{k_6s_6}c_{k_5s_5}c^\dagger_{k_1s_1}a^\dagger_{k_2\lambda_2}|0\rangle$

(b) $\langle 0|a_{k_4\lambda_4}a_{k_3\lambda_3}a^\dagger_{k_8\lambda_8}a^\dagger_{k_7\lambda_7}a_{k_6\lambda_6}a_{k_5\lambda_5}a^\dagger_{k_1\lambda_1}a^\dagger_{k_2\lambda_2}|0\rangle$

(c) $\langle 0|a_{k_4\lambda_4}b_{k_3\lambda_3}b^\dagger_{k_8\lambda_8}a^\dagger_{k_7\lambda_7}b_{k_6\lambda_6}a_{k_5\lambda_5}b^\dagger_{k_1\lambda_1}a^\dagger_{k_2\lambda_2}|0\rangle$

Solution to Problem 7.16

(a) First, we can move like operators together, since they commute with the unlike operators, and we can explicitly indicate the appropriate vacuum

$$\langle 0|a_{k_4\lambda_4}c_{k_3s_3}a^\dagger_{k_8\lambda_8}a_{k_7\lambda_7}c^\dagger_{k_6s_6}c_{k_5s_5}c^\dagger_{k_1s_1}a^\dagger_{k_2\lambda_2}|0\rangle =$$
$$\langle 0|a_{k_4\lambda_4}a^\dagger_{k_8\lambda_8}a_{k_7\lambda_7}a^\dagger_{k_2\lambda_2}|0\rangle\langle 0|c_{k_3s_3}c^\dagger_{k_6s_6}c_{k_5s_5}c^\dagger_{k_1s_1}|0\rangle$$

Now use the (anti-) commutation relations to move the destruction operators over to the vacuum, where they give zero

$$\langle 0|a_{k_4\lambda_4}c_{k_3s_3}a^\dagger_{k_8\lambda_8}a_{k_7\lambda_7}c^\dagger_{k_6s_6}c_{k_5s_5}c^\dagger_{k_1s_1}a^\dagger_{k_2\lambda_2}|0\rangle =$$
$$[\delta_{k_2\lambda_2,\,k_7\lambda_7}\,\delta_{k_4\lambda_4,\,k_8\lambda_8}]\,[\delta_{k_1s_1,\,k_5s_5}\,\delta_{k_3s_3,\,k_6s_6}]$$

(b) With the use of the anticommutation relations for the Dirac operators, we can move the destruction operators to the right, where they annihilate the vacuum, and move the creation operators to the left, where the similarly annihilate the vacuum

$$\langle 0|a_{k_4\lambda_4}a_{k_3\lambda_3}a^\dagger_{k_8\lambda_8}a^\dagger_{k_7\lambda_7}a_{k_6\lambda_6}a_{k_5\lambda_5}a^\dagger_{k_1\lambda_1}a^\dagger_{k_2\lambda_2}|0\rangle$$
$$= \langle 0|a_{k_4\lambda_4}[\delta_{k_3\lambda_3,\,k_8\lambda_8} - a^\dagger_{k_8\lambda_8}a_{k_3\lambda_3}]a^\dagger_{k_7\lambda_7}\times$$
$$a_{k_6\lambda_6}[\delta_{k_1\lambda_1,\,k_5\lambda_5} - a^\dagger_{k_1\lambda_1}a_{k_5\lambda_5}]a^\dagger_{k_2\lambda_2}|0\rangle$$
$$= [\delta_{k_3\lambda_3,\,k_8\lambda_8}\delta_{k_4\lambda_4,\,k_7\lambda_7} - \delta_{k_4\lambda_4,\,k_8\lambda_8}\delta_{k_3\lambda_3,\,k_7\lambda_7}]\times$$
$$[\delta_{k_1\lambda_1,\,k_5\lambda_5}\delta_{k_2\lambda_2,\,k_6\lambda_6} - \delta_{k_1\lambda_1,\,k_6\lambda_6}\delta_{k_2\lambda_2,\,k_5\lambda_5}]$$

Notice the exchange contribution here, coming in with opposite sign.

(c) We can do this one as in (a), keeping track of the fact that all the fermion operators anticommute

$$\langle 0|a_{k_4\lambda_4}b_{k_3\lambda_3}b^\dagger_{k_8\lambda_8}a^\dagger_{k_7\lambda_7}b_{k_6\lambda_6}a_{k_5\lambda_5}b^\dagger_{k_1\lambda_1}a^\dagger_{k_2\lambda_2}|0\rangle$$
$$= -\langle 0|b_{k_3\lambda_3}b^\dagger_{k_8\lambda_8}b_{k_6\lambda_6}b^\dagger_{k_1\lambda_1}|0\rangle\langle 0|a_{k_4\lambda_4}a^\dagger_{k_7\lambda_7}a_{k_5\lambda_5}a^\dagger_{k_2\lambda_2}|0\rangle$$
$$= -[\delta_{k_3\lambda_3,\,k_8\lambda_8}\,\delta_{k_1\lambda_1,\,k_6\lambda_6}]\,[\delta_{k_2\lambda_2,\,k_5\lambda_5}\,\delta_{k_4\lambda_4,\,k_7\lambda_7}]$$

Chapter 8

Quantum Electrodynamics (QED)

Problem 8.1 (a) Verify that the interaction-picture Dirac current is conserved in Eqs. (8.33);

(b) Show that the equal-time interaction-picture commutator of the charge and current densities vanishes in Eq. (8.38).

Solution to Problem 8.1

(a) The Dirac field in the interaction picture is given in Eqs. (5.102)

$$\hat{\psi}(\mathbf{x}, t) = \frac{1}{\sqrt{\Omega}} \sum_{\mathbf{k}} \sum_{\lambda} \left[a_{\mathbf{k}\lambda} u_\lambda(\mathbf{k}) e^{i(\mathbf{k}\cdot\mathbf{x} - \omega_k t)} + b^\dagger_{\mathbf{k}\lambda} v_\lambda(-\mathbf{k}) e^{-i(\mathbf{k}\cdot\mathbf{x} - \omega_k t)} \right]$$
$$\omega_k = c\sqrt{\mathbf{k}^2 + M^2}$$

It satisfies the free Dirac Eqs. (5.94)

$$\left(\gamma_\mu \frac{\partial}{\partial x_\mu} + M \right) \hat{\psi} = 0$$
$$\hat{\bar{\psi}} \left(\gamma_\mu \frac{\overleftarrow{\partial}}{\partial x_\mu} - M \right) = 0$$

The Dirac current in the interaction picture is given in Eq. (8.12)

$$\hat{j}_\mu(x) = i : \hat{\bar{\psi}}(x) \gamma_\mu \hat{\psi}(x) : \, = i \hat{\bar{\psi}}(x) \gamma_\mu \hat{\psi}(x) - \langle 0 | i \hat{\bar{\psi}}(x) \gamma_\mu \hat{\psi}(x) | 0 \rangle$$

Take the four-divergence of the operator, and use the Dirac equation,

$$\frac{\partial}{\partial x_\mu} i \hat{\bar{\psi}}(x) \gamma_\mu \hat{\psi}(x) = i \frac{\partial \hat{\bar{\psi}}(x)}{\partial x_\mu} \gamma_\mu \hat{\psi}(x) + i \hat{\bar{\psi}}(x) \gamma_\mu \frac{\partial \hat{\psi}(x)}{\partial x_\mu}$$
$$= i \hat{\bar{\psi}}(x) \left(M - M \right) \hat{\psi}(x) = 0$$

Hence the interaction-picture Dirac current is conserved

$$\frac{\partial \hat{j}_\mu(x)}{\partial x_\mu} = 0$$

(b) The equal-time anticommutator of the Dirac fields in the interaction picture is given in Eq. (5.104)

$$\left\{ \hat{\psi}_\alpha(\mathbf{x}, t), \hat{\psi}_\beta^\dagger(\mathbf{x}', t') \right\}_{t=t'} = \delta_{\alpha\beta}\, \delta^{(3)}(\mathbf{x} - \mathbf{x}')$$

with all other anticommutators vanishing.

We are asked to evaluate the commutator in Eq. (8.38)

$$[\hat{\rho}(x), \hat{j}_\lambda(y)]_{t_x = t_y} = i[\hat{\psi}^\dagger(\mathbf{x}, t)\hat{\psi}(\mathbf{x}, t),\ \hat{\psi}^\dagger(\mathbf{y}, t)\gamma_4\gamma_\lambda\hat{\psi}(\mathbf{y}, t)]_{t=t_x}$$

If all the Dirac indices are supplied, this is

$$C_\lambda(x, y) \equiv i \sum_\alpha \sum_\rho \sum_\sigma \left\{ \hat{\psi}_\alpha^\dagger(\mathbf{x}, t)\hat{\psi}_\alpha(\mathbf{x}, t)\hat{\psi}_\rho^\dagger(\mathbf{y}, t)\, [\gamma_4\gamma_\lambda]_{\rho\sigma}\, \hat{\psi}_\sigma(\mathbf{y}, t) - \right.$$

$$\left. \hat{\psi}_\rho^\dagger(\mathbf{y}, t)\, [\gamma_4\gamma_\lambda]_{\rho\sigma}\, \hat{\psi}_\sigma(\mathbf{y}, t)\hat{\psi}_\alpha^\dagger(\mathbf{x}, t)\hat{\psi}_\alpha(\mathbf{x}, t) \right\}$$

Now use the anticommutation relations on the first term to re-order it so that it cancels the second. The terms left over in this re-ordering are

$$C_\lambda(x, y) \equiv i \sum_\alpha \sum_\rho \sum_\sigma \left\{ \hat{\psi}_\alpha^\dagger(\mathbf{x}, t)\delta_{\alpha\rho}\delta^{(3)}(\mathbf{x} - \mathbf{y})\, [\gamma_4\gamma_\lambda]_{\rho\sigma}\, \hat{\psi}_\sigma(\mathbf{y}, t) - \right.$$

$$\left. \hat{\psi}_\rho^\dagger(\mathbf{y}, t)\delta_{\alpha\sigma}\delta^{(3)}(\mathbf{x} - \mathbf{y})\, [\gamma_4\gamma_\lambda]_{\rho\sigma}\, \hat{\psi}_\alpha(\mathbf{x}, t) \right\}$$

If the Dirac indices are again suppressed, this is

$$C_\lambda(x, y) = i\delta^{(3)}(\mathbf{x} - \mathbf{y})\, \hat{\psi}^\dagger(\mathbf{x}, t)\, [\gamma_4\gamma_\lambda - \gamma_4\gamma_\lambda]\, \hat{\psi}(\mathbf{x}, t)$$

$$= 0$$

Hence the equal-time interaction-picture commutator of the charge and current densities in Eq. (8.38) vanishes

$$i[\hat{\psi}^\dagger(\mathbf{x}, t)\hat{\psi}(\mathbf{x}, t),\ \hat{\psi}^\dagger(\mathbf{y}, t)\gamma_4\gamma_\lambda\hat{\psi}(\mathbf{y}, t)]_{t=t_x} = 0$$

Problem 8.2 Carry out the four-dimensional Fourier transform in Minkowski space, and show that the Coulomb interaction can be written as in Eq. (8.18).[1]

---

[1] *Hint:* see Vol. I.

Solution to Problem 8.2

Consider the function

$$D^C(x_1 - x_2) = \int \frac{d^4k}{(2\pi)^4} \frac{1}{\mathbf{k}^2} e^{ik\cdot(x_1 - x_2)}$$

Here $x_\mu = (\mathbf{x}, ict)$, $k_\mu = (\mathbf{k}, ik_0)$, and $d^4k = d^3k dk_0$. It follows that

$$\int \frac{d^4k}{(2\pi)^4} \frac{1}{\mathbf{k}^2} e^{ik\cdot(x_1 - x_2)} = \int \frac{d^3k}{(2\pi)^3} \frac{1}{\mathbf{k}^2} e^{i\mathbf{k}\cdot(\mathbf{x}_1 - \mathbf{x}_2)} \int_{-\infty}^{\infty} \frac{dk_0}{2\pi} e^{-ik_0 c(t_1 - t_2)}$$

The integral over k_0 gives a Dirac delta-function

$$\int_{-\infty}^{\infty} \frac{dk_0}{2\pi} e^{-ik_0 c(t_1 - t_2)} = \delta(ct_1 - ct_2)$$

The remaining Fourier transform was evaluated in Vol. I

$$\int \frac{d^3k}{(2\pi)^3} \frac{1}{\mathbf{k}^2} e^{i\mathbf{k}\cdot(\mathbf{x}_1 - \mathbf{x}_2)} = \frac{1}{4\pi|\mathbf{x}_1 - \mathbf{x}_2|}$$

The result is Eq. (8.18)

$$D^C(x_1 - x_2) = \frac{\delta(ct_1 - ct_2)}{4\pi|\mathbf{x}_1 - \mathbf{x}_2|}$$

Problem 8.3 Prove the following relations involving the Dirac matrices[2]

$$\not{a}\not{b} + \not{b}\not{a} = 2(a \cdot b)$$
$$\gamma_\mu \not{a} \gamma_\mu = -2\not{a}$$
$$\gamma_\mu \not{a} \not{b} \gamma_\mu = 2(\not{a}\not{b} + \not{b}\not{a}) = 4(a \cdot b)$$
$$\gamma_\mu \not{a} \not{b} \not{c} \gamma_\mu = -2\not{c}\not{b}\not{a}$$

Solution to Problem 8.3

The Dirac matrices obey the Clifford algebra in Eq. (8.75)

$$\gamma_\mu \gamma_\nu + \gamma_\nu \gamma_\mu = 2\delta_{\mu\nu}$$

where the unit matrix on the r.h.s. is suppressed. They also satisfy[3]

$$\gamma_\mu \gamma_\mu = 4$$

---

[2] The Feynman notation is $\not{a} = a_\mu \gamma_\mu$, where the four-vector $a_\mu = (\mathbf{a}, ia_0)$.
[3] Recall that repeated Greek indices are summed from 1 to 4.

The first relation is then obtained from

$$\not{a}\not{b} + \not{b}\not{a} = a_\mu b_\nu \left(\gamma_\mu \gamma_\nu + \gamma_\nu \gamma_\mu\right)$$
$$= 2\delta_{\mu\nu} a_\mu b_\nu = 2(a \cdot b)$$

The second relation is derived as

$$\gamma_\mu \not{a} \gamma_\mu = a_\lambda \gamma_\mu \gamma_\lambda \gamma_\mu$$
$$= a_\lambda \gamma_\mu \left(2\delta_{\lambda\mu} - \gamma_\mu \gamma_\lambda\right)$$
$$= 2\not{a} - 4\not{a} = -2\not{a}$$

The third relation follows by moving the final γ_μ through to the left

$$\gamma_\mu \not{a} \not{b} \gamma_\mu = a_\lambda b_\sigma \left(\gamma_\mu \gamma_\lambda \gamma_\sigma \gamma_\mu\right)$$
$$= a_\lambda b_\sigma \left(2\delta_{\sigma\mu}\gamma_\mu\gamma_\lambda - 2\delta_{\lambda\mu}\gamma_\mu\gamma_\sigma + 4\gamma_\lambda\gamma_\sigma\right)$$
$$= 2(\not{a}\not{b} + \not{b}\not{a}) = 4(a \cdot b)$$

The final relation follows in the same manner

$$\gamma_\mu \not{a} \not{b} \not{c} \gamma_\mu = a_\lambda b_\sigma c_\rho \left(\gamma_\mu \gamma_\lambda \gamma_\sigma \gamma_\rho \gamma_\mu\right)$$
$$= a_\lambda b_\sigma c_\rho \left(2\delta_{\mu\rho}\gamma_\mu\gamma_\lambda\gamma_\sigma - 2\delta_{\mu\sigma}\gamma_\mu\gamma_\lambda\gamma_\rho + 2\delta_{\mu\lambda}\gamma_\mu\gamma_\sigma\gamma_\rho - 4\gamma_\lambda\gamma_\sigma\gamma_\rho\right)$$
$$= a_\lambda b_\sigma c_\rho \left(2\gamma_\rho\gamma_\lambda\gamma_\sigma - 2\gamma_\sigma\gamma_\lambda\gamma_\rho - 2\gamma_\lambda\gamma_\sigma\gamma_\rho\right)$$

Now rearrange the gamma matrices into the order $\gamma_\rho \gamma_\sigma \gamma_\lambda$

$$\gamma_\mu \not{a} \not{b} \not{c} \gamma_\mu = a_\lambda b_\sigma c_\rho \left(-2\gamma_\rho\gamma_\sigma\gamma_\lambda + 4\delta_{\lambda\sigma}\gamma_\rho - 2\gamma_\rho\gamma_\sigma\gamma_\lambda - 4\delta_{\lambda\rho}\gamma_\sigma + 4\delta_{\sigma\rho}\gamma_\lambda + \right.$$
$$\left. 2\gamma_\rho\gamma_\sigma\gamma_\lambda - 4\delta_{\sigma\rho}\gamma_\lambda + 4\delta_{\lambda\rho}\gamma_\sigma - 4\delta_{\lambda\sigma}\gamma_\rho\right)$$
$$= a_\lambda b_\sigma c_\rho \left(-2\gamma_\rho\gamma_\sigma\gamma_\lambda\right)$$
$$= -2\not{c}\not{b}\not{a}$$

Problem 8.4 Consider the cross section for Compton scattering in the *laboratory* frame obtained from the amplitude $S_{fi}^{(2)}$ in Eq. (8.44). Denote the initial and final three-momenta of the photons in this frame by $(\mathbf{l}_1, \mathbf{l}_2)$. An essential component of this calculation is the calculation of

$$\left(\frac{\partial W_f}{\partial l_2}\right)_\theta = (\hbar c)^2 \frac{M l_1}{E_2 l_2}$$

Here the partial derivative is a reminder that the change in variables from l_2 to the total final energy W_f in Eq. (8.58) is to be carried out at constant angle, and the result is evaluated at energy conservation $W_i = W_f$. Here E_2 is the final electron energy. Prove this relation.

Solution to Problem 8.4

The statements of momentum and energy conservation in the laboratory frame are

$$\mathbf{l}_1 = \mathbf{l}_2 + \mathbf{k}_2$$
$$l_1 + M = l_2 + \sqrt{\mathbf{k}_2^2 + M^2}$$

where $l = \sqrt{\mathbf{l}^2}$.[4] The final total energy W_f is given by

$$\frac{W_f}{\hbar c} = l_2 + \sqrt{\mathbf{k}_2^2 + M^2} = l_2 + \sqrt{(\mathbf{l}_1 - \mathbf{l}_2)^2 + M^2}$$
$$= l_2 + \sqrt{l_1^2 + l_2^2 - 2l_1 l_2 \cos\theta + M^2}$$

At fixed incident energy l_1, and fixed scattering angle θ, one has

$$\frac{1}{\hbar c}\left(\frac{\partial W_f}{\partial l_2}\right)_\theta = 1 + \frac{l_2 - l_1 \cos\theta}{E_2/\hbar c}$$
$$= \frac{E_2/\hbar c + l_2 - l_1 \cos\theta}{E_2/\hbar c}$$
$$= \frac{M + l_1 - l_1 \cos\theta}{E_2/\hbar c}$$

Now use the Compton formula, derived from the above as

$$l_1 + M - l_2 = \sqrt{l_1^2 + l_2^2 - 2l_1 l_2 \cos\theta + M^2}$$
$$l_1^2 + l_2^2 + M^2 - 2l_1 l_2 + 2Ml_1 - 2Ml_2 = l_1^2 + l_2^2 - 2l_1 l_2 \cos\theta + M^2$$
$$Ml_1 = l_2(M + l_1 - l_1 \cos\theta)$$

We thus arrive at the stated answer

$$\left(\frac{\partial W_f}{\partial l_2}\right)_\theta = (\hbar c)^2 \frac{Ml_1}{E_2 l_2}$$

Problem 8.5 Consider the sum over the transverse polarizations of an external photon, as required, for example, in calculating the Compton cross section. Write the polarization four-vector as [see Eqs. (8.45)] $\varepsilon_\mu(\mathbf{k}s) =$

---

[4]Consistent with our usage in this volume, l_μ is an inverse length; the actual photon three-momentum is $\hbar\mathbf{l}$, and the photon energy is $\hbar c l$.

$(\mathbf{e_{ks}}, 0)$. Assume that in calculating the cross section one has to evaluate an expression of the form $\sum_{s=1}^{2} \varepsilon_\mu(\mathbf{ks}) M_{\mu\nu} \varepsilon_\nu(\mathbf{ks})$ where the response tensor satisfies current conservation in momentum space[5]

$$k_\mu M_{\mu\nu} = M_{\mu\nu} k_\nu = 0 \qquad ; \text{ current conservation}$$

Derive the *covariant polarization sum*

$$\sum_{s=1}^{2} \varepsilon_\mu(\mathbf{ks}) M_{\mu\nu} \varepsilon_\nu(\mathbf{ks}) = M_{\mu\mu}$$

Solution to Problem 8.5

The sum over the transverse polarization vectors is given in Eqs. (C.42)

$$\sum_{s=1}^{2} (\mathbf{e_{ks}})_i (\mathbf{e_{ks}})_j = \delta_{ij} - \frac{k_i k_j}{\mathbf{k}^2}$$

Therefore[6]

$$\sum_{s=1}^{2} (\mathbf{e_{ks}})_i M_{ij} (\mathbf{e_{ks}})_j = M_{ii} - \frac{k_i k_j}{\mathbf{k}^2} M_{ij}$$

From $k_\mu M_{\mu\nu} = 0$, we have for $\nu = j$

$$k_\mu M_{\mu j} = k_i M_{ij} + i k_0 M_{4j} = 0$$
$$\implies \quad k_i M_{ij} = -i k_0 M_{4j}$$

Here the photon four-momentum is $k_\mu = (\mathbf{k}, i k_0)$, with $k_0 = |\mathbf{k}|$.

From $M_{\mu\nu} k_\nu = 0$ with $\mu = 4$, we obtain

$$M_{4\nu} k_\nu = M_{4j} k_j + i k_0 M_{44} = 0$$
$$\implies \quad M_{4j} k_j = -i k_0 M_{44}$$

Hence

$$k_i k_j M_{ij} = -i k_0 k_j M_{4j} = -k_0^2 M_{44}$$
$$= -\mathbf{k}^2 M_{44}$$

---

[5]This can, and should, be checked in any application; this relation generally holds only for the full amplitude.

[6]As usual, repeated Latin subscripts are here summed from 1 to 3, and repeated Greek subscripts from 1 to 4.

This yields the stated result

$$\sum_{s=1}^{2} \varepsilon_\mu(\mathbf{k}s) M_{\mu\nu}\, \varepsilon_\nu(\mathbf{k}s) = M_{ii} + M_{44} = M_{\mu\mu}$$

Problem 8.6 The cross section for Compton scattering in the laboratory frame is given by the Klein-Nishina formula

$$\frac{d\sigma}{d\Omega} = \frac{1}{2} r_0^2 \left(\frac{l_2}{l_1}\right)^2 \left[\left(\frac{l_2}{l_1}\right) + \left(\frac{l_1}{l_2}\right) - \sin^2\theta\right] \qquad ; \; r_0 \equiv \frac{e^2}{4\pi\hbar c\varepsilon_0}\frac{1}{M}$$

Use the results in Probs. 8.3–8.5 to calculate the contribution to this result coming from the square of the first Feynman diagram in Fig. 8.1 in the text. Make sure you understand where all the factors come from.

Solution to Problem 8.6

This solution follows closely the analysis in Sec. 8.7. The S-matrix element corresponding to the two Feynman diagrams in Fig. 8.1 in the text is given in Eq. (8.44), where we now label the initial and final photon four-momenta by (l_1, l_2) and electron four-momenta by (k_1, k_2)[7]

$$S_{fi}^{(2)} = (2\pi)^4 \delta^{(4)}(k_1 + l_1 - k_2 - l_2)\frac{1}{\Omega^2}\frac{\hbar c^2}{\sqrt{4\omega_1\omega_2}}\left(\frac{-e}{\hbar c\sqrt{\varepsilon_0}}\right)^2\left(\frac{1}{i}\right) \times$$

$$\bar{u}(\mathbf{k}_2\lambda')\varepsilon_\nu^f\left[\gamma_\nu\frac{1}{i(\not{k}_1 + \not{l}_1) + M}\gamma_\mu + \gamma_\mu\frac{1}{i(\not{k}_1 - \not{l}_2) + M}\gamma_\nu\right]\varepsilon_\mu^i u(\mathbf{k}_1\lambda)$$

The T-matrix is identified as in Eq. (8.52)

$$\frac{-iT_{fi}^{(2)}}{\hbar c} = \frac{\hbar c^2}{\sqrt{4\omega_1\omega_2}}\left(\frac{-e}{\hbar c\sqrt{\varepsilon_0}}\right)^2\left(\frac{1}{i}\right) \times$$

$$\bar{u}(\mathbf{k}_2\lambda')\varepsilon_\nu^f\left[\gamma_\nu\frac{1}{i(\not{k}_1 + \not{l}_1) + M}\gamma_\mu + \gamma_\mu\frac{1}{i(\not{k}_1 - \not{l}_2) + M}\gamma_\nu\right]\varepsilon_\mu^i u(\mathbf{k}_1\lambda)$$

The cross section then follows as in Eq. (8.53)

$$d\sigma_{fi} = \frac{2\pi}{\hbar}\delta(W_f - W_i)\frac{1}{\Omega^2}|T_{fi}|^2\frac{\Omega d^3 l_2}{(2\pi)^3}\frac{1}{I_{\text{inc}}}$$

We now observe:

---

[7]Here the four-momentum $l_\mu = (\mathbf{l}, i\omega_l/c)$ with $\omega_l = c|\mathbf{l}|$. As usual, the initial and final states are labelled by i and f.

- Initially, in the laboratory frame, we have one photon in the volume Ω incident on an electron at rest. Hence the incident flux is

$$I_{\text{inc}} = \rho v = \frac{c}{\Omega}$$

The artificial quantization volume Ω now cancels from the cross section;

- The density of final states is given by the analysis in Prob. 8.4

$$\int \delta(W_f - W_i)dl_2 = \int \delta(W_f - W_i) \left(\frac{\partial l_2}{\partial W_f}\right)_\theta dW_f = \left(\frac{\partial W_f}{\partial l_2}\right)_\theta^{-1}$$

$$= \frac{E_2 l_2}{(\hbar c)^2 M l_1}$$

- We can identify the classical electron radius

$$r_0 \equiv \frac{e^2}{4\pi\hbar c \varepsilon_0} \frac{1}{M}$$

Hence, with $\omega_l/c = l$, the laboratory cross section becomes

$$\frac{d\sigma}{d\Omega} = r_0^2 \left(\frac{l_2}{l_1}\right)^2 \left[\frac{E_2 M}{\hbar c}\right] \times$$

$$\left| \bar{u}(\mathbf{k}_2\lambda')\varepsilon_\nu^f \left[\gamma_\nu \frac{1}{i(\not{k}_1 + \not{l}_1) + M}\gamma_\mu + \gamma_\mu \frac{1}{i(\not{k}_1 - \not{l}_2) + M}\gamma_\nu \right] \varepsilon_\mu^i u(\mathbf{k}_1\lambda) \right|^2$$

Now for the unpolarized cross section, we must sum over the final electron spin, and average over the initial spin. This is done exactly as in Eqs. (8.61)–(8.70), and introduces the positive-energy projections for the Dirac particles.[8] If we retain just the contribution from the first Feynman diagram in Fig. 8.1 in the text, as instructed in the problem, the result is

$$\frac{d\sigma}{d\Omega} \doteq r_0^2 \left(\frac{l_2}{l_1}\right)^2 \frac{1}{8}\mathcal{M}^2$$

$$\mathcal{M}^2 = \text{Tr}\left[\not{\varepsilon}^i \frac{1}{i(\not{k}_1 + \not{l}_1) + M}\not{\varepsilon}^f (M - i\not{k}_2)\not{\varepsilon}^f \frac{1}{i(\not{k}_1 + \not{l}_1) + M}\not{\varepsilon}^i (M - i\not{k}_1) \right]$$

---

[8]Note that

$$(\gamma_4\, i\not{a}\, i\not{b}\, i\not{c} \cdots)^\dagger = \gamma_4 (\cdots i\not{c}\, i\not{b}\, i\not{a})$$

This follows from the fact that the adjoint is the product of adjoints in reverse order, and since $a_\mu = (\mathbf{a}, ia_0)$, one has $(i\not{a})^\dagger \gamma_4 = \gamma_4 (i\not{a})$, *etc.*

At this point, we have all the factors straight. We observe that $\mathcal{M}^2$ is both dimensionless and Lorentz-invariant.

The trace is invariant under cyclic permutations, so the last relation can be re-written as

$$\mathcal{M}^2 = \text{Tr}\left[(M - i\not{k}_2)\not{\epsilon}^f \frac{1}{i(\not{k}_1 + \not{l}_1) + M}\not{\epsilon}^i(M - i\not{k}_1)\not{\epsilon}^i\frac{1}{i(\not{k}_1 + \not{l}_1) + M}\not{\epsilon}^f\right]$$

The denominators can be rationalized

$$\frac{1}{i(\not{k}_1 + \not{l}_1) + M} = \frac{M - i(\not{k}_1 + \not{l}_1)}{(k_1 + l_1)^2 + M^2} = \frac{M - i(\not{k}_1 + \not{l}_1)}{2k_1 \cdot l_1}$$

where we have used $k_1 \cdot k_1 = -M^2$ and $l_1 \cdot l_1 = 0$.[9] Thus

$$\mathcal{M}^2 = \frac{1}{4(k_1 \cdot l_1)^2} \times$$
$$\text{Tr}\left[(M - i\not{k}_2)\not{\epsilon}^f[M - i(\not{k}_1 + \not{l}_1)]\not{\epsilon}^i(M - i\not{k}_1)\not{\epsilon}^i[M - i(\not{k}_1 + \not{l}_1)]\not{\epsilon}^f\right]$$

To evaluate the trace, we can work from the inside out. Use

$$\not{\epsilon}^i\not{\epsilon}^i = \varepsilon^i \cdot \varepsilon^i = 1 \qquad ; \not{k}_1\not{\epsilon}^i = -\not{\epsilon}^i\not{k}_1$$

where the second relation follows from the fact that $\varepsilon^i \cdot k_1 = 0$.[10] Therefore

$$\not{\epsilon}^i(M - i\not{k}_1)\not{\epsilon}^i = M + i\not{k}_1$$

Now use the Dirac equation

$$(M + i\not{k}_1)(M - i\not{k}_1) = M^2 + k_1^2 = 0$$

Therefore

$$\mathcal{M}^2 = \frac{-1}{4(k_1 \cdot l_1)^2}\text{Tr}\left[(M - i\not{k}_2)\not{\epsilon}^f\not{l}_1(M + i\not{k}_1)\not{l}_1\not{\epsilon}^f\right]$$

Use

$$\not{l}_1\not{l}_1 = l_1 \cdot l_1 = 0 \qquad ; \not{l}_1\not{k}_1\not{l}_1 = 2(k_1 \cdot l_1)\not{l}_1$$

Thus

$$\mathcal{M}^2 = \frac{-i}{2(k_1 \cdot l_1)}\text{Tr}\left[(M - i\not{k}_2)\not{\epsilon}^f\not{l}_1\not{\epsilon}^f\right]$$

For the unpolarized cross section, we are supposed to average over the initial photon polarization ε^i (which has now disappeared from this term),

---

[9]These are now all four-vectors.
[10]This is clear, since in the lab frame $k_1 = (\mathbf{0}, iM)$ and $\varepsilon^i = (\boldsymbol{\varepsilon}^i, 0)$.

and sum over the final polarizations ε^f. Since the entire amplitude satisfies current conservation, we can make use of the covariant polarization sum in Prob. 8.5. Then, from Prob. 8.3

$$\gamma_\mu \rlap{/}{l}_1 \gamma_\mu = -2\rlap{/}{l}_1$$

and

$$\mathcal{M}^2 = \frac{i}{(k_1 \cdot l_1)} \text{Tr} \left[(M - i\rlap{/}{k}_2)\rlap{/}{l}_1 \right]$$

$$= 4\frac{(k_2 \cdot l_1)}{(k_1 \cdot l_1)} = 4\frac{(k_1 \cdot l_2)}{(k_1 \cdot l_1)}$$

Here the last relation follows from squaring the four-momentum relation $k_1 - l_2 = k_2 - l_1$. Thus we finally arrive at a nice, simple, Lorentz-invariant expression for $\mathcal{M}^2$.

In the laboratory frame, the four-vector $k_1 = (\mathbf{0}, iM)$. Therefore, in the lab, $\mathcal{M}^2 = 4\omega_2/\omega_1 = 4l_2/l_1$, where now $l = \omega_l/c$. Hence, the contribution of the first Feynman diagram in Fig. 8.1 in the text to the cross section is

$$\frac{d\sigma}{d\Omega} \doteq \frac{1}{2} r_0^2 \left(\frac{l_2}{l_1} \right)^2 \left[\left(\frac{l_2}{l_1} \right) \right] \qquad ; l = \frac{\omega_l}{c}$$

This is the first term in the Klein-Nishina formula.

Problem 8.7 Take the limit of the result in Prob. 8.6 for a heavy target, and derive the Thomson cross section for photon scattering

$$\frac{d\sigma}{d\Omega} = r_0^2 \frac{1}{2}(1 + \cos^2 \theta) \qquad ; \text{ Thomson cross section}$$

Solution to Problem 8.7

The Compton formula follows from the solution to Prob. 8.4

$$l_1 - l_2 = \frac{l_1 l_2}{M}(1 - \cos\theta)$$

If the target mass $M \to \infty$, there is no shift in the scattered photon energy

$$l_1 = l_2 \qquad ; M \to \infty$$

In this case, the Klein-Nishina formula in Prob. 8.7 reduces to the Thomson cross section

$$\frac{d\sigma}{d\Omega} = r_0^2 \frac{1}{2}(1 + \cos^2 \theta) \qquad ; M \to \infty$$

Problem 8.8 (a) Construct $S_{fi}^{(2)}$ for the Bhabha scattering process $e^-(k_1) + e^+(k_2) \to e^-(k_3) + e^+(k_4)$;

(b) Draw the two momentum-space Feynman diagrams.

Solution to Problem 8.8

(a) The initial and final states for Bhabha scattering are

$$|i\rangle = a_{\mathbf{k}_1\lambda_1}^\dagger b_{\mathbf{k}_2\lambda_2}^\dagger |0\rangle \qquad\qquad ; \; |f\rangle = a_{\mathbf{k}_3\lambda_3}^\dagger b_{\mathbf{k}_4\lambda_4}^\dagger |0\rangle$$

The relevant term in the scattering operator in order e^2 is given in Eq. (8.42)

$$\hat{S}^{(2)} \doteq \left(\frac{-e}{\hbar c}\right)^2 \frac{1}{2!} \int d^4x_1 \int d^4x_2 : \hat{\bar{\psi}}(x_1)\gamma_\mu\hat{\psi}(x_1)\hat{\bar{\psi}}(x_2)\gamma_\nu\hat{\psi}(x_2): \times$$
$$\frac{\hbar c}{i\varepsilon_0} D_{\mu\nu}^F(x_1 - x_2)$$

where we have retained the contributing term in the application of Wick's theorem to the Dirac currents. The matrix element of the scattering operator then follows directly, where we proceed right to the momentum-space result, as in Eq. (8.46)[11]

$$S_{fi}^{(2)} = (2\pi)^4\delta^{(4)}(k_1 + k_2 - k_3 - k_4)\frac{1}{\Omega^2}\left(\frac{-e}{\hbar c\sqrt{\varepsilon_0}}\right)^2\left(\frac{\hbar c}{i}\right) \times$$
$$\left[\bar{u}(k_3\lambda_3)\gamma_\mu v(-\mathbf{k}_4,\lambda_4)\frac{1}{(k_1 + k_2)^2}\bar{v}(-\mathbf{k}_2,\lambda_2)\gamma_\mu u(k_1\lambda_1) - \right.$$
$$\left. \bar{u}(k_3\lambda_3)\gamma_\mu u(k_1\lambda_1)\frac{1}{(k_1 - k_3)^2}\bar{v}(-\mathbf{k}_2,\lambda_2)\gamma_\mu v(-\mathbf{k}_4,\lambda_4)\right]$$

(b) The corresponding Feynman diagrams are shown in Fig. 8.1.

Problem 8.9 Make use of the Dirac equation on the external lepton legs, and show explicitly that the terms proportional to q_μ or q_ν in the photon propagator $\tilde{D}_{\mu\nu}^F(q)$ in Eq. (8.28) do not contribute to the scattering amplitudes in Eqs. (8.52) and (8.96).

Solution to Problem 8.9

The photon propagator in Eq. (8.28), written in terms of q, is

$$\tilde{D}_{\mu\nu}^F(q) = \frac{1}{q^2}\delta_{\mu\nu} - \frac{1}{q^2}\frac{1}{q^2 + (q\cdot\eta)^2}\left[q_\mu q_\nu + (q\cdot\eta)(q_\mu\eta_\nu + q_\nu\eta_\mu)\right]$$

---

[11]The required matrix element of the creation and destruction operators is evaluated as in Prob. 7.16.

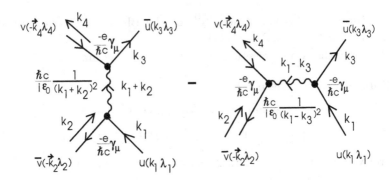

Fig. 8.1   Feynman diagrams for Bhabha scattering.

The lowest-order T-matrix for the $e^- + \mu^- \to e^- + \mu^-$ scattering amplitude described by the Feynman diagram in Fig. 8.4 in the text is given in Eq. (8.52)[12]

$$\frac{-iT^{(2)}_{fi}}{\hbar} = \left(\frac{-e}{\hbar c\sqrt{\varepsilon_0}}\right)^2 \left(\frac{\hbar c}{i}\right) [\bar{u}(\mathbf{k}_3\lambda_3)\gamma_\mu u(\mathbf{k}_1\lambda_1)]_{e^-}\frac{1}{q^2}[\bar{u}(\mathbf{k}_4\lambda_4)\gamma_\mu u(\mathbf{k}_2\lambda_2)]_{\mu^-}$$

Here $q = k_4 - k_2 = k_1 - k_3$. The terms proportional to q_μ or q_ν in the photon propagator vanish in this amplitude by the use of the Dirac equation on the external legs.[13] First, we have

$$\bar{u}(\mathbf{k}_3\lambda_3)\slashed{q}\,u(\mathbf{k}_1\lambda_1) = -i\bar{u}(\mathbf{k}_3\lambda_3)(i\slashed{k}_1 - i\slashed{k}_3)u(\mathbf{k}_1\lambda_1)$$
$$= -i\bar{u}(\mathbf{k}_3\lambda_3)(-M_e + M_e)u(\mathbf{k}_1\lambda_1) = 0$$

Then we also have

$$\bar{u}(\mathbf{k}_4\lambda_4)\slashed{q}\,u(\mathbf{k}_2\lambda_2) = -i\bar{u}(\mathbf{k}_4\lambda_4)(i\slashed{k}_4 - i\slashed{k}_2)u(\mathbf{k}_2\lambda_2)$$
$$= -i\bar{u}(\mathbf{k}_4\lambda_4)(-M_\mu + M_\mu)u(\mathbf{k}_2\lambda_2) = 0$$

The lowest-order T-matrix for the $e^+ + e^- \to \mu^+ + \mu^-$ scattering amplitude described by the Feynman diagram in Fig. 8.7 in the text is given

---

[12]Recall that in this book, as in the first volume, we are using SI units.
[13]The Dirac equations are $(i\slashed{k} + M)u(\mathbf{k}) = \bar{u}(\mathbf{k})(i\slashed{k} + M) = 0$, with $k_\mu = (\mathbf{k}, iE_k/\hbar c)$. Also, $(i\slashed{k} - M)v(-\mathbf{k}) = \bar{v}(-\mathbf{k})(i\slashed{k} - M) = 0$.

in Eq. (8.96)

$$\frac{-iT_{fi}^{(2)}}{\hbar c} = \left(\frac{-e}{\hbar c\sqrt{\varepsilon_0}}\right)^2 \left(\frac{\hbar c}{i}\right) \times$$

$$[\bar{u}(\mathbf{k}_4\lambda_4)\gamma_\mu v(-\mathbf{k}_2, \lambda_2)]_\mu - \frac{1}{(k_1+k_3)^2}[\bar{v}(-\mathbf{k}_3, \lambda_3)\gamma_\mu u(\mathbf{k}_1\lambda_1)]_{e^-}$$

Here $q = k_2 + k_4 = k_1 + k_3$. Again, the terms proportional to q_μ or q_ν in the photon propagator vanish in this amplitude by the use of the Dirac equation on the external legs. First, we have

$$\bar{v}(-\mathbf{k}_3, \lambda_3)\displaystyle{\not}q u(\mathbf{k}_1\lambda_1) = -i\bar{v}(-\mathbf{k}_3, \lambda_3)(i\displaystyle{\not}k_3 + i\displaystyle{\not}k_1)u(\mathbf{k}_1\lambda_1)$$
$$= -i\bar{v}(-\mathbf{k}_3, \lambda_3)(M_e - M_e)u(\mathbf{k}_1\lambda_1) = 0$$

Then we also have

$$\bar{u}(\mathbf{k}_4\lambda_4)\displaystyle{\not}q v(-\mathbf{k}_2, \lambda_2) = -i\bar{u}(\mathbf{k}_4\lambda_4)(i\displaystyle{\not}k_4 + i\displaystyle{\not}k_2)v(-\mathbf{k}_2, \lambda_2)$$
$$= -i\bar{u}(\mathbf{k}_4\lambda_4)(-M_\mu + M_\mu)v(-\mathbf{k}_2, \lambda_2) = 0$$

Problem 8.10 Show that Eq. (8.93) is the Rutherford cross section (see Vol. I).

Solution to Problem 8.10

The Rutherford cross section is given in Vol. I as

$$\sigma_{\text{Ruth}} = \left|\frac{\hbar c\,\alpha}{4E\sin^2\theta/2}\right|^2 \qquad ; \text{Rutherford}$$

where $E = \hbar^2 k^2/2m$ is the incident energy.

The NRL limit of the cross section for (e^-, μ^-) scattering in the C-M system is given in Eq. (8.93) as

$$\frac{d\sigma}{d\Omega} = \frac{\alpha^2}{4\mathbf{k}^4\sin^4(\theta/2)}\left(\frac{M_e M_\mu}{M_\mu + M_e}\right)^2 \qquad ; \text{NRL}$$

Use the reduced mass μ, and write the incident energy in the C-M system as

$$E_{\text{inc}} = \frac{(\hbar\mathbf{k})^2}{2\mu} = (\hbar\mathbf{k})^2\frac{m_e + m_\mu}{2m_e m_\mu}$$
$$= \hbar c\,\mathbf{k}^2\frac{M_e + M_\mu}{2M_e M_\mu}$$

One then recovers the Rutherford cross section

$$\frac{d\sigma}{d\Omega} = \left| \frac{\hbar c \alpha}{4 E_{\text{inc}} \sin^2 \theta/2} \right|^2$$

Problem 8.11 (a) Start from the scattering operator in Eq. (8.120), and derive the expression for the bremsstrahlung amplitude in Eq. (8.126); (b) Repeat part (a) for the pair-production amplitude in Eq. (8.128).

Solution to Problem 8.11

(a) The scattering operator for scattering in an external field in Eq. (8.120) is

$$e^2 \hat{S}_{12}^{\text{ext}} = \left(\frac{-e}{\hbar c} \right)^2 \int d^4 x_1 \int d^4 x_2 \, P \left[: \hat{\bar{\psi}}(x_1) \gamma_\mu \hat{\psi}(x_1) : : \hat{\bar{\psi}}(x_2) \gamma_\nu \hat{\psi}(x_2) : \right] \times$$
$$c^2 \hat{A}_\mu(x_1) A_\nu^{\text{ext}}(x_2) \qquad\qquad ; \, \hat{A}_4(x) \equiv 0$$

Here the vector potentials for the photon field and external Coulomb field are given in Eqs. (8.10) and (8.122) as

$$c\hat{\mathbf{A}}(\mathbf{x}, t) = \sum_{\mathbf{k}} \sum_{s=1}^{2} \left(\frac{\hbar c^2}{2 \omega_k \varepsilon_0 \Omega} \right)^{1/2} \left[c_{\mathbf{k}s} \mathbf{e}_{\mathbf{k}s} e^{i(\mathbf{k}\cdot\mathbf{x} - \omega_k t)} + c_{\mathbf{k}s}^\dagger \mathbf{e}_{\mathbf{k}s} e^{-i(\mathbf{k}\cdot\mathbf{x} - \omega_k t)} \right]$$
$$c A_\mu^{\text{ext}}(x) \equiv \int \frac{d^4 q}{(2\pi)^4} \left[\frac{1}{\sqrt{\varepsilon_0}} a_\mu(q) \right] e^{iq \cdot x}$$

An application of Wick's theorem reduces the scattering operator to the following form for the processes of interest

$$e^2 \hat{S}_{12}^{\text{ext}} \doteq \left(\frac{-e}{\hbar c} \right)^2 \int d^4 x_1 \int d^4 x_2 \left[: \hat{\bar{\psi}}(x_1) \gamma_\mu \, i S_F(x_1 - x_2) \gamma_\nu \hat{\psi}(x_2) : + \right.$$
$$\left. : \hat{\bar{\psi}}(x_2) \gamma_\nu \, i S_F(x_2 - x_1) \gamma_\mu \hat{\psi}(x_1) : \right] c^2 \hat{A}_\mu(x_1) A_\nu^{\text{ext}}(x_2)$$

The initial and final states for the bremsstrahlung process are those in Eq. (8.125)

$$|i\rangle = a_{\mathbf{k}_1 \lambda_1}^\dagger |0\rangle \qquad\qquad ; \, |f\rangle = c_{\mathbf{1}s}^\dagger a_{\mathbf{k}_2 \lambda_2}^\dagger |0\rangle$$

The matrix element of the scattering operator then follows immediately, and we can proceed directly to momentum space as in chapter 7 to arrive at

$$\left[e^2 \hat{S}_{12}^{\text{ext}}\right]_{fi} = \frac{-i}{\sqrt{\Omega^3}} \left(\frac{-e}{\hbar c \sqrt{\varepsilon_0}}\right)^2 \left(\frac{\hbar c^2}{2\omega_l}\right)^{1/2} \bar{u}(\mathbf{k}_2 \lambda_2) \times$$

$$\left[\not{\epsilon} \frac{1}{i(\not{k}_2 + \not{l}) + M} \not{A}(k_2 + l - k_1) + \not{A}(k_2 + l - k_1) \frac{1}{i(\not{k}_1 - \not{l}) + M} \not{\epsilon}\right] u(\mathbf{k}_1 \lambda_1)$$

This is the expression in Eq. (8.126). The two corresponding Feynman diagrams are those shown in Fig. 8.9 in the text.

(b) For pair production, the initial and final states are given by

$$|i\rangle = c_{1s}^\dagger |0\rangle \qquad\qquad ; \; |f\rangle = a_{\mathbf{k}_2 \lambda_2}^\dagger b_{\mathbf{k}_2' \lambda_2'}^\dagger |0\rangle$$

A calculation analogous to that in part (a) then produces the result in Eq. (8.128)

$$\left[e^2 \hat{S}_{12}^{\text{ext}}\right]_{fi} = \frac{-i}{\sqrt{\Omega^3}} \left(\frac{-e}{\hbar c \sqrt{\varepsilon_0}}\right)^2 \left(\frac{\hbar c^2}{2\omega_l}\right)^{1/2} \bar{u}(\mathbf{k}_2 \lambda_2) \times$$

$$\left[\not{\epsilon} \frac{1}{i(\not{k}_2 - \not{l}) + M} \not{A}(k_2 + k_2' - l) + \not{A}(k_2 + k_2' - l) \frac{1}{i(\not{l} - \not{k}_2') + M} \not{\epsilon}\right] v(-\mathbf{k}_2', \lambda_2')$$

The two corresponding Feynman diagrams are those shown in Fig. 8.10 in the text.

Problem 8.12 It is a theorem that there is only one inequivalent irreducible representation of dimension-4 of the Clifford algebra of the Dirac gamma matrices $\gamma_\mu \gamma_\nu + \gamma_\nu \gamma_\mu = 2\delta_{\mu\nu}$. This implies that any other representation must be related to the standard representation by a similarity transformation $s\gamma_\mu s^{-1}$ where s is a non-singular 4×4 matrix.

(a) Show $-\gamma_\mu^T$ satisfies the same Clifford algebra;
(b) Show $s_c = \gamma_2 \gamma_4$ gives $s_c \gamma_\mu s_c^{-1} = -\gamma_\mu^T$.

Solution to Problem 8.12

(a) The Dirac gamma matrices satisfy the Clifford algebra

$$\gamma_\mu \gamma_\nu + \gamma_\nu \gamma_\mu = 2\delta_{\mu\nu}$$

Take the transpose of this matrix relation

$$\gamma_\nu^T \gamma_\mu^T + \gamma_\mu^T \gamma_\nu^T = 2\delta_{\mu\nu}$$

Now let $\gamma_\mu^T \to -\gamma_\mu^T$

$$(-\gamma_\nu^T)(-\gamma_\mu^T) + (-\gamma_\mu^T)(-\gamma_\nu^T) = 2\delta_{\mu\nu}$$

Hence $-\gamma_\mu^T$ satisfies the same Clifford algebra.

(b) The standard representation of the γ-matrices is

$$\gamma_1 = \begin{bmatrix} & & & -i \\ & & -i & \\ & i & & \\ i & & & \end{bmatrix} \quad \gamma_2 = \begin{bmatrix} & & & -1 \\ & & 1 & \\ & 1 & & \\ -1 & & & \end{bmatrix} \quad \gamma_3 = \begin{bmatrix} & & -i & \\ & & & i \\ i & & & \\ & -i & & \end{bmatrix} \quad \gamma_4 = \begin{bmatrix} 1 & & & \\ & 1 & & \\ & & -1 & \\ & & & -1 \end{bmatrix}$$

Define the transformation matrix s_c by

$$s_c \equiv \gamma_2\gamma_4 \qquad ; \; s_c^{-1} = \gamma_4\gamma_2$$

It follows that the gamma matrices are then transformed according to

$$s_c\gamma_\mu s_c^{-1} = \gamma_\mu \qquad ; \; \mu = 1, 3$$
$$= -\gamma_\mu \qquad ; \; \mu = 2, 4$$

A comparison with the above gives

$$s_c\gamma_\mu s_c^{-1} = -\gamma_\mu^T$$

Problem 8.13 *Furry's theorem* states that a closed fermion loop with an odd number of electromagnetic vertices makes no contribution to the scattering operator in QED.

(a) Consider a loop with three vertices (the proof is readily extended). Convince yourself from Wick's theorem that there will be two Feynman diagrams with the fermion line running in opposite directions around the loop. Show that the sum of these contributions appears as

$$S = -\text{Tr}\,\gamma_\mu iS_F(x_1 - x_2)\gamma_\nu iS_F(x_2 - x_3)\gamma_\lambda iS_F(x_3 - x_1)$$
$$-\text{Tr}\,\gamma_\nu iS_F(x_2 - x_1)\gamma_\mu iS_F(x_1 - x_3)\gamma_\lambda iS_F(x_3 - x_2)$$

where $S_F(x_i - x_j) = -\int d^4q(2\pi)^{-4}[i\slashed{q} + M]^{-1} e^{iq\cdot(x_i - x_j)}$;

(b) Insert $s_c^{-1}s_c$ everywhere, where s_c is the "charge conjugation" matrix from Prob. 8.12, and use invariance of the trace under cyclic permutations;

(c) Show $s_c S_F(x_i - x_j)s_c^{-1} = S_F^T(x_j - x_i)$;

(d) Show $\text{Tr}\,[a^T b^T \cdots y^T z^T] = \text{Tr}\,[zy \cdots ba]$;

(e) Hence show $\mathcal{S} = (-1)^3 \mathcal{S}$ and conclude $\mathcal{S} = 0$. This is Furry's theorem.

Solution to Problem 8.13

(a) The third-order S-matrix will have in the integrand a factor arising from the Dirac currents of

$$\mathcal{S} = P\left[:\hat{\bar{\psi}}(x_1)\gamma_\mu\hat{\psi}(x_1)::\hat{\bar{\psi}}(x_2)\gamma_\nu\hat{\psi}(x_2)::\hat{\bar{\psi}}(x_3)\gamma_\lambda\hat{\psi}(x_3):\right]$$

Wick's theorem then contains two distinct fully-contracted terms

$$\mathcal{S} \doteq \hat{\bar{\psi}}(x_1)\cdots\gamma_\mu\hat{\psi}(x_1)\cdot\hat{\bar{\psi}}(x_2)\cdot\gamma_\nu\hat{\psi}(x_2)\cdot\cdot\hat{\bar{\psi}}(x_3)\cdots\gamma_\lambda\hat{\psi}(x_3)\cdots +$$
$$\hat{\bar{\psi}}(x_1)\cdots\gamma_\mu\hat{\psi}(x_1)\cdot\hat{\bar{\psi}}(x_3)\cdot\gamma_\lambda\hat{\psi}(x_3)\cdot\cdot\hat{\bar{\psi}}(x_2)\cdot\cdot\gamma_\nu\hat{\psi}(x_2)\cdots$$

The contractions are given in Eq. (7.16), and the result is

$$\mathcal{S} = -\text{Tr}\,\gamma_\mu iS_F(x_1-x_2)\gamma_\nu iS_F(x_2-x_3)\gamma_\lambda iS_F(x_3-x_1)$$
$$-\text{Tr}\,\gamma_\nu iS_F(x_2-x_1)\gamma_\mu iS_F(x_1-x_3)\gamma_\lambda iS_F(x_3-x_2)$$

This produces two Feynman diagrams with the fermion line running in opposite directions around the loop.

(b) If we insert $s_c^{-1}s_c$ everywhere, where s_c is the charge conjugation matrix from Prob. 8.12, and use the invariance of the trace under cyclic permutations to cancel the first s_c^{-1} and last s_c, the expression for $\mathcal{S}$ becomes

$$\mathcal{S} = +\text{Tr}\,\gamma_\mu^T is_cS_F(x_1-x_2)s_c^{-1}\gamma_\nu^T is_cS_F(x_2-x_3)s_c^{-1}\gamma_\lambda^T is_cS_F(x_3-x_1)s_c^{-1}$$
$$+\text{Tr}\,\gamma_\nu^T is_cS_F(x_2-x_1)s_c^{-1}\gamma_\mu^T is_cS_F(x_1-x_3)s_c^{-1}\gamma_\lambda^T is_cS_F(x_3-x_2)s_c^{-1}$$

The rationalized propagator is given by

$$S_F(x_j-x_k) = -\int\frac{d^4q}{(2\pi)^4}\frac{M-i\not{q}}{q^2+M^2}e^{iq\cdot(x_j-x_k)}$$

Hence,

$$s_cS_F(x_j-x_k)s_c^{-1} = -\int\frac{d^4q}{(2\pi)^4}\frac{M+i\not{q}^T}{q^2+M^2}e^{iq\cdot(x_j-x_k)}$$

With a change of integration variable $q \to -q$, this becomes

$$s_cS_F(x_j-x_k)s_c^{-1} = -\int\frac{d^4q}{(2\pi)^4}\frac{M-i\not{q}^T}{q^2+M^2}e^{-iq\cdot(x_j-x_k)}$$
$$= S_F^T(x_k-x_j)$$

Thus $\mathcal{S}$ takes the form

$$\mathcal{S} = +\,\mathrm{Tr}\,\gamma_\mu^T i S_F^T(x_2-x_1)\gamma_\nu^T i S_F^T(x_3-x_2)\gamma_\lambda^T i S_F^T(x_1-x_3)$$
$$+\,\mathrm{Tr}\,\gamma_\nu^T i S_F^T(x_1-x_2)\gamma_\mu^T i S_F^T(x_3-x_1)\gamma_\lambda^T i S_F^T(x_2-x_3)$$

(d) If we supply the indices, with repeated indices summed, then it is easy to see that the trace of a product of transposes is the trace of the product in reverse order

$$\mathrm{Tr}\,[a^T b^T \cdots y^T z^T] = (a^T)_{ij}(b^T)_{jk}\cdots(y^T)_{lm}(z^T)_{mi}$$
$$= z_{im}y_{ml}\cdots b_{kj}a_{ji}$$
$$= \mathrm{Tr}\,[zy\cdots ba]$$

Therefore

$$\mathcal{S} = +\,\mathrm{Tr}\,i S_F(x_1-x_3)\gamma_\lambda i S_F(x_3-x_2)\gamma_\nu i S_F(x_2-x_1)\gamma_\mu$$
$$+\,\mathrm{Tr}\,i S_F(x_2-x_3)\gamma_\lambda i S_F(x_3-x_1)\gamma_\mu i S_F(x_1-x_2)\gamma_\nu$$

(e) Since the trace is invariant under cyclic permutations, we observe that this is just the *negative* of the expression we started with in part (a).[14] Therefore

$$\mathcal{S} = (-1)^3 \mathcal{S}$$
$$\implies \qquad \mathcal{S} = 0$$

This is Furry's theorem, at least in this simple case. The result is immediately extended to closed loops containing any odd number of photon interactions.

Problem 8.14 Assume a big box with p.b.c., and work in the Coulomb gauge.

(a) Suppose that for some reason the interaction-picture current were to be augmented by a term $\hat{\mathbf{j}}(x) \to \hat{\mathbf{j}}(x) - \nabla\hat{\vartheta}(x)$. Show the scattering operator $\hat{S}$ is unchanged;

(b) How would current conservation then be maintained? (*Hint:* Recall $\hat{\rho}(x) = (i/\hbar)[\hat{H}_0, \hat{\rho}(x)]$.)

Solution to Problem 8.14

(a) The interaction hamiltonian in the interaction picture for QED is given in Eqs. (8.14). Now imagine that the current $\hat{\mathbf{j}}(x)$ is modified in some

---

[14]Note the role of the two terms has been interchanged.

fashion. If this is the only change, then only $\hat{H}_I^\gamma(t)$ is affected

$$\hat{H}_I^\gamma(t) = -e \int_\Omega d^3x \, \hat{\mathbf{j}}(x) \cdot \hat{\mathbf{A}}(x)$$

Suppose that the modification is of the form

$$\hat{\mathbf{j}}(x) \to \hat{\mathbf{j}}(x) - \boldsymbol{\nabla}\hat{\vartheta}(x)$$

Insert this expression in the above and perform a partial integration on the second term. In a big box with p.b.c. the surface term vanishes, and the change in $\hat{H}_I^\gamma(t)$ is

$$\delta\hat{H}_I^\gamma(t) = -e \int_\Omega d^3x \, \hat{\vartheta}(x) \, \boldsymbol{\nabla} \cdot \hat{\mathbf{A}}(x)$$

This vanishes since $\boldsymbol{\nabla} \cdot \hat{\mathbf{A}}(x) = 0$ in the Coulomb gauge

$$\delta\hat{H}_I^\gamma(t) = 0 \qquad \text{; Coulomb gauge}$$

Hence the scattering operator $\hat{S}^{(2)}$ in Eq. (8.16) is *unchanged*.

(b) Suppose $\nabla^2\hat{\vartheta}(x) \neq 0$; then the current is no longer conserved in the presence of the additional term. The time derivative of a quantity at the spatial point $\mathbf{x}$ is given by Ehrenfest's theorem, and hence in the absence of $\hat{\vartheta}(x)$, current conservation in the interaction picture reads

$$\frac{i}{\hbar}[\hat{H}_0, \hat{\rho}(x)] + \boldsymbol{\nabla} \cdot \hat{\mathbf{j}}(x) = 0$$

Without further instruction, the modification required to restore current conservation is model-dependent. One possibility is to find an additional term in the charge density $\hat{\rho} \to \hat{\rho} + \delta\hat{\rho}^\vartheta$ that satisfies

$$\frac{i}{\hbar}[\hat{H}_0, \delta\hat{\rho}^\vartheta(x)] = \nabla^2\hat{\vartheta}(x)$$

The interaction-picture current is then conserved

$$\frac{i}{\hbar}[\hat{H}_0, \hat{\rho}(x)] + \boldsymbol{\nabla} \cdot \hat{\mathbf{j}}(x) = 0$$

The interaction hamiltonian in Eqs. (8.14) now retains the same form; however, the Coulomb interaction is modified.

Problem 8.15 Work in a big box with p.b.c., where the spatial integrals are of the form $\int_\Omega d^3x \int_\Omega d^3y$. Assume that for some reason the equal-time commutator in Eq. (8.37) did not vanish, but had a non-vanishing remainder of the form[15]

$$\delta(t_x - t_y)[\hat{\rho}(x), \hat{\mathbf{j}}(y)]_{t_x=t_y} \doteq i\delta(t_x - t_y)\delta^{(3)}(\mathbf{x} - \mathbf{y})\,\boldsymbol{\nabla}_y\,\hat{\varphi}(y)$$

Show that $\delta\hat{S}^{(2)}$ in Eq. (8.31) still vanishes.

Solution to Problem 8.15

Suppose the last term in Eq. (8.37) is not zero. Then the following term remains in $\delta\hat{S}^{(2)}$ in Eq. (8.32)

$$\delta\hat{S}^{(2)} = \left(\frac{ie}{\hbar c}\right)^2 \left(\frac{\hbar c}{i\varepsilon_0}\right)\frac{1}{2!}\int d^4x \int d^4y\,\frac{1}{c}\,\delta(t_x - t_y)[\hat{\rho}(x),\,\hat{j}_\nu(y)]_{t_x=t_y} \times$$
$$i\int \frac{d^4k}{(2\pi)^4}F_\nu(k,\eta)e^{ik\cdot(x-y)}$$

We are given the assumed non-vanishing expression

$$\delta(t_x - t_y)[\hat{\rho}(x), \hat{\mathbf{j}}(y)]_{t_x=t_y} \doteq i\delta(t_x - t_y)\delta^{(3)}(\mathbf{x} - \mathbf{y})\,\boldsymbol{\nabla}_y\,\hat{\varphi}(y)$$

Substitution into the above gives[16]

$$\delta\hat{S}^{(2)} = -\left(\frac{ie}{\hbar c}\right)^2 \left(\frac{\hbar}{i\varepsilon_0}\right)\frac{1}{2!}\int_\Omega d^4x\,\boldsymbol{\nabla}\,\hat{\varphi}(x)\cdot\int \frac{d^4k}{(2\pi)^4}\,\mathbf{F}(k,\eta)$$

The gradient can now be integrated out to the surface of the box, where the p.b.c. ensure that the integral indeed vanishes

$$\int_\Omega d^4x\,\boldsymbol{\nabla}\,\hat{\varphi}(x) = 0 \qquad\qquad ;\text{ p.b.c.}$$

Hence $\delta\hat{S}^{(2)}$ still vanishes

$$\delta\hat{S}^{(2)} = 0$$

---

[15]Such a contribution is known as a "Schwinger term".
[16]Recall from Eq. (8.12) that $j_\mu = (\mathbf{j}/c,\, i\rho)$.

Chapter 9

Higher-Order Processes

Problem 9.1 Assume the mass counter-term δM has a power-series expansion in e^2, and show that the additional interaction $\hat{H}_I^{\text{mass}}(t)$ in Eq. (9.6) gives rise to the additional terms in $[e^3 \hat{S}_{13}^{\text{ext}}]_{fi}$ in Eq. (9.7).

Solution to Problem 9.1

Assume the mass counter term δM in Eq. (9.6) has an expansion

$$\delta M = e^2 \, \delta M^{(2)} + e^4 \, \delta M^{(4)} + \cdots$$

Now add the additional term $\hat{H}_I^{\text{mass}}(t)$ in Eq. (9.6) to the hamiltonian describing the scattering of the Dirac particle in an external field

$$\hat{H}_I^{\text{mass}}(t) = -\hbar c \, \delta M \int d^3x : \hat{\bar{\psi}}(x)\hat{\psi}(x) :$$

To order e^3, there will then be two additional momentum-space Feynman diagrams corresponding to Figs. 9.1(d,e) in the text, where we replace the self-energy insertion on the external legs in these diagrams by a cross, indicating the insertion of the term $e^2 \, \delta M^{(2)}$.[1] The additional factors over the lowest-order S-matrix element $\left[e\hat{S}_{11}^{\text{ext}}\right]_{fi}$ in Eq. (9.2) are then, from the diagram analogous to Fig. 9.1(e),

$$\left(\frac{-i}{\hbar c}\right) \frac{1}{i} \frac{1}{i\slashed{k} + M} [-\hbar c e^2 \, \delta M^{(2)}]$$

---

[1] There will be two identical, additional cross terms in the scattering operator, cancelling the factor $1/2$!

The additional factors from the diagram analogous to Fig. 9.1(d) are

$$\left(\frac{-i}{\hbar c}\right)[-\hbar c e^2\,\delta M^{(2)}]\frac{1}{i}\frac{1}{i\not{k}' + M}$$

Hence $\left[e^3 \hat{S}_{13}^{\text{ext}}\right]_{fi}$ gets an additional contribution

$$\delta\left[e^3 \hat{S}_{13}^{\text{ext}}\right]_{fi} = \frac{1}{\Omega}\left(\frac{-e}{\hbar c\sqrt{\varepsilon_0}}\right)\bar{u}(\mathbf{k}'\lambda') \times$$

$$\left\{\gamma_\mu \frac{1}{i\not{k} + M}[e^2\delta M^{(2)}] + [e^2\delta M^{(2)}]\frac{1}{i\not{k}' + M}\gamma_\mu\right\} u(\mathbf{k}\lambda) a_\mu(q)$$

This now reproduces the expression for the self-energy insertions in Eq. (9.7)

$$\left[e^3 \hat{S}_{13}^{\text{ext}}\right]_{fi} = \frac{1}{\Omega}\left(\frac{-e}{\hbar c\sqrt{\varepsilon_0}}\right)\bar{u}(\mathbf{k}'\lambda') \times$$

$$\left\{\gamma_\mu \frac{1}{i\not{k} + M}[\delta M - \Sigma(k)] + [\delta M - \Sigma(k')]\frac{1}{i\not{k}' + M}\gamma_\mu\right\} u(\mathbf{k}\lambda) a_\mu(q)$$

Problem 9.2 (a) Evaluate the integral, and prove the Feynman parameterization relation in Eq. (9.25) for the product of two factors in the form $1/ab$;

(b) Repeat part (a) for the product of three factors $1/abc$ in Eq. (9.55);

(c) Use these results to generalize the Feynman parameterization for the product of n factors.

Solution to Problem 9.2

(a) We can just do the integral in Eq. (9.25)

$$\int_0^1 \frac{dx}{[ax + b(1-x)]^2} = \frac{1}{b-a}\left[\frac{1}{ax + b(1-x)}\right]_0^1 = \frac{1}{b-a}\left[\frac{1}{a} - \frac{1}{b}\right] = \frac{1}{ab}$$

(b) Again, with a little more effort, we can do the integrals

$$2!\int_0^1 dx \int_0^x dy\, \frac{1}{[ay + b(x-y) + c(1-x)]^3}$$

$$= \frac{1}{b-a}\int_0^1 dx\left[\frac{1}{[ay + b(x-y) + c(1-x)]^2}\right]_0^x$$

$$= \frac{1}{b-a}\int_0^1 dx\left[\frac{1}{[ax + c(1-x)]^2} - \frac{1}{[bx + c(1-x)]^2}\right]$$

Now make use of the result in part (a)

$$2! \int_0^1 dx \int_0^x dy \, \frac{1}{[ay + b(x - y) + c(1 - x)]^3} = \frac{1}{b - a} \left[\frac{1}{ac} - \frac{1}{bc} \right] = \frac{1}{abc}$$

(c) The generalization of this result is

$$I_n \equiv (n - 1)! \int_0^1 dx_1 \int_0^{x_1} dx_2 \cdots \int_0^{x_{n-2}} dx_{n-1} \times$$

$$\frac{1}{[a_n x_{n-1} + a_{n-1}(x_{n-2} - x_{n-1}) + \cdots + a_1(1 - x_1)]^n}$$

$$= \frac{1}{a_n a_{n-1} \cdots a_1}$$

The proof is by induction. The result clearly holds for I_2, I_3 by parts (a,b). Assume it holds for I_n, and consider I_{n+1}. The new factors are

$$n \int_0^{x_{n-1}} dx_n \, \frac{1}{[a_{n+1} x_n + a_n(x_{n-1} - x_n) + \cdots + a_1(1 - x_1)]^{n+1}}$$

$$= \frac{1}{a_n - a_{n+1}} \left\{ \frac{1}{[a_{n+1} x_{n-1} + a_{n-1}(x_{n-2} - x_{n-1}) + \cdots + a_1(1 - x_1)]^n} - \right.$$

$$\left. \frac{1}{[a_n x_{n-1} + a_{n-1}(x_{n-2} - x_{n-1}) + \cdots + a_1(1 - x_1)]^n} \right\}$$

With the assumption that the result holds for I_n, this gives

$$I_{n+1} = \frac{1}{a_n - a_{n+1}} \left[\frac{1}{a_{n+1} a_{n-1} \cdots a_1} - \frac{1}{a_n a_{n-1} \cdots a_1} \right]$$

$$= \frac{1}{a_{n+1} a_n a_{n-1} \cdots a_1}$$

Thus the result is established.

Problem 9.3 Write $n = 4 - \epsilon$, make a consistent expansion in ϵ as $\epsilon \to 0$, and display the additional finite terms present when passing from Eq. (9.32) to its singular part in Eq. (9.33) [compare Eq. (9.73)].

Solution to Problem 9.3

The constant A appearing in the discussion of the electron self-energy is given in Eq. (9.32) as

$$A = \frac{\alpha M}{4\pi^3} \Gamma \left(2 - \frac{n}{2} \right) \int_0^1 dx \frac{(\pi a^2)^{n/2}}{a^4} [2(1 - x) + nx] \qquad ; \, a^2 \equiv M^2 x^2$$

We write the number of dimensions as $n = 4 - \epsilon$ and investigate the limit as $\epsilon \to 0$. Use $z\Gamma(z) = \Gamma(z+1)$ to obtain[2]

$$\Gamma\left(2 - \frac{n}{2}\right) = \Gamma\left(\frac{\epsilon}{2}\right) = \frac{2}{\epsilon}\Gamma\left(1 + \frac{\epsilon}{2}\right) = \frac{2}{\epsilon} + \Gamma'(1) + O(\epsilon)$$

From an expansion of the exponential, one has

$$\frac{(\pi a^2)^{n/2}}{a^4} = \pi^2(\pi a^2)^{-\epsilon/2} = \pi^2 e^{-(\epsilon/2)\ln(\pi a^2)} = \pi^2\left[1 - \frac{\epsilon}{2}\ln(\pi a^2) + O(\epsilon^2)\right]$$

Hence, with the retention of the constant term as $\epsilon \to 0$,

$$A = \frac{\alpha M}{\pi\epsilon}\int_0^1 dx\,(1+x) + \frac{\alpha M}{2\pi}\int_0^1 dx\,\left\{(1+x)\left[\Gamma'(1) - \ln(\pi a^2)\right] - x\right\}$$

$$= \frac{3\alpha M}{2\pi\epsilon} + \frac{\alpha M}{4\pi}\left[3\,\Gamma'(1) - 1 - 2\int_0^1 dx\,(1+x)\ln(\pi a^2)\right] + O(\epsilon)$$

The divergent term is that given in Eqs. (9.33), and we leave one contribution in the constant term as a definite integral.

Problem 9.4 Use the Dirac equation on both sides $(i\not{k} + M)u(k) = \bar{u}(k)(i\not{k} + M) = 0$, together with the relations $2k_\mu = \not{k}\gamma_\mu + \gamma_\mu\not{k}$ and $\not{k}\not{k} = k^2 = -M^2$, to prove Eq. (9.48).

Solution to Problem 9.4

From Eqs. (9.45)–(9.48), the expression we are interested in is

$$\Lambda'_{c\mu}(k,k) = \frac{ie^2}{\hbar c\varepsilon_0}\int\frac{d^n l}{(2\pi)^4}\times$$

$$\left\{\frac{1}{l^2}\gamma_\lambda\frac{1}{i(\not{k}-\not{l}) + M}\gamma_\mu\frac{1}{i(\not{k}-\not{l}) + M}\gamma_\lambda - (k=0)\right\}$$

Rationalize the denominators. From Lorentz invariance, the resulting integral must have the form

$$\Lambda'_{c\mu}(k,k) = A(k^2,\not{k})\gamma_\mu + B(k^2,\not{k})k_\mu$$

where we observe that $\not{k}\not{k} = k^2$. Consider the matrix element

$$\bar{u}(k)\Lambda'_{c\mu}(k,k)u(k) = \bar{u}(k)\left\{A(k^2,\not{k})\gamma_\mu + B(k^2,\not{k})k_\mu\right\}u(k)$$

This can be analyzed as follows:

---

[2]Note $\Gamma'(1) = -\gamma$ where $\gamma = 0.5772\ldots$ is Euler's constant.

- Use the Dirac equation on both sides

$$(i\not{k} + M)u(k) = 0 \qquad ; \quad \bar{u}(k)(i\not{k} + M) = 0$$

Hence we can replace $\not{k} \to iM$;
- Use $k^2 = -M^2$;
- Use $2k_\mu = \not{k}\gamma_\mu + \gamma_\mu\not{k}$;
- Then use the Dirac equation again.

The result is Eq. (9.48)[3]

$$\bar{u}(k)\Lambda'_{c\mu}(k,k)u(k) = \bar{u}(k)L''\gamma_\mu u(k)$$

Problem 9.5 This problem concerns an explicit proof that the terms proportional to l_μ or l_ν in the photon propagator in the Coulomb gauge $\tilde{D}^F_{\mu\nu}(l)$ in Eq. (8.133) do not contribute to the S-matrix element $[e^3\hat{S}^{\text{ext}}_{13}]_{fi}$ for the scattering of an electron in an external field.

(a) Use the fact that the photon propagator is symmetric in $\mu\nu$ and an even function of l to rewrite the integrand $M_{\mu\nu}(k,q;l)$ into which the photon propagator $\tilde{D}^F_{\mu\nu}(l)$ is contracted in the sum of the vertex and electron self-energy diagrams as (see Fig. 9.1)

$$M_{\mu\nu} = M^{(i)}_{\mu\nu} + M^{(f)}_{\mu\nu}$$

$$M^{(i)}_{\mu\nu} = \frac{1}{2}\left[\gamma_\nu \frac{1}{i(\not{k} - \not{l} + \not{q}) + M}\not{a}\frac{1}{i(\not{k} - \not{l}) + M}\gamma_\mu + \right.$$
$$\left. \not{a}\frac{1}{i\not{k} + M}\gamma_\nu\frac{1}{i(\not{k} - \not{l}) + M}\gamma_\mu + \not{a}\frac{1}{i\not{k} + M}\gamma_\mu\frac{1}{i(\not{k} + \not{l}) + M}\gamma_\nu\right]$$

$$M^{(f)}_{\mu\nu} = \frac{1}{2}\left[\gamma_\mu \frac{1}{i(\not{k} + \not{l} + \not{q}) + M}\not{a}\frac{1}{i(\not{k} + \not{l}) + M}\gamma_\nu + \right.$$
$$\gamma_\nu\frac{1}{i(\not{k} + \not{q} - \not{l}) + M}\gamma_\mu\frac{1}{i(\not{k} + \not{q}) + M}\not{a} + $$
$$\left. \gamma_\mu\frac{1}{i(\not{k} + \not{q} + \not{l}) + M}\gamma_\nu\frac{1}{i(\not{k} + \not{q}) + M}\not{a}\right]$$

In this fashion, the amplitude is separated into contributions where the virtual photon originating at the vertex γ_μ ends up at the vertex γ_ν *located in all possible positions along the electron line.*

---

[3]Compare Eq. (9.65) with $q = 0$.

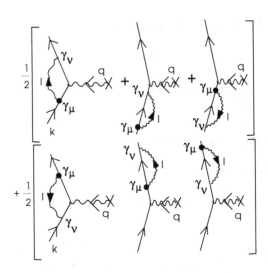

Fig. 9.1　Rewriting of the integrand $M_{\mu\nu}(k, q; l)$ in $[e^3 \hat{S}_{13}^{\text{ext}}]_{fi}$ for the diagrams in Fig. 9.1(b,d,e) in the text.

(b) Now multiply by il_ν, and use the Dirac equation on the initial and final electron legs $\bar{u}(k + q)[i(\not{k} + \not{q}) + M] = (i\not{k} + M)u(k) = 0$. Show

$$M_{\mu\nu}^{(i)} l_\nu = M_{\mu\nu}^{(f)} l_\nu = 0$$

The term in l_μ is handled in a similar fashion.

This is a very useful result since it demonstrates just what class of diagrams must be considered to satisfy current conservation. The extension of the argument to an arbitrary set of Feynman diagrams can be found in [Feynman (1949)] and [Bjorken and Drell (1964); Bjorken and Drell (1965)]. The key is that the charged electron lines run continuously through a diagram.

Solution to Problem 9.5

(a) The contribution from the vertex and self-energy graphs in Fig. 9.1(b,d,e) in the text, to the third-order S-matrix element for scattering an electron in an external field $\left[e^3 \hat{S}_{13}^{\text{ext}} \right]_{fi}$, is obtained exactly as in Eqs. (9.3) and (9.5)

$$\left[e^3 \hat{S}_{13}^{\text{ext}}\right]_{fi} = \frac{1}{\Omega}\left(\frac{-e}{\hbar c\sqrt{\varepsilon_0}}\right)\bar{u}(\mathbf{k}'\lambda')\left\{\frac{ie^2}{\hbar c\varepsilon_0}\int\frac{d^4l}{(2\pi)^4}\tilde{D}_{\mu\nu}^F(l)\,M_{\mu\nu}\right\}u(\mathbf{k}\lambda)$$

$$M_{\mu\nu} = \gamma_\mu\frac{1}{i(\not{k}+\not{q}-\not{l})+M}\not{d}\frac{1}{i(\not{k}-\not{l})+M}\gamma_\nu +$$

$$\not{d}\frac{1}{i\not{k}+M}\gamma_\mu\frac{1}{i(\not{k}-\not{l})+M}\gamma_\nu + \gamma_\mu\frac{1}{i(\not{k}+\not{q}-\not{l})+M}\gamma_\nu\frac{1}{i(\not{k}+\not{q})+M}\not{d}$$

Now use the fact that the fact that the photon propagator $\tilde{D}_{\mu\nu}^F(l)$ is unchanged under $\mu \rightleftharpoons \nu$ and $l \to -l$ to rewrite the integrand $M_{\mu\nu}$ as

$$M_{\mu\nu} = M_{\mu\nu}^{(i)} + M_{\mu\nu}^{(f)}$$

where $M_{\mu\nu}^{(i)}$ and $M_{\mu\nu}^{(f)}$ are given in the statement of the problem.

(b) Multiply $M_{\mu\nu}^{(i)}$ by il_ν

$$iM_{\mu\nu}^{(i)}l_\nu = \frac{1}{2}\left[i\not{l}\frac{1}{i(\not{k}-\not{l}+\not{q})+M}\not{d}\frac{1}{i(\not{k}-\not{l})+M}\gamma_\mu + \right.$$

$$\left.\not{d}\frac{1}{i\not{k}+M}i\not{l}\frac{1}{i(\not{k}-\not{l})+M}\gamma_\mu + \not{d}\frac{1}{i\not{k}+M}\gamma_\mu\frac{1}{i(\not{k}+\not{l})+M}i\not{l}\right]$$

Use the Dirac equation on the external legs

$$\bar{u}(k+q)[i(\not{k}+\not{q})+M] = 0 \qquad\qquad ; \quad (i\not{k}+M)u(k) = 0$$

This reduces the above to

$$iM_{\mu\nu}^{(i)}l_\nu = \frac{1}{2}\left[-\not{d}\frac{1}{i(\not{k}-\not{l})+M}\gamma_\mu + \not{d}\frac{1}{i\not{k}+M}i\not{l}\frac{1}{i(\not{k}-\not{l})+M}\gamma_\mu + \right.$$

$$\left.\not{d}\frac{1}{i\not{k}+M}\gamma_\mu\right]$$

Now decompose the expression appearing in the second term according to

$$\frac{1}{i\not{k}+M}i\not{l}\frac{1}{i(\not{k}-\not{l})+M} = -\frac{1}{i\not{k}+M} + \frac{1}{i(\not{k}-\not{l})+M}$$

Hence

$$iM_{\mu\nu}^{(i)}l_\nu = 0$$

Consider $iM_{\mu\nu}^{(f)}l_\nu$

$$iM_{\mu\nu}^{(f)}l_\nu = \frac{1}{2}\left[\gamma_\mu \frac{1}{i(\slashed{k} + \slashed{l} + \slashed{q}) + M}\slashed{d}\frac{1}{i(\slashed{k} + \slashed{l}) + M}i\slashed{l} + \right.$$
$$i\slashed{l}\frac{1}{i(\slashed{k} + \slashed{q} - \slashed{l}) + M}\gamma_\mu\frac{1}{i(\slashed{k} + \slashed{q}) + M}\slashed{d} +$$
$$\left.\gamma_\mu\frac{1}{i(\slashed{k} + \slashed{q} + \slashed{l}) + M}i\slashed{l}\frac{1}{i(\slashed{k} + \slashed{q}) + M}\slashed{d}\right]$$

With the use of the Dirac equation on the external legs, this is

$$iM_{\mu\nu}^{(f)}l_\nu = \frac{1}{2}\left[\gamma_\mu\frac{1}{i(\slashed{k} + \slashed{l} + \slashed{q}) + M}\slashed{d} - \gamma_\mu\frac{1}{i(\slashed{k} + \slashed{q}) + M}\slashed{d} + \right.$$
$$\left.\gamma_\mu\frac{1}{i(\slashed{k} + \slashed{q} + \slashed{l}) + M}i\slashed{l}\frac{1}{i(\slashed{k} + \slashed{q}) + M}\slashed{d}\right]$$

Now decompose the expression appearing in the last term

$$\frac{1}{i(\slashed{k} + \slashed{q} + \slashed{l}) + M}i\slashed{l}\frac{1}{i(\slashed{k} + \slashed{q}) + M} = -\frac{1}{i(\slashed{k} + \slashed{q} + \slashed{l}) + M} + \frac{1}{i(\slashed{k} + \slashed{q}) + M}$$

Therefore

$$iM_{\mu\nu}^{(f)}l_\nu = 0$$

It is readily verified that the other results follow in exactly the same fashion

$$il_\mu M_{\mu\nu}^{(i)} = il_\mu M_{\mu\nu}^{(f)} = 0$$

The key to using this current-conservation relation, due to Feynman, is to gather the contributions where the virtual photon originating at the vertex γ_μ ends up at the vertex γ_ν *located in all possible positions along the charged electron line.*

Problem 9.6 Invert the Fourier transform of the adiabatic damping factor $e^{-\epsilon|t|}$ in Eq. (9.96), and show $G(\Gamma_0) = \epsilon/\pi(\Gamma_0^2 + \epsilon^2)$. Verify that this $G(\Gamma_0)$ has all the properties in Eqs. (9.97).

Solution to Problem 9.6

The expression for the adiabatic damping factor in Eq. (9.96) is

$$e^{-\epsilon|t|} = \int_{-\infty}^{\infty} G(\Gamma_0)e^{-i\Gamma_0 t}\,d\Gamma_0 \qquad ; \epsilon > 0$$

This Fourier transform is inverted according to

$$G(\Gamma_0) = \frac{1}{2\pi} \int_{-\infty}^{\infty} dt\, e^{-\epsilon|t|} e^{i\Gamma_0 t}$$

$$= \frac{1}{2\pi} \left[\int_{-\infty}^{0} dt\, e^{\epsilon t} e^{i\Gamma_0 t} + \int_{0}^{\infty} dt\, e^{-\epsilon t} e^{i\Gamma_0 t} \right]$$

The integrals are evaluated as

$$G(\Gamma_0) = \frac{1}{2\pi} \left[\frac{1}{\epsilon + i\Gamma_0} + \frac{1}{\epsilon - i\Gamma_0} \right]$$

$$= \frac{1}{\pi} \frac{\epsilon}{\epsilon^2 + \Gamma_0^2}$$

Now calculate the integral over Γ_0 for all ϵ

$$\int_{-\infty}^{\infty} d\Gamma_0\, G(\Gamma_0) = \frac{1}{\pi} \int_{-\infty}^{\infty} \frac{dx}{1 + x^2} = 1 \qquad ;\, x \equiv \frac{\Gamma_0}{\epsilon}$$

The properties of $G(\Gamma_0)$ in the limit $\epsilon \to 0$ follow immediately

$$G(\Gamma_0) \to 0 \qquad ;\, \epsilon \to 0 \qquad ;\, \Gamma_0 \neq 0$$

$$G(\Gamma_0) \to \infty \qquad ;\, \epsilon \to 0 \qquad ;\, \Gamma_0 = 0$$

$$\int_{-\infty}^{\infty} d\Gamma_0\, G(\Gamma_0) = 1$$

These are just the properties of the Dirac delta-function

$$\text{Lim}_{\epsilon \to 0}\, G(\Gamma_0) = \text{Lim}_{\epsilon \to 0} \frac{1}{\pi} \frac{\epsilon}{\epsilon^2 + \Gamma_0^2} = \delta(\Gamma_0)$$

This result provides an explicit representation of the Dirac delta-function.

Problem 9.7 (a) Consider the scattering operator $\hat{S}_0$ in Eq. (8.135) in order e^4. Use Wick's theorem, and compute the S-matrix element $[\hat{S}_{04}]_{fi}$ for photon-photon scattering $\gamma(l_1) + \gamma(l_2) \to \gamma(l_3) + \gamma(l_4)$;
(b) Draw the Feynman diagrams;
(c) Use the Feynman rules to reproduce the result in part (a);
(d) Use the analysis in Prob. 8.13 to relate the contributions where the electron runs in opposite directions around the loop.

Note that while simple power counting would imply that the amplitude for this process is logarithmically divergent, it is actually convergent.

Solution to Problem 9.7

(a) The initial and final states in this case are

$$|i\rangle = c_{l_1 s_1}^\dagger c_{l_2 s_2}^\dagger |0\rangle \qquad ; \ |f\rangle = c_{l_3 s_3}^\dagger c_{l_4 s_4}^\dagger |0\rangle$$

One can now take the matrix element of the scattering operator $\hat{S}_0^{(4)}$ in Eq. (8.135) in order e^4. There are $6 \times 4!$ contributions, 6 sets with different topologies, and $4!$ within each set which are equal by a re-labeling of the dummy vertices $(x_1, \cdots, x_4)$. We proceed directly to the momentum space result as in chapter 7

$$[S_{04}]_{fi} = \frac{(2\pi)^4}{\Omega^2} \delta(l_1 + l_2 - l_3 - l_4) \frac{-(\hbar c^2)^2}{i^4 \sqrt{2^4 \omega_1 \omega_2 \omega_3 \omega_4}} \left(\frac{-e}{\hbar c \sqrt{\varepsilon_0}}\right)^4 \int \frac{d^4 p}{(2\pi)^4} \times$$

$$\mathrm{Tr} \left\{ \frac{1}{i(\not{p} + \not{l}_2 - \not{l}_3 - \not{l}_4) + M} \left[\not{\varepsilon}_4 \frac{1}{i(\not{p} + \not{l}_2 - \not{l}_3) + M} \not{\varepsilon}_3 \frac{1}{i(\not{p} + \not{l}_2) + M} \not{\varepsilon}_2 \right. \right.$$

$$+ \not{\varepsilon}_3 \frac{1}{i(\not{p} + \not{l}_2 - \not{l}_4) + M} \not{\varepsilon}_4 \frac{1}{i(\not{p} + \not{l}_2) + M} \not{\varepsilon}_2$$

$$+ \not{\varepsilon}_4 \frac{1}{i(\not{p} + \not{l}_2 - \not{l}_3) + M} \not{\varepsilon}_2 \frac{1}{i(\not{p} - \not{l}_3) + M} \not{\varepsilon}_3$$

$$+ \not{\varepsilon}_2 \frac{1}{i(\not{p} - \not{l}_3 - \not{l}_4) + M} \not{\varepsilon}_3 \frac{1}{i(\not{p} - \not{l}_4) + M} \not{\varepsilon}_4$$

$$+ \not{\varepsilon}_2 \frac{1}{i(\not{p} - \not{l}_3 - \not{l}_4) + M} \not{\varepsilon}_4 \frac{1}{i(\not{p} - \not{l}_3) + M} \not{\varepsilon}_3$$

$$\left. \left. + \not{\varepsilon}_3 \frac{1}{i(\not{p} + \not{l}_2 - \not{l}_4) + M} \not{\varepsilon}_2 \frac{1}{i(\not{p} - \not{l}_4) + M} \not{\varepsilon}_4 \right] \frac{1}{i\not{p} + M} \not{\varepsilon}_1 \right\}$$

(b) This amplitude is the sum of the six Feynman diagrams shown in Fig. 9.2. At each vertex there is a factor $\not{\varepsilon} = \gamma_\mu \varepsilon_\mu$, and in the last three diagrams, the internal four-momentum p starts in the opposite direction.

(c) The momentum-space Feynman rules for QED are given in section 8.10. It is readily verified that they reproduce the result in part (a).

(d) With four vertices, the analysis in the proof of Furry's theorem in Prob. 8.13 shows that, instead of cancelling, the contribution of the loop running in one direction is *equal* to the contribution where the loop runs in the opposite direction. Hence, the contribution of the last three diagrams in Fig. 9.2 is *equal* to the contribution of the first three.

Problem 9.8 (a) Consider a theory where a Dirac field $\hat{\psi}(x)$ is coupled to the massive vector field $\hat{V}_\mu(x)$ of Prob. C.6 with an interaction hamilto-

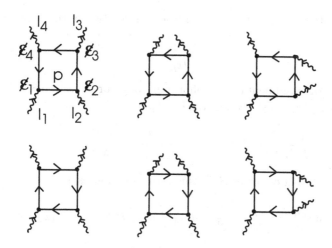

Fig. 9.2 Feynman diagrams for photon-photon scattering $\gamma(l_1) + \gamma(l_2) \rightarrow \gamma(l_3) + \gamma(l_4)$ through order e^4.

nian density $\hat{\mathcal{H}}_I = -g_v \hat{D}_\mu(x) \hat{V}_\mu(x)$ where $\hat{D}_\mu(x) = i :\hat{\bar{\psi}}(x) \gamma_\mu \hat{\psi}(x):$ is the conserved Dirac current. Construct the scattering operator $\hat{S}$ in order g_v^2;

(b) The Fourier transform of the vector propagator was calculated in Prob. F.5 to be $\tilde{\Delta}_{\mu\nu}^F(k) = [k^2 + m_v^2]^{-1}[\delta_{\mu\nu} + k_\mu k_\nu/m_v^2]$. Show that one can use an *effective* vector propagator $\tilde{\Delta}_{\mu\nu}^F(k) \doteq [k^2 + m_v^2]^{-1}\delta_{\mu\nu}$ in this theory.

Solution to Problem 9.8

(a) The second-order scattering operator for this theory follows immediately from the results in Prob. 7.8(c,d)

$$\hat{S}^{(2)} = \left(\frac{ig_v}{\hbar c}\right)^2 \frac{1}{2!} \int d^4x_1 \int d^4x_2 \, P[\hat{D}_\mu(x_1)\hat{D}_\nu(x_2)] \times$$
$$\left[:\hat{V}_\mu(x_1)\hat{V}_\nu(x_2): + \frac{\hbar c}{i}\Delta_{\mu\nu}^F(x_1 - x_2)\right]$$
$$\hat{D}_\mu(x) = i :\hat{\bar{\psi}}(x)\gamma_\mu\hat{\psi}(x):$$

From Prob. F.5, the vector-meson propagator is

$$\Delta_{\mu\nu}^F(x_1 - x_2) = \int \frac{d^4k}{(2\pi)^4} \frac{e^{ik\cdot(x_1-x_2)}}{k^2 + m_v^2}\left(\delta_{\mu\nu} + \frac{k_\mu k_\nu}{m^2}\right)$$

(b) The demonstration that the term in $k_\mu k_\nu/m_v^2$ does not contribute to the scattering operator relies on the conservation of the Dirac current

$$\frac{\partial \hat{D}_\mu(x)}{\partial x_\mu} = 0$$

It follows the argument on the photon propagator in section 8.4. Write the term in $\Delta^F_{\mu\nu}(x_1 - x_2)$ proportional to k_μ as

$$\int \frac{d^4k}{(2\pi)^4} k_\mu F_\nu(k) e^{ik\cdot x} = \frac{1}{i}\frac{\partial}{\partial x_\mu} \int \frac{d^4k}{(2\pi)^4} F_\nu(k) e^{ik\cdot x}$$

In the scattering operator $\hat{S}^{(2)}$, this term contributes as

$$\delta \hat{S}^{(2)} = \left(\frac{ig_v}{\hbar c}\right)^2 \frac{1}{2!} \int d^4x_1 \int d^4x_2\, P[\hat{D}_\mu(x_1)\hat{D}_\nu(x_2)] \times$$

$$\frac{1}{i}\frac{\partial}{\partial x_{1\mu}} \int \frac{d^4k}{(2\pi)^4} F_\nu(k) e^{ik\cdot(x_1-x_2)}$$

Carry out a partial integration on the four-dimensional gradient, getting it over onto the current. This partial integration is justified exactly as in the text. Thus, provided one can take the time derivative through the time-ordering symbol, the above expression becomes

$$\delta \hat{S}^{(2)} = \left(\frac{ig_v}{\hbar c}\right)^2 \frac{1}{2!} \int d^4x_1 \int d^4x_2\, P\left[\frac{\partial \hat{D}_\mu(x_1)}{\partial x_{1\mu}} \hat{D}_\nu(x_2)\right] \times$$

$$i \int \frac{d^4k}{(2\pi)^4} F_\nu(k,\eta) e^{ik\cdot(x_1-x_2)}$$

Now use the conservation of the interaction-picture Dirac current. This establishes the result.

It remains to show that the time derivative can be taken through the time-ordering symbol. Write the analytic expression

$$P[\hat{\rho}(x)\,\hat{D}_\lambda(y)] = \hat{\rho}(x)\,\hat{D}_\lambda(y)\theta(t_x - t_y) + \hat{D}_\lambda(y)\,\hat{\rho}(x)\theta(t_y - t_x)$$

Then, exactly as in the text,

$$\frac{\partial}{\partial t_x} P[\hat{\rho}(x)\,\hat{D}_\lambda(y)] = P\left[\frac{\partial \hat{\rho}(x)}{\partial t_x} \hat{D}_\lambda(y)\right] + \delta(t_x - t_y)[\hat{\rho}(x), \hat{D}_\lambda(y)]_{t_x=t_y}$$

It follows from the solution to Prob. 8.1(b) that the additional commutator on the r.h.s. *vanishes*

$$[\hat{\rho}(x), \hat{D}_\lambda(y)]_{t_x=t_y} = i[\hat{\psi}^\dagger(\mathbf{x},t)\hat{\psi}(\mathbf{x},t),\, \hat{\psi}^\dagger(\mathbf{y},t)\gamma_4\gamma_\lambda\hat{\psi}(\mathbf{y},t)]_{t=t_x} = 0$$

Thus the time derivative can indeed be taken through the time-ordering symbol in this case, and the proof is complete.

We have thus shown, at least to this order, that one can use the effective vector meson propagator $\tilde{\Delta}^F_{\mu\nu}(k) \doteq [k^2 + m_v^2]^{-1}\delta_{\mu\nu}$ in this theory.

Problem 9.9 (a) Verify the n-dimensional gamma-matrix relations used in Eq. (9.112);

(b) Verify that Eq. (9.122) follows from Eq. (9.120) in the limit $n \to 4$.

Solution to Problem 9.9

(a) The gamma-matrix algebra to be used with dimensional regularization is summarized in Eq. (G.25) and the accompanying discussion

- Do the γ-matrix algebra as if in discrete integer-n dimensions;
- Use $\gamma_\mu = (\gamma_1, \gamma_2, \cdots, \gamma_n)$ with

$$\{\gamma_\mu, \gamma_\nu\} = 2\delta_{\mu\nu}$$

All the γ-matrix algebra is based on this relation;[4]

- Do the γ-matrix algebra first, and reduce the problem to k-integrals.

The relations we are asked to prove in Eqs. (9.112) are

$$\gamma_\mu \slashed{d} \gamma_\mu = (2 - n)\slashed{d} \qquad ; \ \gamma_\mu\gamma_\mu = n$$
$$\mathrm{Tr}\,\slashed{d}\slashed{b} = \tau(a \cdot b) \qquad ; \ \mathrm{Tr}\,\slashed{d} = 0$$

We start from the Clifford algebra $\{\gamma_\mu, \gamma_\nu\} = 2\delta_{\mu\nu}$. Then

$$\gamma_\mu\gamma_\mu = \gamma_1\gamma_1 + \gamma_2\gamma_2 + \cdots \gamma_n\gamma_n = n$$

Next, we have

$$\gamma_\mu \slashed{d} \gamma_\mu = a_\lambda\gamma_\mu\gamma_\lambda\gamma_\mu = a_\lambda\gamma_\mu[2\delta_{\mu\lambda} - \gamma_\mu\gamma_\lambda] = (2 - n)\slashed{d}$$

Let $j \neq \mu$, then

$$\gamma_j\gamma_\mu\gamma_j = -\gamma_\mu \qquad ; \ j \neq \mu, \text{ no sum on } j$$

Now take the trace, using the properties of the trace in Eqs. (8.72)

$$\mathrm{Tr}\,\gamma_j\gamma_\mu\gamma_j = \mathrm{Tr}\,\gamma_j\gamma_j\gamma_\mu = \mathrm{Tr}\,\gamma_\mu = -\mathrm{Tr}\,\gamma_\mu = 0$$

---

[4]One can use representations of different *dimension*. With these, $\mathrm{Tr}\,\underline{1} = \tau$; however, since this will always appear as an overall factor in the calculations, one can just set $\tau = 4$ at the outset.

Therefore

$$\text{Tr}\,\rlap{/}{a} = a_\mu \text{Tr}\,\gamma_\mu = 0$$

Finally

$$\text{Tr}\,\rlap{/}{a}\rlap{/}{b} = a_\mu b_\nu \text{Tr}\,\gamma_\mu \gamma_\nu = a_\mu b_\nu \frac{1}{2}\text{Tr}\,(\gamma_\mu \gamma_\nu + \gamma_\nu \gamma_\mu)$$
$$= (a \cdot b)\,\text{Tr}\,\underline{1} = \tau(a \cdot b)$$

This establishes Eqs. (9.112).

(b) Equation (9.120) reads

$$q^2 \Pi_f(q^2) = \frac{\alpha}{2\pi^3}\tau \int_0^1 x(1-x)dx\,\Gamma\left(2 - \frac{n}{2}\right)\left[\frac{(\pi M^2)^{n/2}}{M^4} - \frac{(\pi a^2)^{n/2}}{a^4}\right]$$
$$;\ a^2 = M^2 + q^2 x(1-x)$$

Write $n = 4 - \epsilon$ and use

$$\Gamma\left(\frac{\epsilon}{2}\right) = \frac{2}{\epsilon}\Gamma\left(1 + \frac{\epsilon}{2}\right) = \frac{2}{\epsilon} + \Gamma'(1) + O(\epsilon)$$

Also

$$(M^2)^{-\epsilon/2} = e^{-\epsilon \ln M} = 1 - \epsilon \ln M + +O(\epsilon^2)$$
$$(a^2)^{-\epsilon/2} = e^{-\epsilon \ln a} = 1 - \epsilon \ln a + +O(\epsilon^2)$$

Hence

$$q^2 \Pi_f(q^2) = \frac{\alpha}{2\pi}\tau \int_0^1 x(1-x)dx\,\ln\frac{a^2}{M^2} + O(\epsilon)$$

With $\tau = 4$, this is just Eq. (9.122)

$$q^2 \Pi_f(q^2) = \frac{2\alpha}{\pi}\int_0^1 dx\,x(1-x)\ln\left[1 + x(1-x)\frac{q^2}{M^2}\right]$$

Problem 9.10 It is evident from Eq. (9.133) that the charge renormalization is given to $O(\alpha)$ by $e^2 = e_0^2[1 - C(0)]$. The sign here might be a cause for concern since bad things happen if $e^2 < 0$.[5] Use Eq. (9.118) to estimate the cutoff Λ for which $C(0) = 1$. Discuss.

---

[5]In addition to having $e = i|e|$, there are the difficulties discussed in [Dyson (1952)].

Solution to Problem 9.10

Equation (9.133) implies that in QED the charge renormalization is given to $O(\alpha)$ by

$$e^2 = e_0^2[1 - C(0)]$$

The constant $C(0)$ is actually divergent, but the cut-off value is given in Eq. (9.118)

$$C(0) = \frac{2\alpha}{3\pi} \ln \frac{\Lambda}{M}$$

The cut-off mass for $C(0) = 1$ is[6]

$$\frac{\Lambda}{M} = e^{3\pi/2\alpha} \approx e^{646} \approx 10^{281} \qquad\qquad ; C(0) = 1$$

$$\frac{\hbar\Lambda}{c} \equiv m_\Lambda \approx 10^{281} m_e \approx 10^{254} \text{ gm}$$

The equivalent cut-off distance is

$$\frac{1}{\Lambda} \approx 10^{-281} \frac{\hbar}{m_e c} \approx 4 \times 10^{-292} \text{ cm}$$

Several comments:

- This is a ridiculously large cut-off mass. For comparison, the mass of the sun is $m_\odot = 2 \times 10^{33}$ gm. The cut-off distance is similarly unbelievably small. The very successful theory of QED has actually only been verified down to distance scales of order 10^{-16} cm;
- It is naive, and incorrect, to assume that QED holds as a "stand-alone theory" down to this distance scale. In fact we now know there is a whole world of strong and electroweak interactions at shorter distances (and *who-knows-what* beyond that!);
- Within the calculation itself, one has no business using perturbation theory when the first correction is the same size as the leading term, so it should be looked on as simply giving a qualitative result.

Problem 9.11 One might be concerned that the same renormalized charge e should appear at a vertex involving a real photon.

(a) Argue in analogy to Eq. (9.95) that the correct wave function renormalization for the external photon is to use a factor $-C(0)/2$;

(b) Hence show the external photon vertex gets modified to $e_0[1 - e_0^2 \bar{C}(0)/2] = \sqrt{e^2}$ through $O(e_0^3)$. Discuss.

---

[6]Compare the discussion of Eqs. (9.40)–(9.43).

Solution to Problem 9.11

(a) It is evident that in the case of a real photon, the expression for the vacuum-polarization insertion in Eq. (9.125) is the $q^2 \to 0$ limit of[7]

$$[e^3 \hat{S}_{13}^{\text{ext}}]_{fi} \to \frac{1}{\Omega} \left(\frac{-e}{\hbar c \sqrt{\varepsilon_0}} \right) \bar{u}(k') \gamma_\mu \frac{1}{q^2} \left[-q^2 C(q^2) \right] u(k) a_\mu^\gamma(q) \qquad ; q^2 \to 0$$

As in Eq. (9.91), the limit 0/0 is again ambiguous. The photon *propagator* is modified by the vacuum-polarization insertion insertion according to Eq. (9.130)

$$D'_F(q^2) = D_F(q^2)[1 - C(0)] + \Pi_f(q^2)$$

The argument following Eq. (9.94) suggests that the correct wave-function renormalization for the real photon should be $[1 - C(0)]^{1/2} \approx 1 - C(0)/2$ to this order.

(b) For a real photon, Eq. (9.133) should then read

$$[\hat{S}_3^\gamma]_{fi} = \frac{1}{\Omega} \left(\frac{-e_0}{\hbar c \sqrt{\varepsilon_0}} \right) \bar{u}(k) \gamma_\mu \left[1 - \frac{e_0^2}{2} \bar{C}(0) \right] u(k) a_\mu^\gamma \qquad ; \text{ through } O(e_0^3)$$

The renormalized charge is given in Eq. (9.134)

$$e^2 \equiv e_0^2 [1 - e_0^2 \bar{C}(0)] \qquad ; \text{ through } O(e_0^4)$$

Hence, the same renormalized charge e does indeed appear at a vertex involving a real photon

$$e = e_0 \left[1 - \frac{e_0^2}{2} \bar{C}(0) \right] \qquad ; \text{ through } O(e_0^3)$$

Problem 9.12 This problem concerns power-counting in the momentum integrals for an arbitrary Feynman diagram in QED. Assume the contours have all been rotated so that everything is in the euclidean metric.

---

[7]Here, in contrast to Eqs. (9.131)–(9.132), one would have $a_\mu^\gamma(q) = [\mathbf{a}^\gamma(q), 0]$ with $\mathbf{a}^\gamma(q) = (2\pi)^4 \delta^{(4)}(k - k' + q) (\hbar c^2 / 2\omega_q \Omega)^{1/2} \mathbf{e}_{qs}$ for the absorption of a real photon.

Let

$$\kappa = \# \text{ of extra powers of momenta in denominator}$$
$$F_e = \# \text{ of internal electron lines}$$
$$F_p = \# \text{ of internal photon lines}$$
$$E_e = \# \text{ of external electron lines}$$
$$E_p = \# \text{ of external photon lines}$$
$$n = \# \text{ of vertices (the order)}$$
$$n_s = \# \text{of vertices without photons (mass counter-terms)}$$

(a) Since there are $F_e + F_p$ virtual momenta, with $n - 1$ $\delta^{(4)}$-functions at the vertices (one is overall), and $d^4q = q^3 dq$ counts as 4 powers in the numerator, show

$$\kappa = F_e + 2F_p - 4[(F_e + F_p) - (n - 1)]$$

(b) Since every vertex has an electron line in and out, and an internal electron line counts as two vertices, show

$$2F_e + E_e = 2n$$

(c) Since every vertex (except for n_s) has a photon line, show

$$2F_p + E_p = n - n_s$$

(d) Hence show

$$\kappa = E_p + \frac{3}{2}E_e + n_s - 4$$

This is an important result. Note that κ is *independent of the order n!*

Solution to Problem 9.12

(a) Let κ be the number of extra powers of four-momenta in the denominator of an arbitrary momentum-space Feynman diagram of order n. There are $F_e + F_p$ virtual momenta, which, through the propagators, contribute $F_e + 2F_p$ to κ. There are $4(F_e + F_p)$ powers of four-momenta in the numerator coming from the integration over virtual momenta $q^3 dq$; however, these virtual momenta are not all independent since they are limited by the $(n - 1)$ internal delta-functions at the vertices (one delta-function just expresses overall energy-momentum conservation). The integrations over virtual momenta thus contribute $-4[(F_e + F_p) - (n - 1)]$ to κ. Therefore

$$\kappa = F_e + 2F_p - 4[(F_e + F_p) - (n - 1)]$$

(b) Each vertex connects to two electron lines. Hence the total number of ends of electron lines is twice the number of vertices. The internal electron lines have two ends, while the external lines have just one. Thus

$$2F_e + E_e = 2n$$

(c) All n vertices, except for the n_s which represent the insertion of the mass counter-term, are connected to a photon line. Hence the number of ends of the photon lines is $n - n_s$. Again, the internal photon lines have two ends, while the external photon lines have just one. Thus

$$2F_p + E_p = n - n_s$$

(d) These results can be combined according to

$$\kappa = \frac{1}{2}(2n - E_e) + (n - n_s - E_p) - 4\left[\frac{1}{2}(3n - n_s - E_e - E_p) - (n - 1)\right]$$
$$= E_p + \frac{3}{2}E_e + n_s - 4$$

This is an important result. Note that κ is *independent of the order n!*

Problem 9.13 Apply the result in Prob. 9.12. A necessary condition for the convergence of the momentum integrals in an arbitrary Feynman diagram is $\kappa \geq 1$.[8] If $\kappa \leq 0$, the graph is *primitively divergent*. In QED there are four primitively divergent sets of graphs. What are they? Discuss.[9]

Solution to Problem 9.13

Recall that the following integral is convergent

$$\int^\infty \frac{q^m \, dq}{q^{m+2}} < \infty$$

Thus a *necessary* condition for convergence is

$$\kappa \geq 1 \qquad ; \text{ necessary for convergence}$$

---

[8]The integral $\int q^p dq/q^{p+2}$ is convergent; remember that we are assuming that the contours have all been rotated so that everything is in the euclidean metric.

[9]Recall Probs. 8.13 and 9.7.

where κ is the number of extra powers of four-momenta in the denominator of an arbitrary momentum-space Feynman diagram of order n as defined in Prob. 9.12. If $\kappa \geq 1$, the graph *may still diverge because of the insertion of divergent subunits.*

If $\kappa \leq 0$, the graph is said to be *primitively divergent*

$$\kappa \leq 0 \qquad ; \text{ primitive divergence}$$

It is evident from the result in Prob. 9.12(d) that there are only four sets of primitively divergent diagrams in QED, those with $(E_e, E_p) = (2,0), (0,2), (2,1), (0,4)$. They are shown in Fig. 9.3.

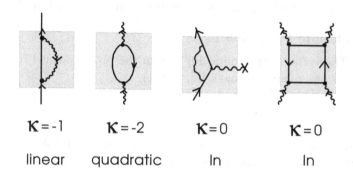

$$K = -1 \qquad\qquad K = -2 \qquad\qquad K = 0 \qquad\qquad K = 0$$

$$\text{linear} \qquad \text{quadratic} \qquad \text{ln} \qquad\qquad \text{ln}$$

Fig. 9.3 Primitively divergent graphs with $\kappa \leq 0$ in QED. The lowest-order contribution in each set is illustrated in the shaded box.

Several comments:

- These graphs all have $n_s = 0$, since the insertion of the mass counterterm increases convergence;
- There are no graphs with $(E_e, E_p) = (0,3)$ due to Furry's theorem (see Prob. 8.13);
- Actually, by Lorentz invariance, the first set of graphs in Fig. 9.3 is only ln divergent, as we saw explicitly with the calculation of the lowest-order electron self-energy in section 9.3;
- From gauge invariance, the second set of graphs in Fig. 9.3 is again only ln divergent, as we saw explicitly with the calculation of the lowest-order photon self-energy in section 9.6;
- From gauge invariance, the last set of diagrams in Fig. 9.3 describing the scattering of light by light is actually *finite* in QED (see Prob. 9.7);

- All other divergences in QED arise from the *insertion of these sub-units*. Hence, we only have to deal with these primitive divergences in order to define the theory;
- It is crucial here that κ *is independent of n*. If this were not the case, then one would have to deal with new types of divergences as one goes to higher and higher order.

Problem 9.14 (a) Iterate Dyson's equation for the vertex $\Gamma_\mu(k_2, k_1)$ in Fig. 9.6 in the text through $O(e_0^2)$, and reproduce the analytic expression for $\Lambda_\mu(k_2, k_1)$ in Eq. (9.3);

(b) Show that an iteration of Dyson's equation for the vertex $\Gamma_\mu(k_2, k_1)$ in Fig. 9.6 in the text through $O(e_0^4)$, produces all the appropriate diagrams.

Solution to Problem 9.14

(a) The vertex correction in Fig. 9.6 in the text is defined relative to the lowest order vertex through Eq. (9.141)

$$\Gamma_\mu(k_2, k_1) = \gamma_\mu + \Lambda_\mu(k_2, k_1)$$

The additional factors appearing in the e_0^2 correction follow from the momentum-space Feynman rules in section 8.10.3:

- Two additional vertices with a factor of $e_0^2/(\hbar c)^2 \varepsilon_0$ and a γ_λ each;
- Three internal lines with two four-momentum conserving vertices providing $\int d^4l/(2\pi)^4$;
- Two additional electron propagators, each with iS_F;
- One additional photon propagator with $(\hbar c/i)D_F$;

If the multiplying factors are combined, one has

$$i^2 \frac{\hbar c}{i} \frac{e_0^2}{(\hbar c)^2 \varepsilon_0} = \frac{ie_0^2}{\hbar c \varepsilon_0}$$

This verifies the statement made in the caption to Fig. 9.6 in the text:

To convert to the present conventions with S_F, D_F, Γ_μ, the Feynman rules require a factor of $(i/\hbar c \varepsilon_0)$ for each pair of charged vertices in the graphs.

A combination of these results then gives the expression for the $O(e_0^2)$ vertex correction in Eq. (9.3)

$$\Lambda_\mu^{(2)}(k',k) = \frac{ie_0^2}{\hbar c \varepsilon_0} \int \frac{d^n l}{(2\pi)^4}\frac{1}{l^2}\, \gamma_\lambda \frac{1}{i(\not{k}'-\not{l})+M}\gamma_\mu \frac{1}{i(\not{k}-\not{l})+M}\gamma_\lambda$$

where the integral is again defined through dimensional regularization

(b) To compute the $O(e_0^4)$ corrections to the vertex in Fig. 9.6 in the text, we must start by substituting the $O(e_0^2)$ modifications of the photon propagator, electron propagator, and vertex in the first irreducible diagram in Dyson's equations.

- The first insertion in the photon propagator is just the lowest-order vacuum polarization graph in Fig. 9.4 in the text. This produces diagram 8 in Fig. 9.4 below;
- The first insertion in the electron propagator is the self-energy graph in Fig. 9.2 in the text, where we associate the mass counter-term with it. This gives rise to diagrams 6 and 7 in Fig. 9.4 below;
- The lowest-order vertex correction is that shown in Fig. 9.1(b) in the text and calculated in part (a). When inserted in the first irreducible graph, it gives rise to diagrams 3,4, and 5 in Fig. 9.4 below;
- Finally, there is the fourth-order irreducible diagram in Fig. 9.6 in the text, which must be calculated from all the lowest-order components. This is the last diagram in Fig. 9.4 below.

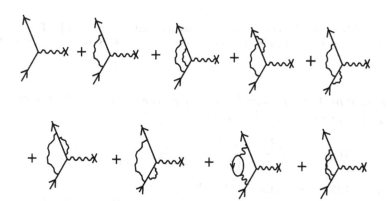

Fig. 9.4   Iteration of Dyson's equation for the vertex $\Gamma_\mu(k_2, k_1)$ in Fig. 9.6 in the text through $O(e_0^4)$. The external lines are indicated for orientation, and factors of e_0 are suppressed.

Problem 9.15 Verify Ward's identity in Eq. (9.142) through $O(e_0^4)$.

Solution to Problem 9.15

We provide a diagrammatic demonstration of Ward's identity in Eq. (9.142) by extending Fig. 9.7 in the text to include all three proper self energies $\Sigma^\star(k)$ of $O(e_0^4)$. The result is shown in Fig. 9.5 below. Once again, the external four-momentum k runs along the electron line, and differentiation with respect to k inserts a zero-momentum-transfer vertex everywhere along that line. It is evident that all seven vertex corrections of $O(e_0^4)$ in Fig. 9.4 above are reproduced.

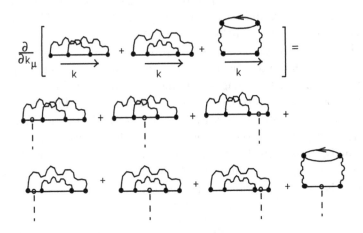

Fig. 9.5   Diagrammatic demonstration of Ward's identity through $O(e_0^4)$. The external four-momentum k is carried by the electron line, and each filled dot has an associated factor of e_0.

Problem 9.16 (a) Start from the following expression for the vacuum polarization tensor in $O(e_0^2)$ [compare Eq. (9.4)]

$$\Pi_{\lambda\sigma}(q) = \frac{e_0^2}{i\hbar c\varepsilon_0} \int \frac{d^n k}{(2\pi)^4} \mathrm{Tr}\, \frac{1}{i\not{k} + M} \gamma_\lambda \frac{1}{i(\not{k} + \not{q}) + M} \gamma_\sigma$$

Use Eq. (9.12) to compute $(\partial/\partial q_\mu)\Pi_{\lambda\sigma}(q)$;

(b) Iterate Dyson's equation for the vertex $W_\mu(q,q)$ in Fig. 9.9 in the text through $O(e_0^2)$ for fermion vertices (λ,σ). Show the $O(e_0^2)$ term $[W_\mu^{(2)}(q,q)]_{\lambda\sigma}$ produces the expression $-(\partial/\partial q_\mu)\Pi_{\lambda\sigma}(q)$;

(c) Use the form of $\Pi_{\lambda\sigma}(q)$ in Eq. (9.109) to solve for $\Pi(q)$, and then use Eq. (9.150) to show

$$(n-1)\Pi(q) = \Pi_{\lambda\lambda}(q)$$
$$i\Delta_\mu(q,q) = \frac{\partial \Pi(q)}{\partial q_\mu} = \frac{1}{(n-1)}\frac{\partial}{\partial q_\mu}\Pi_{\lambda\lambda}(q)$$

(d) Hence conclude from Eqs. (9.154) that $W_\mu^{(2)}(q,q) \equiv (n-1)^{-1}$ $[W_\mu^{(2)}(q,q)]_{\lambda\lambda}$ in part (b) produces the correct $i\Delta_\mu(q,q)$ through $O(e_0^2)$.[10]

Solution to Problem 9.16

(a) The second-order vacuum polarization tensor follows from Eq. (9.4) as

$$\Pi_{\lambda\sigma}^{(2)}(q) \equiv \frac{e_0^2}{i\hbar c \varepsilon_0} \int \frac{d^n k}{(2\pi)^4} \mathrm{Tr}\, \frac{1}{i\not k + M}\gamma_\lambda \frac{1}{i(\not k + \not q) + M}\gamma_\sigma$$

Here we employ the un-renormalized charge e_0, work in n-dimensions, and shift the integration variable so that the external four-momentum q runs through just one leg of the electron loop. Equation (9.12) allows us to differentiate this expression with respect to q_μ

$$\frac{\partial}{\partial q_\mu}\frac{1}{i(\not k + \not q) + M} = -\frac{1}{i(\not k + \not q) + M}i\gamma_\mu \frac{1}{i(\not k + \not q) + M}$$

The result is

$$\frac{\partial}{\partial q_\mu}\Pi_{\lambda\sigma}^{(2)}(q) = \frac{-e_0^2}{\hbar c \varepsilon_0}\int \frac{d^n k}{(2\pi)^4}\mathrm{Tr}\,\frac{1}{i\not k + M}\gamma_\lambda \frac{1}{i(\not k + \not q) + M}\gamma_\mu \frac{1}{i(\not k + \not q) + M}\gamma_\sigma$$

(b) Go to Fig. 9.9 in the text, which presents Dyson's equation for Ward's full vertex construct $W_\mu(q,q)$, and use the rules stated in the figure caption

> *To convert to the present conventions with* $S_F, D_F, \Gamma_\mu, W_\mu$, *the Feynman rules require a factor of* $(i/\hbar c\varepsilon_0)$ *for each pair of charged electron vertices in a graph, and a factor of* $(-i)$ *for the odd one.*

Then, with an additional factor of (-1) for the closed fermion loop, we have

---

[10] Alternatively, one can just select the coefficient of $\delta_{\lambda\sigma}$ in $[W_\mu^{(2)}(q,q)]_{\lambda\sigma}$. Note that from Eq. (9.129) $\Pi_{\lambda\sigma} \doteq \Pi(q)\delta_{\lambda\sigma}$, and in $O(e_0^2)$, $\Pi(q) \equiv \Pi^\star(q)$.

through $O(e_0^2)$

$$[W_\mu(q,q)]_{\lambda\sigma} = 2q_\mu - \left(\frac{ie_0^2}{\hbar c\varepsilon_0}\right)\frac{(-i)}{(-1)^3}\int\frac{d^nk}{(2\pi)^4} \times$$

$$\mathrm{Tr}\,\frac{1}{i\not{k}+M}\gamma_\lambda\frac{1}{i(\not{k}+\not{q})+M}\gamma_\mu\frac{1}{i(\not{k}+\not{q})+M}\gamma_\sigma$$

A comparison with the result in part (a) allows us to express the second-order contribution in this expression as

$$[W_\mu^{(2)}(q,q)]_{\lambda\sigma} = -\frac{\partial}{\partial q_\mu}\Pi_{\lambda\sigma}^{(2)}(q)$$

(c) The general form of $\Pi_{\lambda\sigma}(q)$ is given in Eq. (9.109)

$$\Pi_{\lambda\sigma}(q) = (q_\lambda q_\sigma - q^2\delta_{\lambda\sigma})C(q^2) \equiv \left(\delta_{\lambda\sigma} - \frac{q_\lambda q_\sigma}{q^2}\right)\Pi(q)$$

where the last relation defines $\Pi(q)$. It follows that

$$\Pi_{\lambda\lambda}^{(2)}(q) = (n-1)\Pi^{(2)}(q)$$

Ward's second identity in Eq. (9.150) relates the derivative of the photon proper self-energy $\Pi^\star(q)$ to Ward's constructed vertex $\Delta_\mu(q,q)$ through

$$\frac{\partial\Pi^\star(q)}{\partial q_\mu} \equiv i\Delta_\mu(q,q)$$

We note that to $O(e_0^2)$, the polarization insertion $\Pi(q)$ is identical to the proper polarization $\Pi^\star(q)$. Hence Ward's second identity implies

$$i\Delta_\mu^{(2)}(q,q) = \frac{1}{n-1}\frac{\partial}{\partial q_\mu}\Pi_{\lambda\lambda}^{(2)}(q)$$

(d) In a similar fashion, solve for $W_\mu^{(2)}(q,q)$ as

$$W_\mu^{(2)}(q,q) \equiv \frac{1}{n-1}[W_\mu^{(2)}(q,q)]_{\lambda\lambda}$$

where $[W_\mu^{(2)}(q,q)]_{\lambda\lambda}$ was calculated in part (b). Then, from parts (b,c)

$$W_\mu^{(2)}(q,q) = -\frac{1}{n-1}\frac{\partial}{\partial q_\mu}\Pi_{\lambda\lambda}^{(2)}(q) = -i\Delta_\mu^{(2)}(q,q)$$

Equations (9.154) relate the vertex $\Delta_\mu(q,q)$ appearing Ward's second identity and $W_\mu(q,q)$ calculated from Fig. 9.9 in the text through

$$2q_\mu - i\Delta_\mu(q,q) = W_\mu(q,q)$$

The present calculation explicitly verifies this relation through $O(e_0^2)$.

Problem 9.17 The *skeleton* graphs for any physical process consist of the set of distinct diagrams obtained by removing all self-energy and vertex insertions from the Feynman diagrams for that process. All diagrams are then re-obtained by using the full propagators (S_F', D_F') and full vertex $e_0 \Gamma_\mu$ in the skeletons. Show it then follows from Eqs. (9.157) that the S-matrix for any physical process is given by the use of the *finite functions* $(S_{F1}, D_{F1}, \Gamma_{\mu 1})$ and the *renormalized charge* e_1. You may assume the same wave function renormalizations $(Z_2^{1/2}, Z_3^{1/2})$ for the external electrons and photons that were established in the text to second order.

Solution to Problem 9.17

Consider the entire set of Feynman diagrams describing any physical process. Remove all the self-energy and vertex insertions to produce the set of *skeleton graphs* for that process. Now insert the full self-energy and vertex parts to recover the complete set of graphs. The full self-energy and vertex parts are related to the renormalized parts through Eqs. (9.157)

$$\Gamma_\mu = Z_2^{-1} \Gamma_{\mu 1}(e_1)$$
$$W_\mu = Z_3^{-1} W_{\mu 1}(e_1)$$
$$S_F' = Z_2 S_{F1}(e_1)$$
$$D_F' = Z_3 D_{F1}(e_1) \qquad ; \; e_1^2 \equiv Z_3 e_0^2$$

In the scattering amplitude, each vertex Γ_μ has associated with it a factor of the un-renormalized charge e_0. Every internal photon line ends on two vertices. Associate a factor of $Z_3^{1/2}$ with each vertex. If the photon line is external, it brings in precisely that $Z_3^{1/2}$. Every internal electron line connects to two vertices. Associate a factor of $Z_2^{1/2}$ with each vertex. Again, if the electron line is external, it brings in $Z_2^{1/2}$. Note that there are two electron lines at each vertex. Thus, when building the entire set from the skeletons, we can employ $S_{F1}(e_1), D_{F1}(e_1)$ for the propagators, and use the following vertices

$$e_0 Z_3^{1/2} Z_2^{1/2} Z_2^{1/2} \Gamma_\mu = e_1 \Gamma_{\mu 1}(e_1) \qquad ; \; e_1^2 \equiv Z_3 e_0^2$$

The scattering amplitude is now expressed entirely in terms of renormalized quantities. in particular, the renormalized charge.[11]

---

[11] A lovely result!

It follows from Probs. 9.12–9.13 that this procedure deals with the divergences in the theory, and the remaining integrals over the finite parts should be finite.[12]

Problem 9.18 Suppose the additional term of Prob. 8.15 is actually present in the equal-time commutator, while the external field has the form of Eq. (8.121).

(a) Write out the 3! possible time-orderings of the current operators in Eq. (8.132), and differentiate with respect to t_{x_1} as in Eqs. (8.34)–(8.37);

(b) Verify that the effective photon propagator is still given by Eq. (8.134).[13]

Solution to Problem 9.18

(a) The expression we are interested in is

$$P[\hat{j}_\mu(x)\hat{j}_\nu(y)\hat{j}_\lambda(z)] =$$
$$\hat{j}_\mu(x)\hat{j}_\nu(y)\hat{j}_\lambda(z)\theta(t_x - t_y)\theta(t_y - t_z) + \hat{j}_\mu(x)\hat{j}_\lambda(z)\hat{j}_\nu(y)\theta(t_x - t_z)\theta(t_z - t_y) +$$
$$\hat{j}_\nu(y)\hat{j}_\mu(x)\hat{j}_\lambda(z)\theta(t_y - t_x)\theta(t_x - t_z) + \hat{j}_\nu(y)\hat{j}_\lambda(z)\hat{j}_\mu(x)\theta(t_y - t_z)\theta(t_z - t_x) +$$
$$\hat{j}_\lambda(z)\hat{j}_\mu(x)\hat{j}_\nu(y)\theta(t_z - t_x)\theta(t_x - t_y) + \hat{j}_\lambda(z)\hat{j}_\nu(y)\hat{j}_\mu(x)\theta(t_z - t_y)\theta(t_y - t_x)$$

The terms left over when we differentiate with respect to t_x as in Eqs. (8.34)–(8.37) arise from the differentiation of the θ-functions, and take the form

$$\delta P = \hat{\rho}(x)\hat{j}_\nu(y)\hat{j}_\lambda(z)\frac{\partial}{\partial t_x}\theta(t_x - t_y)\theta(t_y - t_z) +$$

$$\hat{\rho}(x)\hat{j}_\lambda(z)\hat{j}_\nu(y)\frac{\partial}{\partial t_x}\theta(t_x - t_z)\theta(t_z - t_y) +$$

$$\hat{j}_\nu(y)\hat{\rho}(x)\hat{j}_\lambda(z)\frac{\partial}{\partial t_x}\theta(t_y - t_x)\theta(t_x - t_z) +$$

$$\hat{j}_\nu(y)\hat{j}_\lambda(z)\hat{\rho}(x)\frac{\partial}{\partial t_x}\theta(t_y - t_z)\theta(t_z - t_x) +$$

$$\hat{j}_\lambda(z)\hat{\rho}(x)\hat{j}_\nu(y)\frac{\partial}{\partial t_x}\theta(t_z - t_x)\theta(t_x - t_y) +$$

$$\hat{j}_\lambda(z)\hat{j}_\nu(y)\hat{\rho}(x)\frac{\partial}{\partial t_x}\theta(t_z - t_y)\theta(t_y - t_x)$$

---

[12]The proof of this last result, however, lies beyond the scope of the present work (see [Weinberg (1960)]).

[13]Compare Prob. 9.5.

The use of Eqs. (8.36) gives

$$
\begin{aligned}
\delta P = {} & \hat{\rho}(x)\hat{j}_\nu(y)\hat{j}_\lambda(z)\delta(t_x - t_y)\theta(t_y - t_z) + \\
& \hat{\rho}(x)\hat{j}_\lambda(z)\hat{j}_\nu(y)\delta(t_x - t_z)\theta(t_z - t_y) + \\
& \hat{j}_\nu(y)\hat{\rho}(x)\hat{j}_\lambda(z)\left[-\delta(t_y - t_x)\theta(t_x - t_z) + \theta(t_y - t_x)\delta(t_x - t_z)\right] + \\
& \hat{j}_\nu(y)\hat{j}_\lambda(z)\hat{\rho}(x)\left[-\theta(t_y - t_z)\delta(t_z - t_x)\right] + \\
& \hat{j}_\lambda(z)\hat{\rho}(x)\hat{j}_\nu(y)\left[-\delta(t_z - t_x)\theta(t_x - t_y) + \theta(t_z - t_x)\delta(t_x - t_y)\right] + \\
& \hat{j}_\lambda(z)\hat{j}_\nu(y)\hat{\rho}(x)\left[-\theta(t_z - t_y)\delta(t_y - t_x)\right]
\end{aligned}
$$

These terms can be combined to give

$$
\begin{aligned}
\delta P = {} & \delta(t_x - t_y)\Big\{ [\hat{\rho}(x),\, \hat{j}_\nu(y)]\,\hat{j}_\lambda(z)\,\theta(t_y - t_z) + \hat{j}_\lambda(z)\theta(t_z - t_x)[\hat{\rho}(x),\, \hat{j}_\nu(y)] \Big\} \\
& + \delta(t_x - t_z)\Big\{ [\hat{\rho}(x),\, \hat{j}_\lambda(z)]\hat{j}_\nu(y)\theta(t_x - t_y) + \hat{j}_\nu(y)\theta(t_y - t_x)[\hat{\rho}(x),\, \hat{j}_\lambda(z)] \Big\}
\end{aligned}
$$

(b) Now apply this result to the problem at hand:

- Assume that $cA^{\text{ext}}_\lambda(x)$ has the form in Eq. (8.121), so that with only a fourth component $\hat{j}_\lambda(z) \to i\delta_{\mu4}\hat{\rho}(z)$. Furthermore, assume

$$
[\hat{\rho}(x),\, \hat{\rho}(z)]_{t_x = t_z} = 0
$$

- Assume the remaining non-zero commutator has the form in Prob. 8.15

$$
\delta(t_x - t_y)[\hat{\rho}(x),\, \hat{j}(y)]_{t_x = t_y} \doteq i\delta(t_x - t_y)\delta^{(3)}(\mathbf{x} - \mathbf{y})\,\boldsymbol{\nabla}_y\,\hat{\varphi}(y)
$$

It follows that

$$
\begin{aligned}
\delta P \doteq {} & -\frac{1}{c}\delta(t_x - t_y)\delta^{(3)}(\mathbf{x} - \mathbf{y}) \times \\
& \left\{ \boldsymbol{\nabla}_y\,\hat{\varphi}(y)\,\hat{\rho}(z)\,\theta(t_y - t_z) + \hat{\rho}(z)\,\boldsymbol{\nabla}_y\,\hat{\varphi}(y)\,\theta(t_z - t_y) \right\}
\end{aligned}
$$

Now make use of this result in the proof that the extra terms in the photon propagator in Eq. (8.133) do not contribute to the scattering operator in Eq. (8.132). Exactly as in Prob. 8.15, the gradient can again be spatially integrated out to the surface of the box where the p.b.c. ensure that it vanishes. Hence

$$
\delta\left[e^3 \hat{S}^{\text{ext}}_{13}\right] = 0
$$

As a consequence, the effective photon propagator is indeed given by Eq. (8.134).

Problem 9.19 Verify Eqs. (9.137); remember to renormalize the charge.

Solution to Problem 9.19

Suppose one had a world where there were only muons and no electrons. One could go through the calculation in chapter 9 of the contributions of all the diagrams in Fig. 9.1 in the text, and end up with the additional terms in Eqs. (9.106), (9.126), and (9.135)

$$[\hat{S}_1^{ext}]_{fi} = \frac{4\pi\alpha}{\Omega}\bar{u}(k')\left\{\gamma_\sigma\left[1 + \alpha q^2\bar{\Pi}_f^{(\mu)}(q^2)\right] + \alpha\bar{\Lambda}_{c\sigma}^{(\mu)}(k',k)\right\}u(k)\bar{a}_\sigma(q)$$

where we label the corrections with the superscript (μ). Here the fine-structure constant α is computed with the renormalized charge

$$e^2 = e_0^2[1 - e_0^2\bar{C}^{(\mu)}(0)]$$

The above result holds through $O(e_0^4)$.

As explained in the text, one must then extend the calculation by including the vacuum polarization contribution in Fig. 9.1(c) in the text coming from the actual lightest-mass intermediate pair $(e^\pm)$. The result for $q^2/M_e^2 \gg 1$ is to add the following term to the S-matrix

$$\delta[\hat{S}_1^{ext}]_{fi} = \frac{4\pi\alpha_0}{\Omega}\bar{u}(k')\left\{\gamma_\sigma\left[-e_0^2\bar{C}^{(e)}(0) + \frac{\alpha_0}{3\pi}\ln\frac{q^2}{M_e^2}\right]\right\}u(k)\bar{a}_\sigma(q)$$

where this fine-structure constant is

$$\alpha_0 = \frac{e_0^2}{4\pi\hbar c\varepsilon_0}$$

Now add the two contributions together. The result can be written

$$[\hat{S}_1^{ext}]_{fi} = \frac{4\pi\alpha(q^2)}{\Omega}\bar{u}(k')\left\{\gamma_\sigma\left[1 + \alpha(q^2)q^2\bar{\Pi}_f(q^2)\right] + \alpha(q^2)\bar{\Lambda}_{c\sigma}(k',k)\right\}u(k)\bar{a}_\sigma(q)$$

$$\alpha(q^2) \equiv \alpha\left[1 + \frac{\alpha}{3\pi}\ln\frac{q^2}{M_e^2}\right] \qquad ; \quad \frac{q^2}{M_e^2} \gg 1$$

where the first line is computed entirely for the μ^-. Here the renormalized charge is now

$$e^2 = e_0^2\left\{1 - e_0^2\left[\bar{C}^{(e)}(0) + \bar{C}^{(\mu)}(0)\right]\right\}$$

and the result again holds through order e_0^4. Once the charge e^2 has been identified through the first term of $O(\alpha)$ in the S-matrix element, this last expression can be inverted

$$e_0^4 = e^4[1 + O(e^2)]$$

and used in the correction terms to give an expression for $[\hat{S}_1^{\text{ext}}]_{fi}$ which holds through $O(\alpha^2)$. This is Eq. (9.137).

Chapter 10

Path Integrals

Problem 10.1 Restore the Dirac indices and make them explicit in Eqs. (10.143)–(10.148). Explain the Dirac matrix structure of Eqs. (10.148).

Solution to Problem 10.1

If we make the Dirac indices explicit, and use a convention that repeated Dirac indices are summed from 1 to 4, then the generating functional in Eq. (10.143) becomes

$$\tilde{W}_0[\bar{\zeta}, \zeta] = \exp\left\{-\frac{i}{(\hbar c)^2} \int d^4x \int d^4y \, \bar{\zeta}_\alpha(x) S^F_{\alpha\beta}(x-y) \zeta_\beta(y)\right\}$$

The bilinear term in Eq. (10.144) is

$$\tilde{W}_0^{(2)}[\bar{\zeta}, \zeta] = -\frac{i}{(\hbar c)^2} \int d^4x \int d^4y \, \bar{\zeta}_\alpha(x) S^F_{\alpha\beta}(x-y) \zeta_\beta(y)$$

Hence the free propagator in Eq. (10.145) is

$$\left(\frac{\hbar c}{i}\right)^2 \left[\frac{\delta}{\delta\bar{\zeta}_\alpha(x)} \tilde{W}_0[\bar{\zeta}, \zeta] \frac{\delta}{\delta\zeta_\beta(y)}\right]_{\bar{\zeta}=\zeta=0} = \langle\psi_0|P[\hat{\psi}_\alpha(x), \hat{\bar{\psi}}_\beta(y)]|\psi_0\rangle$$
$$= iS^F_{\alpha\beta}(x-y)$$

In a similar fashion, Eqs. (10.148) become

$$\left(\frac{\hbar c}{i}\right)^4 \left[\frac{\delta}{\delta\bar{\zeta}_\alpha(x_2)} \frac{\delta}{\delta\bar{\zeta}_\beta(x_1)} \tilde{W}_0[\bar{\zeta}, \zeta] \frac{\delta}{\delta\zeta_\rho(y_2)} \frac{\delta}{\delta\zeta_\sigma(y_1)}\right]_{\bar{\zeta}=\zeta=0} =$$
$$= \langle\psi_0|P[\hat{\psi}_\alpha(x_2)\hat{\psi}_\beta(x_1)\hat{\bar{\psi}}_\rho(y_2)\hat{\bar{\psi}}_\sigma(y_1)]|\psi_0\rangle_C$$
$$= i^2[S^F_{\alpha\sigma}(x_2-y_1)S^F_{\beta\rho}(x_1-y_2) - S^F_{\beta\sigma}(x_1-y_1)S^F_{\alpha\rho}(x_2-y_2)]$$

The matrix indices in the Dirac propagators can thus be associated with the coordinate points in the Feynman diagrams.

Problem 10.2 Verify the following left-variational derivatives

$$\frac{1}{i}\frac{\delta}{\delta\bar{\zeta}(x)}\exp\left\{i\int d^4y\,[\bar{\zeta}(y)\psi(y)]\right\} = \psi(x)\exp\left\{i\int d^4y\,[\bar{\zeta}(y)\psi(y)]\right\}$$

$$\frac{1}{i}\frac{\delta}{\delta\zeta(x)}\exp\left\{i\int d^4y\,[\bar{\psi}(y)\zeta(y)]\right\} = -\bar{\psi}(x)\exp\left\{i\int d^4y\,[\bar{\psi}(y)\zeta(y)]\right\}$$

Solution to Problem 10.2

Let $\bar{\zeta}(y) \to \bar{\zeta}(y) + \lambda\bar{\eta}(y)$, where $\bar{\eta}(y)$ is also a Grassmann variable. Then as $\lambda \to 0$

$$\frac{1}{\lambda}\left[\exp\left\{i\int d^4y\,[\bar{\zeta}(y) + \lambda\bar{\eta}(y)]\,\psi(y)\right\} - \exp\left\{i\int d^4y\,[\bar{\zeta}(y)\psi(y)]\right\}\right] =$$

$$i\int d^4x\,[\bar{\eta}(x)\psi(x)]\exp\left\{i\int d^4y\,[\bar{\zeta}(y)\psi(y)]\right\} \qquad ; \lambda \to 0$$

From the definition of variational derivative in Eq. (10.53), with the correct ordering of Grassmann variables for the left-variational derivative, this is

$$\int dx\,\bar{\eta}(x)\frac{\delta}{\delta\bar{\zeta}(x)}\exp\left\{i\int d^4y\,[\bar{\zeta}(y)\psi(y)]\right\}$$

Hence

$$\frac{1}{i}\frac{\delta}{\delta\bar{\zeta}(x)}\exp\left\{i\int d^4y\,[\bar{\zeta}(y)\psi(y)]\right\} = \psi(x)\exp\left\{i\int d^4y\,[\bar{\zeta}(y)\psi(y)]\right\}$$

In a similar fashion, let $\zeta(y) \to \zeta(y) + \lambda\eta(y)$, where $\eta(y)$ is also a Grassmann variable. Then as $\lambda \to 0$

$$\frac{1}{\lambda}\left[\exp\left\{i\int d^4y\,\bar{\psi}(y)\,[\zeta(y) + \lambda\eta(y)]\right\} - \exp\left\{i\int d^4y\,[\bar{\psi}(y)\zeta(y)]\right\}\right] =$$

$$i\int d^4x\,[\bar{\psi}(x)\eta(x)]\exp\left\{i\int d^4y\,[\bar{\psi}(y)\zeta(y)]\right\} \qquad ; \lambda \to 0$$

From the definition of variational derivative in Eq. (10.53), with the correct ordering of Grassmann variables for the left-variational derivative, this is

$$\int dx\,\eta(x)\frac{\delta}{\delta\zeta(x)}\exp\left\{i\int d^4y\,[\bar{\psi}(y)\zeta(y)]\right\}$$

Hence

$$\frac{1}{i}\frac{\delta}{\delta\zeta(x)}\exp\left\{i\int d^4y\,[\bar\psi(y)\zeta(y)]\right\} = -\bar\psi(x)\exp\left\{i\int d^4y\,[\bar\psi(y)\zeta(y)]\right\}$$

Here we have used the fact that with Grassmann variables, $\bar\psi(x)\eta(x) = -\eta(x)\bar\psi(x)$.

Problem 10.3 Use the scalar interaction in Eq. (10.108). Expand the free functional $W_0(J)$ in Eq. (10.100) and retain $W_0^{(6)}(J)$. Then evaluate the interacting functional $W(J)$ in Eq. (10.112) to $O(\lambda)$.

Solution to Problem 10.3

The starting point is the generating functional of Eq. (10.100)

$$\tilde W_0(J) = \exp\left\{\frac{i}{2\hbar c^3}\int d^4x\int d^4y\, J(x)\Delta_F(x-y)J(y)\right\}$$

With the expansion of the exponential in the generating functional, we have

$$\tilde W_0(J) = 1 + \tilde W_0^{(2)}(J) + \tilde W_0^{(4)}(J) + \tilde W_0^{(6)}(J) + \cdots$$

where the terms $\tilde W_0^{(2)}(J)$ and $\tilde W_0^{(4)}(J)$ are reported in Eqs. (10.102) and (10.106) of the text.

We need to work out the expression for $\tilde W_0^{(6)}(J)$

$$\tilde W_0^{(6)}(J) = \frac{1}{3!}\left(\frac{i}{2\hbar c^3}\right)^3\int d^4x_1\cdots\int d^4x_6 \times$$
$$\Delta_F(x_1-x_2)\Delta_F(x_3-x_4)\Delta_F(x_5-x_6)\,J(x_1)\cdots J(x_6)$$

After a change of dummy integration variables, $\tilde W_0^{(6)}(J)$ can be written as a functional with a completely symmetric kernel

$$\tilde W_0^{(6)}(J) = \left(\frac{i}{\hbar c}\right)^6\int d^4x_1\cdots\int d^4x_6\,\frac{1}{6!}K_6(x_1,\ldots,x_6)\times$$
$$J(x_1)\cdots J(x_6)$$

The symmetric kernel, with 15 terms, is given by

$$K_6(x_1, \ldots, x_6) = \left(\frac{\hbar}{ic}\right)^3 [\Delta_F(x_1 - x_2)\Delta_F(x_3 - x_4)\Delta_F(x_5 - x_6) +$$

$$\Delta_F(x_1 - x_2)\Delta_F(x_3 - x_6)\Delta_F(x_4 - x_5) + \Delta_F(x_1 - x_2)\Delta_F(x_4 - x_6)\Delta_F(x_3 - x_5) +$$

$$\Delta_F(x_1 - x_3)\Delta_F(x_2 - x_4)\Delta_F(x_5 - x_6) + \Delta_F(x_1 - x_3)\Delta_F(x_2 - x_6)\Delta_F(x_4 - x_5) +$$

$$\Delta_F(x_1 - x_3)\Delta_F(x_4 - x_6)\Delta_F(x_2 - x_5) + \Delta_F(x_1 - x_4)\Delta_F(x_3 - x_2)\Delta_F(x_5 - x_6) +$$

$$\Delta_F(x_1 - x_4)\Delta_F(x_3 - x_6)\Delta_F(x_2 - x_5) + \Delta_F(x_1 - x_4)\Delta_F(x_2 - x_6)\Delta_F(x_3 - x_5) +$$

$$\Delta_F(x_1 - x_5)\Delta_F(x_3 - x_4)\Delta_F(x_2 - x_6) + \Delta_F(x_1 - x_5)\Delta_F(x_3 - x_6)\Delta_F(x_2 - x_4) +$$

$$\Delta_F(x_1 - x_5)\Delta_F(x_4 - x_6)\Delta_F(x_2 - x_3) + \Delta_F(x_1 - x_6)\Delta_F(x_3 - x_4)\Delta_F(x_2 - x_5) +$$

$$\Delta_F(x_1 - x_6)\Delta_F(x_2 - x_3)\Delta_F(x_4 - x_5) + \Delta_F(x_1 - x_6)\Delta_F(x_2 - x_4)\Delta_F(x_3 - x_5)]$$

The interacting generating functional is given in Eq. (10.112)

$$\tilde{W}(J) = \left(\exp\left\{\frac{i}{\hbar c}\int d^4x\, \mathcal{L}_1\left[\frac{\hbar c}{i}\frac{\delta}{\delta J(x)}\right]\right\}\tilde{W}_0(J)\right) / (\cdots)_{J=0}$$

where the interaction is displayed in Eq. (10.108)

$$\mathcal{L}_1(\phi) = -\frac{\lambda}{4!}\phi^4$$

Now, as requested, we can evaluate $\tilde{W}(J)$ to $O(\lambda)$ where we approximate

$$\tilde{W}_0(J) \approx 1 + \tilde{W}_0^{(2)}(J) + \tilde{W}_0^{(4)}(J) + \tilde{W}_0^{(6)}(J) + \cdots$$

Let us expand the numerator of the generating functional to $O(\lambda)$

$$\left\{1 - \frac{\lambda}{4!}\frac{i}{\hbar c}\int d^4x\left[\frac{\hbar c}{i}\frac{\delta}{\delta J(x)}\right]^4\right\}\tilde{W}_0(J) \approx 1 + \tilde{W}_0^{(2)}(J) + \tilde{W}_0^{(4)}(J) + \tilde{W}_0^{(6)}(J)$$

$$+ \cdots - \frac{\lambda}{4!}\frac{i}{\hbar c}\int d^4x\left[\frac{\hbar c}{i}\frac{\delta}{\delta J(x)}\right]^4\left[\tilde{W}_0^{(4)}(J) + \tilde{W}_0^{(6)}(J) + \cdots\right]$$

The denominator is simply obtained by setting the scalar sources equal to zero, and therefore it reduces to[1]

$$\left\{1 - \frac{\lambda}{4!}\frac{i}{\hbar c}\int d^4x\left[\frac{\hbar c}{i}\frac{\delta}{\delta J(x)}\right]^4\right\}\tilde{W}_0(J)\Bigg|_{J=0} \approx$$

$$1 - \frac{\lambda}{4!}\frac{i}{\hbar c}\int d^4x\left[\frac{\hbar c}{i}\frac{\delta}{\delta J(x)}\right]^4\tilde{W}_0^{(4)}(J) + \cdots$$

---

[1]Notice that $[(\hbar c/i)\delta/\delta J(x)]^4\tilde{W}_0^{(4)}(J)$ is independent of the sources.

To $O(\lambda)$, the interacting generating functional thus reads

$$\tilde{W}(J) \approx \left\{1 + \tilde{W}_0^{(2)}(J) + \tilde{W}_0^{(4)}(J) + \tilde{W}_0^{(6)}(J) + \cdots \right.$$

$$\left. -\frac{\lambda}{4!}\frac{i}{\hbar c}\int d^4x \left[\frac{\hbar c}{i}\frac{\delta}{\delta J(x)}\right]^4 \left[\tilde{W}_0^{(6)}(J) + \cdots\right]\right\} +$$

$$\left[\tilde{W}_0^{(2)}(J) + \tilde{W}_0^{(4)}(J) + \tilde{W}_0^{(6)}(J) + \cdots\right] \times$$

$$\frac{\lambda}{4!}\frac{i}{\hbar c}\int d^4x \left[\frac{\hbar c}{i}\frac{\delta}{\delta J(x)}\right]^4 \tilde{W}_0^{(4)}(J) + O\left(\lambda^2\right)$$

The last two lines represent the remaining contribution from the expansion of the denominator.

Let us explicitly evaluate the indicated functional derivatives. First, from Eqs. (10.106),

$$\left[\frac{\hbar c}{i}\frac{\delta}{\delta J(x)}\right]^4 \tilde{W}_0^{(4)}(J) = K_4(x,x,x,x) = -3\frac{\hbar^2}{c^2}\Delta_F(x-x)^2$$

Then, from the above,

$$\left[\frac{\hbar c}{i}\frac{\delta}{\delta J(x)}\right]^4 \tilde{W}_0^{(6)}(J) = -\frac{1}{(\hbar c)^2}\frac{6!}{2!}\frac{1}{6!}\int d^4x_1 \int d^4x_2 \times$$

$$K_6(x_1,x_2,x,x,x,x)J(x_1)J(x_2)$$

$$= -3\frac{i\hbar}{2c^5}\int d^4x_1 \int d^4x_2 \times$$

$$\left[4\Delta_F(x_1-x)\Delta_F(x-x_2)\Delta_F(x-x) + \Delta_F(x-x)^2\Delta_F(x_1-x_2)\right]J(x_1)J(x_2)$$

where $6!/2! = 360$ is the number of modes of taking the functional derivatives, keeping in mind that K_6 is completely symmetric in the coordinates.

For simplicity, we focus on the contribution to $\tilde{W}(J)$ of $O(\lambda)$ and bilinear in the scalar sources[2]

$$\frac{\lambda}{4!}\frac{i}{\hbar c}\int d^4x \left\{-\left[\frac{\hbar c}{i}\frac{\delta}{\delta J(x)}\right]^4 \tilde{W}_0^{(6)}(J) + \tilde{W}_0^{(2)}(J)\left[\frac{\hbar c}{i}\frac{\delta}{\delta J(x)}\right]^4 \tilde{W}_0^{(4)}(J)\right\} =$$

$$-\frac{\lambda}{4c^6}\int d^4x \int d^4x_1 \int d^4x_2\, \Delta_F(x_1-x)\Delta_F(x-x)\Delta_F(x-x_2)J(x_1)J(x_2)$$

where the disconnected contributions have canceled out.

---

[2]Clearly, $\tilde{W}(J)$ contains contributions of order $O(\lambda)$ with any even number of scalar sources as well. For $\tilde{W}_0^{(2)}(J)$, see Eqs. (10.102).

Therefore, the scalar propagator to $O(\lambda)$ will be

$$\left(\frac{\hbar c}{i}\right)^2 \frac{\delta^2 \tilde{W}(J)}{\delta J(x_1)\,\delta J(x_2)}\bigg|_{J=0} = \frac{\hbar}{ic}\Delta_F(x_1 - x_2) - $$

$$\frac{i}{2}\frac{\lambda}{\hbar c}\left(\frac{\hbar}{ic}\right)^3 \int d^4x\; \Delta_F(x_1 - x)\Delta_F(x - x)\Delta_F(x - x_2)$$

The additional non-linear interaction $\mathcal{L}_1(\phi) = -\lambda\phi^4/4!$ is not normal-ordered, and it gives rise to an additional self-energy term in the scalar propagator proportional to $\Delta_F(0)$, which is correctly included in the path-integral approach.[3]

Problems 10.4–10.8 show how the partition function of statistical mechanics (see appendix E of Vol. I) can be written as a path integral.

Problem 10.4 Consider a single non-relativistic particle moving in a potential in one dimension (Fig. 10.1 in the text) with the hamiltonian of Eq. (10.1). The partition function in the microcanonical ensemble is defined by

$$Z \equiv \text{Trace}\left[e^{-\beta\hat{H}}\right] \qquad ; \beta \equiv 1/k_{\text{B}}T$$

where "Trace" indicates the sum of the diagonal elements for a complete set of states. For example, if one uses the eigenstates of $\hat{H}$ with $\hat{H}|E_n\rangle = E_n|E_n\rangle$, then $Z = \sum_n e^{-\beta E_n}$. Use completeness to show that Z can also be computed with a complete set of *eigenstates of position*

$$Z = \int dq\, \langle q|e^{-\beta\hat{H}}|q\rangle$$

Solution to Problem 10.4

Start from the expression for the partition function as the sum of the diagonal matrix elements of the eigenstates of $\hat{H}$

$$Z = \sum_n e^{-\beta E_n} = \sum_n \langle E_n|e^{-\beta\hat{H}}|E_n\rangle$$

Now insert the completeness relation for the eigenstates of position (twice)

$$\int dq\, |q\rangle\langle q| = \hat{1}$$

---

[3]Compare Prob. 7.11.

This gives

$$Z = \int dq \int dq' \sum_n \langle E_n|q\rangle \langle q|e^{-\beta \hat{H}}|q'\rangle \langle q'|E_n\rangle$$

Use the completeness relation for the eigenstates of $\hat{H}$

$$\sum_n |E_n\rangle \langle E_n| = \hat{1}$$

This allows one to perform the sum over n

$$\sum_n \langle q'|E_n\rangle \langle E_n|q\rangle = \langle q'|q\rangle = \delta(q' - q)$$

Hence

$$Z = \int dq \, \langle q|e^{-\beta \hat{H}}|q\rangle$$

The "Trace" can thus be calculated from the diagonal elements of *any* complete set of states.

Problem 10.5 Use the arguments in the text to show that as $\varepsilon \to 0$

$$\langle q_l|e^{-(\varepsilon/\hbar)\hat{H}}|q_{l+1}\rangle = \left(\frac{m}{2\pi\varepsilon\hbar}\right)^{1/2} \exp\left\{-\frac{\varepsilon}{\hbar}\left[\frac{m}{2\varepsilon^2}(q_l - q_{l+1})^2 + V(q_l)\right]\right\}$$

Solution to Problem 10.5

The arguments in the text in Eqs. (10.10)–(10.18) show that as $\Delta t \to 0$,

$$\langle q_2|e^{-i\Delta t\hat{H}/\hbar}|q_1\rangle = \left(\frac{me^{-i\pi/2}}{2\pi\hbar\Delta t}\right)^{1/2} \exp\left\{\frac{i}{\hbar}\left[\frac{m(q_2 - q_1)^2}{2\Delta t} - \Delta t\, V(q_1)\right]\right\}$$

$$; \ \Delta t = t_2 - t_1 \to 0$$

Simply make the substitution

$$i\Delta t \to \varepsilon$$

The previous relation then takes the form

$$\langle q_{l+1}|e^{-(\varepsilon/\hbar)\hat{H}}|q_l\rangle = \left(\frac{m}{2\pi\hbar\varepsilon}\right)^{1/2} \exp\left\{-\frac{\varepsilon}{\hbar}\left[\frac{m}{2\varepsilon^2}(q_{l+1} - q_l)^2 + V(q_l)\right]\right\}$$

$$; \ \varepsilon \to 0$$

The arguments in the text imply that for an infinitesimal $\varepsilon \to 0$, this can equivalently be written as[4]

$$\langle q_l | e^{-(\varepsilon/\hbar)\hat{H}} | q_{l+1} \rangle = \left(\frac{m}{2\pi\hbar\varepsilon} \right)^{1/2} \exp\left\{ -\frac{\varepsilon}{\hbar} \left[\frac{m}{2\varepsilon^2}(q_l - q_{l+1})^2 + V(q_l) \right] \right\}$$
$$; \; \varepsilon \to 0$$

Problem 10.6 Let the variable τ run over the interval $[0, \hbar\beta]$. Divide this interval into n-subintervals of length ε, and label the corresponding variables by $(\tau_0, \tau_1, \cdots, \tau_n)$. Introduce coordinates $(q_0, q_1, \cdots, q_n)$ associated with these values, where $q_n \equiv q_0$. Write $e^{-\beta\hat{H}}$ as a product of n factors $e^{-(\varepsilon/\hbar)\hat{H}}$, and insert $n - 1$ complete sets of eigenstates of position between the factors. Use the result in Prob. 10.5 to show

$$Z = \mathrm{Lim}_{\varepsilon \to 0} \left(\frac{m}{2\pi\hbar\varepsilon} \right)^{n/2} \int \cdots \int \prod_{l=0}^{n-1} dq_l \; \times$$

$$\exp\left\{ -\frac{\varepsilon}{\hbar} \sum_{p=0}^{n-1} \left[\frac{m}{2\varepsilon^2}(q_p - q_{p+1})^2 + V(q_p) \right] \right\}$$

Solution to Problem 10.6

Start from the expression for the partition function in Prob. 10.4

$$Z = \int dq_0 \, \langle q_0 | e^{-\beta\hat{H}} | q_0 \rangle$$

Write the exponential as the product of n factors

$$e^{-\beta\hat{H}} = e^{-\varepsilon\hat{H}/\hbar} \cdots e^{-\varepsilon\hat{H}/\hbar} \qquad ; \; n \text{ factors}$$
$$; \; \varepsilon \equiv \hbar\beta/n$$

Now follow the arguments in Eqs. (10.22)–(10.26) in the text, with the substitutions

$$i(t_2 - t_1) \to \hbar\beta$$
$$i\Delta t \to \varepsilon \equiv \frac{\hbar\beta}{n}$$
$$\dot{q}_p \equiv \frac{q_{p+1} - q_p}{\Delta t} \to \frac{i(q_{p+1} - q_p)}{\varepsilon}$$

---

[4]Note the real gaussian in this amplitude driving $q_l \to q_{l+1}$ as $\varepsilon \to 0$.

Equation (10.26) for the propagation amplitude reads

$$\langle q_f | e^{-i(t_f - t_i)\hat{H}/\hbar} | q_i \rangle = \mathrm{Lim}_{n \to \infty} \left[\frac{nme^{-i\pi/2}}{2\pi\hbar(t_f - t_i)} \right]^{n/2} \int \cdots \int \prod_{l=1}^{n-1} dq_l \times$$

$$\exp \left\{ \frac{i}{\hbar} \sum_{p=0}^{n-1} \Delta t \left[\frac{m}{2} \dot{q}_p^2 - V(q_p) \right] \right\}$$

where $q_n \equiv q_f$. The analog of Eq. (10.26) for the partition function then reads

$$Z = \mathrm{Lim}_{\varepsilon \to 0} \left(\frac{m}{2\pi\hbar\varepsilon} \right)^{n/2} \int \cdots \int \prod_{l=0}^{n-1} dq_l \times$$

$$\exp \left\{ -\frac{\varepsilon}{\hbar} \sum_{p=0}^{n-1} \left[\frac{m}{2\varepsilon^2} (q_p - q_{p+1})^2 + V(q_p) \right] \right\}$$

with $\varepsilon = \hbar\beta/n$. Here $q_n \equiv q_0$, and the coordinates are tied together to produce the Trace.

Problem 10.7 The result in Prob. 10.6 is a path integral

$$Z = \int \mathcal{D}(q) \exp \left\{ -\frac{1}{\hbar} \bar{S}(\hbar\beta, 0) \right\}$$

$$\bar{S}(\hbar\beta, 0) = \int_0^{\hbar\beta} d\tau \left[\frac{m}{2} \left(\frac{dq}{d\tau} \right)^2 + V(q) \right]$$

Here $i\bar{S}$ is the classical action computed for *imaginary time*, $t = -i\tau$. Construct the analogue of Fig. 10.3 in the text and use the result in Prob. 10.6 to give the corresponding rules for evaluating this path integral. Note, in particular, the role here of the cyclic boundary condition $q_n = q_0$.

Solution to Problem 10.7

The analog of Fig. 10.3 in the text is shown in the accompanying Fig. 10.1. The corresponding rules for evaluating the partition function are then as follows (see Prob. 10.6):

(1) Divide the τ integration for the action at imaginary times into n intervals of size $\Delta\tau$ with

$$\hbar\beta = n\Delta\tau \equiv n\varepsilon$$

Label the coordinate at the intermediate time τ_p by q_p as indicated in Fig. 10.1;

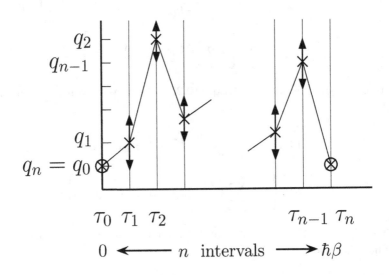

$$q_2$$
$$q_{n-1}$$

$$q_1$$
$$q_n = q_0$$

$$\tau_0 \; \tau_1 \; \tau_2 \qquad\qquad \tau_{n-1} \; \tau_n$$

$$0 \longleftarrow n \;\; \text{intervals} \longrightarrow \hbar\beta$$

Fig. 10.1 Partition function as a path integral. There are n intervals of size $\Delta\tau = \hbar\beta/n \equiv \varepsilon$. Note the cyclic boundary condition $q_n = q_0$ required by the Trace.

(2) Write the action $\bar{S}(\hbar\beta, 0)$ as the corresponding finite sum;

(3) Define the τ derivative appearing in that expression by

$$\frac{dq_p}{d\tau} \equiv \frac{q_{p+1} - q_p}{\Delta\tau}$$

(4) Impose cyclic boundary conditions required by the Trace

$$q_n \equiv q_0$$

(5) Carry out the multiple integral over the coordinates at all intermediate taus (Fig. 10.1); include $\int dq_0$ to recover the Trace

$$\int dq_0 \int dq_1 \cdots \int dq_{n-1}$$

(6) As a measure for the integration, include the following factor for each interval

$$\left(\frac{m}{2\pi\hbar\Delta\tau}\right)^{1/2} \qquad \text{; one for each interval}$$

(7) At the end of the calculation, take the limit $n \to \infty$ (which implies $\Delta\tau = \hbar\beta/n \to 0$).

Problem 10.8 The *thermal average* of an operator $O(\hat{q})$ is defined by

$$\langle\langle O(\hat{q})\rangle\rangle \equiv \frac{\text{Trace}\,[O(\hat{q})e^{-\beta\hat{H}}]}{\text{Trace}\,[e^{-\beta\hat{H}}]}$$

(a) Show this is the ratio of two path integrals of the form in Prob. 10.7, where the path integral in the numerator has an additional factor of $O(q_0)$;

(b) Discuss the relation to the path-integral expression for the Green's functions in the euclidean metric [compare Eq. (10.81)]. What are the similarities? What are the differences?[5]

Solution to Problem 10.8

(a) From the solution to Prob. 10.6, the "Trace" in the numerator is

$$\text{Trace}\,[O(\hat{q})e^{-\beta\hat{H}}] = \int dq_0 \langle q_0|O(\hat{q})e^{-\beta\hat{H}}|q_0\rangle$$

$$= \int dq_0 O(q_0)\langle q_0|e^{-\beta\hat{H}}|q_0\rangle$$

This simply adds a factor $O(q_0)$ to the analysis in Probs. 10.6–10.7 of the path integral in the denominator.

(b) In a comparison of the path-integral expression for the thermal average in Probs. 10.6–10.8 with that for the Green's function in the euclidean metric [see Eq. (10.81)], we observe the following:

- The integrals in both cases run over imaginary times

$$t \to -i\tau$$

- The weighting factor in the path integral in both cases turns out to be $iS/\hbar$ where S is the action computed with the imaginary time [see Prob. 10.7 and Eqs. (10.83)];
- In the case of the Green's function in the euclidean metric, the τ-integrals run over the infinite interval $[-\infty, \infty]$;
- In the case of the thermal averages, the τ-integrals run over the finite interval $[0, \hbar\beta]$;

---

[5]Note that the reduction of any problem to convergent many-component multiple integrals, as in Probs. 10.7–10.8, lends itself to numerical approximation methods (see, for example, lattice gauge theory for QCD as described in [Walecka (2004)]).

- In order to recover the Trace in the case of the thermal averages, one must impose cyclic boundary conditions on the coordinate, $q_n = q_0$.

Problem 10.9 Derive Eq. (10.121) in the case $n = 3$.

Solution to Problem 10.9

We start working on the l.h.s. of Eq. (10.121) for the case $n = 3$

$$\text{l.h.s.} = \int d\bar{c}_3 \int d\bar{c}_2 \int d\bar{c}_1 \int dc_1 \int dc_2 \int dc_3 \exp\left\{ -\sum_{i=1}^{3}\sum_{j=1}^{3} \bar{c}_i N_{ij} c_j \right\}$$

$$= \int d\bar{c}_3 \int d\bar{c}_2 \int d\bar{c}_1 \int dc_1 \int dc_2 \int dc_3 \frac{1}{3!} \left[-\sum_{i=1}^{3}\sum_{i=j}^{3} \bar{c}_i N_{ij} c_j \right]^3$$

where the last line has been obtained by expanding the exponential and retaining only the non-vanishing terms [see Eqs. (10.118) and (10.120)]. We must keep in mind that we are dealing with Grassmann variables, and therefore only factors which contain different variables can contribute.

In particular, we consider the factor

$$\left[\sum_{i=1}^{3}\sum_{i=j}^{3} \bar{c}_i N_{ij} c_j \right]^3 = \bar{c}_1 N_{11} c_1 \left(\bar{c}_2 N_{22} c_2 + \bar{c}_2 N_{23} c_3 + \bar{c}_3 N_{32} c_2 + \bar{c}_3 N_{33} c_3 \right)^2 +$$

$$\bar{c}_1 N_{12} c_2 \left(\bar{c}_2 N_{21} c_1 + \bar{c}_2 N_{23} c_3 + \bar{c}_3 N_{31} c_1 + \bar{c}_3 N_{33} c_3 \right)^2 + \cdots +$$

$$\bar{c}_3 N_{33} c_3 \left(\bar{c}_1 N_{11} c_1 + \bar{c}_1 N_{12} c_2 + \bar{c}_2 N_{21} c_1 + \bar{c}_2 N_{22} c_2 \right)^2$$

This gives

$$\left[\sum_{i=1}^{3}\sum_{i=j}^{3} \bar{c}_i N_{ij} c_j \right]^3 = 2\bar{c}_1 N_{11} c_1 \left(\bar{c}_2 N_{22} c_2 \bar{c}_3 N_{33} c_3 + \bar{c}_2 N_{23} c_3 \bar{c}_3 N_{32} c_2 \right) +$$

$$2\bar{c}_1 N_{12} c_2 \left(\bar{c}_2 N_{21} c_1 \bar{c}_3 N_{33} c_3 + \bar{c}_2 N_{23} c_3 \bar{c}_3 N_{31} c_1 \right) + \cdots +$$

$$2\bar{c}_3 N_{33} c_3 \left(\bar{c}_1 N_{11} c_1 \bar{c}_2 N_{22} c_2 + \bar{c}_1 N_{12} c_2 \bar{c}_2 N_{21} c_1 \right)$$

where we have use the nilpotency of the Grassmann variables to eliminate the vanishing terms.

After re-arranging this expression, keeping in mind the fact that the

Grassmann variables anticommute, we have

$$\left[\sum_{i=1}^{3}\sum_{i=j}^{3}\bar{c}_i N_{ij} c_j\right]^3 = 6\bar{c}_3\bar{c}_2\bar{c}_1 c_1 c_2 c_3 \times$$

$$[-N_{13}N_{31}N_{22} + N_{23}N_{31}N_{12} + N_{13}N_{21}N_{32}$$
$$-N_{11}N_{23}N_{32} - N_{12}N_{21}N_{33} + N_{33}N_{11}N_{22}]$$
$$= 6\,\bar{c}_3\bar{c}_2\bar{c}_1 c_1 c_2 c_3\,\det\underline{N}$$

With the use of the properties in Eqs. (10.118) and (10.120), it is easy to see that

$$\int d\bar{c}_3 \int d\bar{c}_2 \int d\bar{c}_1 \int dc_1 \int dc_2 \int dc_3\ \bar{c}_3\bar{c}_2\bar{c}_1 c_1 c_2 c_3 = -1$$

and therefore

$$\int d\bar{c}_3 \int d\bar{c}_2 \int d\bar{c}_1 \int dc_1 \int dc_2 \int dc_3 \exp\left\{-\sum_{i=1}^{3}\sum_{i=j}^{3}\bar{c}_i N_{ij} c_j\right\} = \det\underline{N}$$

as anticipated.

Problem 10.10 Many physical systems can be described with *effective field theories* where one builds all the symmetries of the problem into an effective lagrangian density and then expands physical quantities in some meaningful small dimensionless ratio. The appropriate lagrangian density can involve nonlinear, derivative couplings, and it may be nonrenormalizable; nevertheless, path integral techniques provide a means for quantizing the theory.

(a) As an example of a model, non-linear, effective field theory consider the following lagrangian density involving nucleon, pion, and scalar fields $(\underline{\psi}, \boldsymbol{\pi}, \varphi)$

$$\mathcal{L}^{\text{eff}} = -\hbar c\left[\bar{\underline{\psi}}_L \gamma_\lambda \frac{\partial}{\partial x_\lambda}\underline{\psi}_L + \bar{\underline{\psi}}_R \gamma_\lambda \frac{\partial}{\partial x_\lambda}\underline{\psi}_R -\right.$$
$$\left. g\sigma_0\left(1 + \frac{\varphi}{\sigma_0}\right)\left(\bar{\underline{\psi}}_R \underline{U}\,\underline{\psi}_L + \bar{\underline{\psi}}_L\,\underline{U}^\dagger\,\underline{\psi}_R\right)\right]$$
$$-\frac{c^2}{2}\left(\frac{\partial\varphi}{\partial x_\lambda}\right)^2 - \frac{c^2\sigma_0^2}{4}\text{tr}\left(\frac{\partial\underline{U}^\dagger}{\partial x_\lambda}\frac{\partial\underline{U}}{\partial x_\lambda}\right) - V\left(\underline{U}, \frac{\partial\underline{U}}{\partial x_\lambda}; \varphi\right)$$
$$\underline{U} \equiv \exp\left\{\frac{i}{\sigma_0}\boldsymbol{\tau}\cdot\boldsymbol{\pi}\right\}$$

Show that the lagrangian density, without $\mathcal{V}$, is invariant under the following transformation

$$\begin{aligned} \underline{\psi}_L &\rightarrow \underline{L}\,\underline{\psi}_L & ;\ \underline{U} &\rightarrow \underline{R}\,\underline{U}\,\underline{L}^\dagger \\ \underline{\psi}_R &\rightarrow \underline{R}\,\underline{\psi}_R & ;\ \varphi &\rightarrow \varphi \end{aligned}$$

where $(\underline{L},\ \underline{R})$ are independent, global $SU(2)$ matrices. Hence conclude that as long as the potential $\mathcal{V}$ is constructed to be invariant, this lagrangian density has an exact $SU(2)_L \otimes SU(2)_R$ symmetry (compare Prob. 6.12);

(b) Take the limit $|\sigma_0| \rightarrow \infty$, assume that in this limit $\mathcal{V} \rightarrow (m_s^2 c^2/2)\varphi^2 + O(1/|\sigma_0|)$, and reproduce the chiral-symmetric σ-model lagrangian density of Eq. (6.166) with $\sigma_0 = -M/g$;

(c) Show that the following additional contribution

$$\mathcal{L}_\pi^{\text{mass}} = \frac{m_\pi^2 c^2 \sigma_0^2}{4}\,\text{tr}\left(\underline{U} + \underline{U}^\dagger - 2\right)$$

reduces to the usual pion mass term in the limit $|\sigma_0| \rightarrow \infty$ and is *not* chiral invariant.[6]

Solution to Problem 10.10

(a) Consider a model non-linear, effective field theory characterized by the following lagrangian density involving nucleon, pion, and scalar fields

$$\begin{aligned} \mathcal{L}^{\text{eff}} = -\hbar c &\left[\underline{\bar{\psi}}_L \gamma_\lambda \frac{\partial}{\partial x_\lambda}\underline{\psi}_L + \underline{\bar{\psi}}_R \gamma_\lambda \frac{\partial}{\partial x_\lambda}\underline{\psi}_R - \right. \\ &\left. g\sigma_0\left(1 + \frac{\varphi}{\sigma_0}\right)\left(\underline{\bar{\psi}}_R\,\underline{U}\,\underline{\psi}_L + \underline{\bar{\psi}}_L\,\underline{U}^\dagger\,\underline{\psi}_R\right)\right] \\ &-\frac{c^2}{2}\left(\frac{\partial \varphi}{\partial x_\lambda}\right)^2 - \frac{c^2\sigma_0^2}{4}\text{tr}\left(\frac{\partial \underline{U}^\dagger}{\partial x_\lambda}\frac{\partial \underline{U}}{\partial x_\lambda}\right) - \mathcal{V}\left(\underline{U}, \frac{\partial \underline{U}}{\partial x_\lambda}; \varphi\right) \end{aligned}$$

$$\underline{U} \equiv \exp\left\{\frac{i}{\sigma_0}\boldsymbol{\tau}\cdot\boldsymbol{\pi}\right\}$$

Define

$$\mathcal{L}^{\text{eff}} = \mathcal{L}_0^{\text{eff}} - \mathcal{V}$$

---

[6]The conventional notation in Prob. 10.10 is $-\sigma_0 = M/g \equiv f_\pi$. The most general form of $\mathcal{L}^{\text{eff}}$ here is the point of departure for a subsequent expansion of scattering amplitudes in powers of $(q_\lambda, m_\pi)/f_\pi$ where q_λ is an external four-momentum; this is the basic idea behind *chiral perturbation theory* (see [Donoghue, Golowich, and Holstein (2014); Walecka (2004)]).

Start with $\mathcal{L}_0^{\text{eff}}$, and consider the following transformation (compare the solution to Prob. 6.12)

$$\underline{\psi}_L \to \underline{L}\,\underline{\psi}_L \qquad ; \underline{U} \to \underline{R}\,\underline{U}\,\underline{L}^\dagger$$
$$\underline{\psi}_R \to \underline{R}\,\underline{\psi}_R \qquad ; \varphi \to \varphi$$

where $(\underline{L},\ \underline{R})$ are independent, global SU(2) matrices . For this transformation

$$\underline{U}^\dagger \to \underline{L}\,\underline{U}^\dagger\,\underline{R}^\dagger$$

Since $(\underline{L},\ \underline{R})$ are global transformations, they can be moved through any derivatives, and

$$\underline{R}^\dagger \underline{R} = \underline{L}^\dagger \underline{L} = \underline{1}$$

Furthermore, the trace "tr" is invariant under cyclic permutations. It then follows by inspection that $\mathcal{L}_0^{\text{eff}}$ is invariant under this transformation.

It follows that as long as the potential V is constructed to also be invariant under this transformation, the lagrangian density $\mathcal{L}^{\text{eff}}$ has an exact $SU(2)_L \otimes SU(2)_R$ symmetry.

(b) Define the parameter σ_0 according to

$$\sigma_0 \equiv -\frac{M}{g}$$

Assume that in the limit $|\sigma_0| \to \infty$, the potential reduces to a simple quadratic form (compare Fig. 6.4 in the text)

$$V \to \frac{1}{2}m_s^2 c^2 \varphi^2 \qquad ; |\sigma_0| \to \infty$$

In this limit, $\mathcal{L}^{\text{eff}}$ then takes the following form

$$\mathcal{L}^{\text{eff}} = -\hbar c\left[\bar{\psi}\gamma_\lambda \frac{\partial}{\partial x_\lambda}\underline{\psi} - g\sigma_0\left(1+\frac{\varphi}{\sigma_0}\right)\bar{\psi}\left(1+\frac{i}{\sigma_0}\boldsymbol{\tau}\cdot\boldsymbol{\pi}\right)\frac{1}{2}(1+\gamma_5)\underline{\psi} - \right.$$
$$\left. g\sigma_0\left(1+\frac{\varphi}{\sigma_0}\right)\bar{\psi}\left(1-\frac{i}{\sigma_0}\boldsymbol{\tau}\cdot\boldsymbol{\pi}\right)\frac{1}{2}(1-\gamma_5)\underline{\psi}\right]$$
$$-\frac{c^2}{2}\left[\left(\frac{\partial\varphi}{\partial x_\lambda}\right)^2 + m_s^2\varphi^2\right] - \frac{c^2\sigma_0^2}{2}\left(\frac{1}{\sigma_0^2}\frac{\partial\boldsymbol{\pi}}{\partial x_\lambda}\cdot\frac{\partial\boldsymbol{\pi}}{\partial x_\lambda}\right) + \cdots$$

which reduces to

$$\mathcal{L}^{\text{eff}} = -\hbar c \bar{\psi} \left[\gamma_\mu \frac{\partial}{\partial x_\mu} + M - g \left(i\gamma_5 \, \boldsymbol{\tau} \cdot \boldsymbol{\pi} + \varphi \right) \right] \psi - \frac{c^2}{2} \left(\frac{\partial \boldsymbol{\pi}}{\partial x_\mu} \cdot \frac{\partial \boldsymbol{\pi}}{\partial x_\mu} \right)$$
$$- \frac{c^2}{2} \left(\frac{\partial \varphi}{\partial x_\mu} \frac{\partial \varphi}{\partial x_\mu} + m_s^2 \varphi^2 \right) \qquad ; \, |\sigma_0| \to \infty$$

This expression coincides with the corresponding limit of the chiral-symmetric lagrangian density of the σ-model in Eq. (6.166).

(c) A pion mass term can be included by augmenting the effective lagrangian density with the following

$$\mathcal{L}_\pi^{\text{mass}} = \frac{m_\pi^2 c^2 \sigma_0^2}{4} \, \text{tr} \left(\underline{U} + \underline{U}^\dagger - 2 \right)$$

In the limit $|\sigma_0| \to \infty$, this term reproduces the usual pion mass contribution

$$\mathcal{L}_\pi^{\text{mass}} = \frac{m_\pi^2 c^2 \sigma_0^2}{4} \, \text{tr} \left[1 + \left(\frac{i}{\sigma_0} \boldsymbol{\tau} \cdot \boldsymbol{\pi} \right) + \frac{1}{2!} \left(\frac{i}{\sigma_0} \boldsymbol{\tau} \cdot \boldsymbol{\pi} \right)^2 + \right.$$
$$\left. 1 + \left(\frac{-i}{\sigma_0} \boldsymbol{\tau} \cdot \boldsymbol{\pi} \right) + \frac{1}{2!} \left(\frac{-i}{\sigma_0} \boldsymbol{\tau} \cdot \boldsymbol{\pi} \right)^2 + \cdots - 2 \right]$$
$$= -\frac{1}{2} m_\pi^2 c^2 \boldsymbol{\pi}^2 \qquad ; \, |\sigma_0| \to \infty$$

The term $\mathcal{L}_\pi^{\text{mass}}$ is evidently *not* invariant under the transformation in part (a); it constitutes an explicit chiral-symmetry breaking contribution.

Chapter 11

Canonical Transformations for Quantum Systems

Problem 11.1 (a) Consider a real, massive, scalar field ϕ interacting with a time-independent, localized, c-number source $\lambda s(\mathbf{x})$; the equation of motion is

$$(\Box - m^2)\phi(x) = \lambda s(\mathbf{x})$$

Construct the lagrangian and hamiltonian for this system.

(b) Quantize this system in the Schrödinger picture (see Chap. 5). Work in a big box with p.b.c. Introduce the normal-mode expansion for $\hat{\phi}(\mathbf{x})$, and show

$$\hat{H} = \hat{H}_0 + \hat{H}_1$$
$$\hat{H}_0 = \sum_{\mathbf{k}} \hbar\omega_k \left(c_{\mathbf{k}}^\dagger c_{\mathbf{k}} + 1/2 \right)$$
$$\hat{H}_1 = \lambda c^2 \sum_{\mathbf{k}} \left(\frac{\hbar}{2\omega_k} \right)^{1/2} \left[c_{\mathbf{k}}\tilde{s}(\mathbf{k}) + c_{\mathbf{k}}^\dagger \tilde{s}^\dagger(\mathbf{k}) \right]$$

where $s(\mathbf{x})$ has the Fourier expansion $s(\mathbf{x}) = (1/\sqrt{\Omega})\sum_{\mathbf{k}} \tilde{s}(\mathbf{k})e^{-i\mathbf{k}\cdot\mathbf{x}}$.

(c) Show the following canonical transformation diagonalizes the hamiltonian

$$C_{\mathbf{k}} = c_{\mathbf{k}} + \lambda c^2 \left(\frac{1}{2\hbar\omega_k^3} \right)^{1/2} \tilde{s}^\dagger(\mathbf{k})$$
$$C_{\mathbf{k}}^\dagger = c_{\mathbf{k}}^\dagger + \lambda c^2 \left(\frac{1}{2\hbar\omega_k^3} \right)^{1/2} \tilde{s}(\mathbf{k})$$

(d) Assume two point sources with $s(\mathbf{x}) = \delta^{(3)}(\mathbf{x} - \mathbf{x}_1) + \delta^{(3)}(\mathbf{x} - \mathbf{x}_2)$. Identify the interaction energy, and show that the result is the *Yukawa potential*

$$E_{\text{int}} = -\frac{\lambda^2 c^2}{4\pi} \frac{e^{-m|\mathbf{x}_1 - \mathbf{x}_2|}}{|\mathbf{x}_1 - \mathbf{x}_2|}$$

to all orders in λ. Note the sign.

(e) Compute the overlap of the exact vacuum $|\underline{0}\rangle$ with the non-interacting occupation-number eigenstates of $\hat{H}_0$, and show that

$$\langle n_1 n_2 \cdots n_\infty | \underline{0} \rangle = \frac{(-1)^{n_1 + n_2 + \cdots}}{\sqrt{n_1! n_2! \cdots}} \left(\frac{\lambda c^2}{\sqrt{2\hbar\omega_1^3}} \tilde{s}^\dagger(\mathbf{k}_1) \right)^{n_1} \left(\frac{\lambda c^2}{\sqrt{2\hbar\omega_2^3}} \tilde{s}^\dagger(\mathbf{k}_2) \right)^{n_2} \times$$
$$\cdots \langle 0 | \underline{0} \rangle$$

(f) Use completeness to show that the overlap of the exact vacuum and the non-interaction vacuum is given by

$$|\langle 0 | \underline{0} \rangle|^2 = \exp \left\{ -\sum_{\mathbf{k}} \frac{\lambda^2 c^4}{2\hbar\omega_k^3} |\tilde{s}(\mathbf{k})|^2 \right\}$$

Solution to Problem 11.1[1]

(a) Take the following as the lagrangian density for this neutral scalar meson theory with a static, c-number source

$$\mathcal{L} = -\frac{c^2}{2} \left[\left(\frac{\partial \phi}{\partial x_\mu} \right)^2 + m^2 \phi^2 \right] - \lambda c^2 s(\mathbf{x}) \phi$$

Lagrange's equation reads

$$\frac{\partial}{\partial x_\mu} \frac{\partial \mathcal{L}}{\partial(\partial \phi / \partial x_\mu)} - \frac{\partial \mathcal{L}}{\partial \phi} = 0$$

This yields the anticipated equation of motion

$$(\Box - m^2)\phi(x) = \lambda s(\mathbf{x})$$

The canonical momentum density is obtained as

$$\pi = \frac{\partial \mathcal{L}}{\partial(\partial \phi / \partial t)} = \frac{\partial \phi}{\partial t} \equiv \dot{\phi}$$

---

[1]Problems 11.1–11.2 are longer, but extremely valuable, as they provide exact solutions for two important problems in quantum field theory.

The hamiltonian density then follows as

$$\mathcal{H} = \pi\dot{\phi} - \mathcal{L}$$
$$= \frac{1}{2}\left[\pi^2 + c^2(\boldsymbol{\nabla}\phi)^2 + m^2c^2\phi^2\right] + \lambda c^2 s(\mathbf{x})\phi$$
$$\equiv \mathcal{H}_0 + \mathcal{H}_1$$

(b) Insert the Schrödinger-picture field expansions[2]

$$\hat{\phi}(\mathbf{x}) = \frac{1}{\sqrt{\Omega}}\sum_{\mathbf{k}}\left(\frac{\hbar}{2\omega_k}\right)^{1/2}\left[c_{\mathbf{k}}e^{i\mathbf{k}\cdot\mathbf{x}} + c_{\mathbf{k}}^{\dagger}e^{-i\mathbf{k}\cdot\mathbf{x}}\right]$$

$$\hat{\pi}(\mathbf{x}) = \frac{1}{\sqrt{\Omega}}\frac{1}{i}\sum_{\mathbf{k}}\left(\frac{\hbar\omega_k}{2}\right)^{1/2}\left[c_{\mathbf{k}}e^{i\mathbf{k}\cdot\mathbf{x}} - c_{\mathbf{k}}^{\dagger}e^{-i\mathbf{k}\cdot\mathbf{x}}\right]$$

$$\omega_k = c\sqrt{\mathbf{k}^2 + m^2}$$

The creation and destruction operators satisfy the familiar expression

$$[c_{\mathbf{k}}, c_{\mathbf{k}'}^{\dagger}] = \delta_{\mathbf{k}\mathbf{k}'}$$

This ensures that the canonical commutation relation is obeyed by the fields[3]

$$[\hat{\phi}(\mathbf{x}), \hat{\pi}(\mathbf{x}')] = i\hbar\delta^{(3)}(\mathbf{x} - \mathbf{x}')$$

The hamiltonian $\hat{H}_0$ is then just the result in Eq. (5.62)

$$\hat{H}_0 = \int_{\text{Box}} d^3x\, \hat{\mathcal{H}}_0 = \sum_{\mathbf{k}}\hbar\omega_k\left(c_{\mathbf{k}}^{\dagger}c_{\mathbf{k}} + \frac{1}{2}\right)$$

Insert the field $\hat{\phi}(\mathbf{x})$ and the Fourier transform of the source

$$s(\mathbf{x}) = \frac{1}{\sqrt{\Omega}}\sum_{\mathbf{k}}\tilde{s}(\mathbf{k})e^{-i\mathbf{k}\cdot\mathbf{x}}$$

$$\tilde{s}(-\mathbf{k}) = \tilde{s}^{\dagger}(\mathbf{k})$$

Here the last relation follows from the reality of $s(\mathbf{x})$. The interaction hamiltonian is then

$$\hat{H}_1 = \int_{\text{Box}} d^3x\, \hat{\mathcal{H}}_1 = \lambda c^2\sum_{\mathbf{k}}\left(\frac{\hbar}{2\omega_k}\right)^{1/2}\left[c_{\mathbf{k}}\,\tilde{s}(\mathbf{k}) + c_{\mathbf{k}}^{\dagger}\,\tilde{s}^{\dagger}(\mathbf{k})\right]$$

---

[2]Compare the argument in Eqs. (5.60)–(5.62) and (5.65); the Schrödinger picture coincides with the interaction picture at $t = 0$.

[3]Even when the interaction $\hat{H}_1$ is present.

(c) The hamiltonian $\hat{H} = \hat{H}_0 + \hat{H}_1$ is a quadratic form in the creation and destruction operators, and it can be diagonalized with a canonical transformation. Define

$$C_{\mathbf{k}} \equiv c_{\mathbf{k}} + \lambda c^2 \left(\frac{1}{2\hbar\omega_k^3} \right)^{1/2} \tilde{s}^\dagger(\mathbf{k})$$

$$C_{\mathbf{k}}^\dagger \equiv c_{\mathbf{k}}^\dagger + \lambda c^2 \left(\frac{1}{2\hbar\omega_k^3} \right)^{1/2} \tilde{s}(\mathbf{k})$$

In terms of these new operators, the hamiltonian takes the form

$$\hat{H} = \sum_{\mathbf{k}} \hbar\omega_k \left(C_{\mathbf{k}}^\dagger C_{\mathbf{k}} + \frac{1}{2} \right) + H'$$

$$H' = -(\lambda c^2)^2 \sum_{\mathbf{k}} \frac{1}{2\omega_k^2} |\tilde{s}(\mathbf{k})|^2$$

The new operators satisfy the same commutation relations as the original operators[4]

$$[C_{\mathbf{k}}, C_{\mathbf{k}'}^\dagger] = \delta_{\mathbf{k}\mathbf{k}'}$$

Therefore we know all their properties. The operator $C_{\mathbf{k}}^\dagger C_{\mathbf{k}}$ is again the number operator, with a spectrum of the positive integers and zero, and the hamiltonian $\hat{H}$ has been diagonalized.

(d) There is a c-number shift in the energy of the system given by H'. An inversion of the Fourier transform gives

$$\tilde{s}(\mathbf{k}) = \frac{1}{\sqrt{\Omega}} \int d^3x \, e^{i\mathbf{k}\cdot\mathbf{x}} s(\mathbf{x})$$

The interaction H' can therefore be re-written as

$$H' = -\frac{(\lambda c^2)^2}{\Omega} \sum_{\mathbf{k}} \frac{1}{2\omega_k^2} \int d^3x \int d^3x' \, s(\mathbf{x})s(\mathbf{x}') \, e^{i\mathbf{k}\cdot(\mathbf{x}-\mathbf{x}')}$$

Suppose one has *two point sources* located at $\mathbf{x}_1$ and $\mathbf{x}_2$, so that

$$s(\mathbf{x}) = \delta^{(3)}(\mathbf{x} - \mathbf{x}_1) + \delta^{(3)}(\mathbf{x} - \mathbf{x}_2)$$

---

[4]Also $[C_{\mathbf{k}}, C_{\mathbf{k}'}] = [C_{\mathbf{k}}^\dagger, C_{\mathbf{k}'}^\dagger] = 0$.

One can then identify an *interaction energy* between the sources as[5]

$$H'_{\text{int}} = -\frac{\lambda^2 c^2}{\Omega} \sum_{\mathbf{k}} \frac{e^{i\mathbf{k}\cdot(\mathbf{x}_1 - \mathbf{x}_2)}}{\mathbf{k}^2 + m^2}$$

The Yukawa potential has a Fourier series representation

$$\frac{e^{-mr}}{4\pi r} = \frac{1}{\Omega} \sum_{\mathbf{k}} \frac{e^{i\mathbf{k}\cdot\mathbf{r}}}{\mathbf{k}^2 + m^2}$$

Hence

$$H'_{\text{int}} = -\frac{\lambda^2 c^2}{4\pi} \frac{e^{-m|\mathbf{x}_1 - \mathbf{x}_2|}}{|\mathbf{x}_1 - \mathbf{x}_2|}$$

The attractive Yukawa interaction between two point sources is an *exact result* for this neutral scalar meson theory.

(e) The ground state of the interacting system is defined by

$$C_{\mathbf{k}}|\underline{0}\rangle = 0 \qquad ; \text{ all } \mathbf{k}$$

Consider the normalized state

$$|n_1 n_2 \cdots\rangle = \frac{1}{\sqrt{n_1! n_2! \cdots}} (c_1^\dagger)^{n_1} (c_2^\dagger)^{n_2} \cdots |0\rangle$$

The overlap of this state with the new vacuum is given by

$$\langle n_1 n_2 \cdots |\underline{0}\rangle = \frac{1}{\sqrt{n_1! n_2! \cdots}} \langle 0|(c_1)^{n_1} (c_2)^{n_2} \cdots |\underline{0}\rangle$$

Now substitute the transformed operator from part (c)

$$c_{\mathbf{k}} \equiv C_{\mathbf{k}} - \lambda c^2 \left(\frac{1}{2\hbar\omega_k^3}\right)^{1/2} \tilde{s}^\dagger(\mathbf{k})$$

Make use of the above relation defining $|\underline{0}\rangle$. All the $C_{\mathbf{k}}$ annihilate this new vacuum, and hence they disappear in the overlap. Thus

$$\langle n_1 n_2 \cdots |\underline{0}\rangle = \frac{(-1)^{n_1 + n_2 \cdots}}{\sqrt{n_1! n_2! \cdots}} \left[\frac{\lambda c^2}{\sqrt{2\hbar\omega_1^3}} \tilde{s}^\dagger(\mathbf{k}_1)\right]^{n_1} \left[\frac{\lambda c^2}{\sqrt{2\hbar\omega_2^3}} \tilde{s}^\dagger(\mathbf{k}_2)\right]^{n_2} \cdots \langle 0|\underline{0}\rangle$$

The completeness relation states that

$$\sum_{n_1 n_2 \cdots} |n_1 n_2 \cdots\rangle\langle n_1 n_2 \cdots| = \hat{1}$$

---

[5]There is also a self-energy for each source, whose effects we do not pursue here.

Hence the normalization condition on $|\underline{0}\rangle$ can be written as

$$\sum_{n_1 n_2 \cdots} |\langle n_1 n_2 \cdots |\underline{0}\rangle|^2 = \langle \underline{0}|\underline{0}\rangle = 1$$

Substitution of the above amplitudes, and identification of the exponential for each mode, gives

$$\exp\left\{ \sum_{\mathbf{k}} \frac{\lambda^2 c^4}{2\hbar\omega_k^3} |\tilde{s}(\mathbf{k})|^2 \right\} |\langle 0|\underline{0}\rangle|^2 = 1$$

Thus the overlap of the interacting and non-interacting vacuua is given by[6]

$$|\langle 0|\underline{0}\rangle|^2 = \exp\left\{ -\sum_{\mathbf{k}} \frac{\lambda^2 c^4}{2\hbar\omega_k^3} |\tilde{s}(\mathbf{k})|^2 \right\}$$

Problem 11.2 Consider the Bloch-Nordsieck problem of the quantized radiation field interacting with a time-independent, classical external current source $j_\mu^{\text{ext}}(\mathbf{x}) = [\mathbf{j}^{\text{ext}}(\mathbf{x})/c, \, i\rho^{\text{ext}}(\mathbf{x})]$.[7]

(a) The hamiltonian for this system in the interaction picture was derived in Eq. (C.73). Go back to the Schrödinger picture (see Chap. 5). Introduce the expansion $\mathbf{j}^{\text{ext}}(\mathbf{x}) = (c/\sqrt{\Omega}) \sum_{\mathbf{q}} \tilde{\mathbf{j}}(\mathbf{q}) e^{-i\mathbf{q}\cdot\mathbf{x}}$ and show that

$$\hat{H} = \sum_{\mathbf{k},\lambda} \hbar\omega_k \left(a_{\mathbf{k}\lambda}^\dagger a_{\mathbf{k}\lambda} + 1/2 \right) - e \sum_{\mathbf{k},\lambda} \left(\frac{\hbar c^2}{2\omega_k \varepsilon_0} \right)^{1/2} \left(j(\mathbf{k}\lambda) a_{\mathbf{k}\lambda} + j^\dagger(\mathbf{k}\lambda) a_{\mathbf{k}\lambda}^\dagger \right)$$
$$+ \frac{e^2}{8\pi\varepsilon_0} \int d^3x \int d^3x' \, \frac{\rho^{\text{ext}}(\mathbf{x})\rho^{\text{ext}}(\mathbf{x}')}{|\mathbf{x}-\mathbf{x}'|}$$

where $j(\mathbf{k}\lambda) \equiv \mathbf{e}_{\mathbf{k}\lambda} \cdot \tilde{\mathbf{j}}(\mathbf{k})$. Here λ is the photon helicity;

(b) Show the following canonical transformation *diagonalizes* the hamiltonian

$$A_{\mathbf{k}\lambda} = a_{\mathbf{k}\lambda} - \left(\frac{c^2}{2\hbar\omega_k^3 \varepsilon_0} \right)^{1/2} ej^\dagger(\mathbf{k}\lambda)$$

$$A_{\mathbf{k}\lambda}^\dagger = a_{\mathbf{k}\lambda}^\dagger - \left(\frac{c^2}{2\hbar\omega_k^3 \varepsilon_0} \right)^{1/2} ej(\mathbf{k}\lambda)$$

(c) Expand the exact ground state $|\underline{0}\rangle$ of the diagonalized hamiltonian in terms of the eigenstates of the free photon field, and show that the

---

[6]Note that this result is non-perturbative in λ, and correctly lies between 0 and 1.
[7]The reference is [Bloch and Nordsieck (1937)].

expansion coefficients are given by

$$\langle n_1 n_2 \cdots n_\infty | \underline{0} \rangle = \frac{1}{\sqrt{n_1! n_2! \cdots}} \times$$

$$\left(\sqrt{\frac{c^2}{2\hbar \omega_1^3 \varepsilon_0}} \, e j^\dagger(\mathbf{k}_1 \lambda_1) \right)^{n_1} \left(\sqrt{\frac{c^2}{2\hbar \omega_2^3 \varepsilon_0}} \, e j^\dagger(\mathbf{k}_2 \lambda_2) \right)^{n_2} \cdots \langle 0 | \underline{0} \rangle$$

(d) Use completeness to demonstrate that

$$|\langle 0 | \underline{0} \rangle|^2 = \exp \left\{ -\frac{e^2 c^2}{\varepsilon_0} \sum_{\mathbf{k}, \lambda} \frac{|j(\mathbf{k}\lambda)|^2}{2\hbar \omega_k^3} \right\}$$

(e) Show the number of photons with $(\mathbf{k}\lambda)$ in the true ground state is

$$\langle \underline{0} | a_{\mathbf{k}\lambda}^\dagger a_{\mathbf{k}\lambda} | \underline{0} \rangle = \frac{e^2 c^2}{\varepsilon_0} \frac{|j(\mathbf{k}\lambda)|^2}{2\hbar \omega_k^3}$$

Hence conclude that the result in part (d) can be written

$$|\langle 0 | \underline{0} \rangle|^2 = e^{-N}$$

(f) Convert $\sum_{\mathbf{k}} \to \Omega(2\pi)^{-3} \int d^3 k$, and show that if $|j(\mathbf{0}\lambda)|^2 \neq 0$, then the resulting integral in part (d) *diverges* at long wavelength. Hence conclude that there are an *infinite number* of long-wavelength photons in the true ground state. Show that the *energy* carried by these photons is finite.[8]

Solution to Problem 11.2

(a) The hamiltonian for the quantized radiation field interacting with a specified, time-independent, classical external current $j_\mu^{\text{ext}}(\mathbf{x}) = [\mathbf{j}^{\text{ext}}(\mathbf{x})/c, \, i\rho^{\text{ext}}(\mathbf{x})]$ is derived in Eq. (C.73)[9]

$$\hat{H} = \hat{H}_0 - e \int_\Omega d^3 x \, \mathbf{j}^{\text{ext}}(\mathbf{x}) \cdot \hat{\mathbf{A}}(\mathbf{x})$$

$$+ \frac{e^2}{8\pi\varepsilon_0} \int_\Omega d^3 x \int_\Omega d^3 x' \frac{\rho^{\text{ext}}(\mathbf{x})\rho^{\text{ext}}(\mathbf{x}')}{|\mathbf{x} - \mathbf{x}'|}$$

$$\hat{H}_0 = \sum_{\mathbf{k}} \sum_{\lambda = \pm 1} \hbar \omega_k \left(a_{\mathbf{k}\lambda}^\dagger a_{\mathbf{k}\lambda} + \frac{1}{2} \right)$$

---

[8] As discussed in the text, this *infrared divergence* is usually handled by giving the photon a tiny mass M_γ, so that $\omega_k/c = \sqrt{k^2 + M_\gamma^2}$, and then taking the limit $M_\gamma \to 0$ at the end of the calculation of any physical quantity (the energy, for example).

[9] Recall that here we work in SI units, in the Coulomb gauge.

We work in the Schrödinger picture, where the radiation field is given by the time-independent expression

$$\hat{\mathbf{A}}(\mathbf{x}) = \sum_{\mathbf{k}} \sum_{\lambda=\pm 1} \left(\frac{\hbar}{2\omega_k \varepsilon_0 \Omega} \right)^{1/2} \left[a_{\mathbf{k}\lambda} \mathbf{e}_{\mathbf{k}\lambda} e^{i\mathbf{k}\cdot\mathbf{x}} + a^\dagger_{\mathbf{k}s\lambda} \mathbf{e}^\dagger_{\mathbf{k}\lambda} e^{-i\mathbf{k}\cdot\mathbf{x}} \right]$$

Here we employ the helicity states of Eqs. (A.22). The creation and destruction operators satisfy the familiar commutation relations of Eqs. (A.21)

$$[a_{\mathbf{k}\lambda}, a^\dagger_{\mathbf{k}'\lambda'}] = \delta_{\mathbf{k}\mathbf{k}'}\delta_{\lambda\lambda'}$$

Make a Fourier series expansion of the external current

$$\mathbf{j}^{\text{ext}}(\mathbf{x}) = \frac{c}{\sqrt{\Omega}} \sum_{\mathbf{q}} \tilde{\mathbf{j}}(\mathbf{q}) e^{-i\mathbf{q}\cdot\mathbf{x}}$$

$$\tilde{\mathbf{j}}(-\mathbf{q}) = \tilde{\mathbf{j}}^\dagger(\mathbf{q})$$

where the second relation follows from the reality of $\mathbf{j}^{\text{ext}}(\mathbf{x})$. Also, define

$$\mathbf{e}_{\mathbf{k}\lambda} \cdot \tilde{\mathbf{j}}(\mathbf{k}) \equiv j(\mathbf{k}\lambda)$$

The introduction of the expansions of the vector potential and the current then allow the hamiltonian to be re-written as[10]

$$\hat{H} = \hat{H}_0 - e \sum_{\mathbf{k}} \sum_{\lambda=\pm 1} \left(\frac{\hbar c^2}{2\omega_k \varepsilon_0} \right)^{1/2} \left[j(\mathbf{k}\lambda) a_{\mathbf{k}\lambda} + j^\dagger(\mathbf{k}\lambda) a^\dagger_{\mathbf{k}\lambda} \right]$$

$$+ \frac{e^2}{8\pi\varepsilon_0} \int d^3x \int d^3x' \frac{\rho^{\text{ext}}(\mathbf{x})\rho^{\text{ext}}(\mathbf{x}')}{|\mathbf{x} - \mathbf{x}'|}$$

(b) The hamiltonian is now a quadratic form in the creation and destruction operators, and it is diagonalized with the following transformation

$$A_{\mathbf{k}\lambda} \equiv a_{\mathbf{k}\lambda} - \left(\frac{c^2}{2\hbar\omega_k^3 \varepsilon_0} \right)^{1/2} e j^\dagger(\mathbf{k}\lambda)$$

$$A^\dagger_{\mathbf{k}\lambda} \equiv a^\dagger_{\mathbf{k}\lambda} - \left(\frac{c^2}{2\hbar\omega_k^3 \varepsilon_0} \right)^{1/2} e j(\mathbf{k}\lambda)$$

The transformation is canonical, since the new creation and destruction operators satisfy the same commutation relations

$$[A_{\mathbf{k}\lambda}, A^\dagger_{\mathbf{k}'\lambda'}] = \delta_{\mathbf{k}\mathbf{k}'}\delta_{\lambda\lambda'}$$

---

[10] Use $\Omega^{-1} \int_\Omega d^3x \, e^{i(\mathbf{k}-\mathbf{q})\cdot\mathbf{x}} = \delta_{\mathbf{k},\mathbf{q}}$.

As a consequence, we again know all their properties.[11] The transformed hamiltonian takes the form

$$\hat{H} = \sum_{\mathbf{k}} \sum_{\lambda=\pm 1} \hbar\omega_k \left(A_{\mathbf{k}\lambda}^{\dagger} A_{\mathbf{k}\lambda} + \frac{1}{2} \right) + H'$$

$$H' = -e^2 \sum_{\mathbf{k}} \sum_{\lambda=\pm 1} \left(\frac{c^2}{2\omega_k^2 \varepsilon_0} \right) |j(\mathbf{k}\lambda)|^2 + \frac{e^2}{8\pi\varepsilon_0} \int_{\Omega} d^3x \int_{\Omega} d^3x' \frac{\rho^{\text{ext}}(\mathbf{x})\rho^{\text{ext}}(\mathbf{x}')}{|\mathbf{x} - \mathbf{x}'|}$$

H' is a c-number, and since $A_{\mathbf{k}\lambda}^{\dagger} A_{\mathbf{k}\lambda}$ is again the number operator, with a spectrum of the positive integers and zero, the hamiltonian has been diagonalized.

(c) The ground state of the interacting system is defined by

$$A_{\mathbf{k}\lambda}|\underline{0}\rangle = 0 \qquad ; \text{ all } (\mathbf{k}, \lambda)$$

Introduce the normalized multi-photon state

$$|n_1 n_2 \cdots\rangle = \frac{1}{\sqrt{n_1! n_2! \cdots}} (a_1^{\dagger})^{n_1} (a_2^{\dagger})^{n_2} \cdots |0\rangle$$

The overlap of this state with the new vacuum is given by

$$\langle n_1 n_2 \cdots |\underline{0}\rangle = \frac{1}{\sqrt{n_1! n_2! \cdots}} \langle 0|(a_1)^{n_1} (a_2)^{n_2} \cdots |\underline{0}\rangle$$

Now substitute the transformed operator from part (b)

$$a_{\mathbf{k}\lambda} \equiv A_{\mathbf{k}\lambda} + \left(\frac{c^2}{2\hbar\omega_k^3 \varepsilon_0} \right)^{1/2} ej^{\dagger}(\mathbf{k}\lambda)$$

Make use of the above relation defining $|\underline{0}\rangle$. All the $A_{\mathbf{k}\lambda}$ annihilate the new vacuum, and hence they disappear in the overlap. Thus

$$\langle n_1 n_2 \cdots |\underline{0}\rangle = \frac{1}{\sqrt{n_1! n_2! \cdots}} \left[e \left(\frac{c^2}{2\hbar\omega_1^3 \varepsilon_0} \right)^{1/2} j^{\dagger}(\mathbf{k}_1\lambda_1) \right]^{n_1} \times$$

$$\left[e \left(\frac{c^2}{2\hbar\omega_2^3 \varepsilon_0} \right)^{1/2} j^{\dagger}(\mathbf{k}_2\lambda_2) \right]^{n_2} \cdots \langle 0|\underline{0}\rangle$$

(d) The completeness relation states that

$$\sum_{n_1 n_2 \cdots} |n_1 n_2 \cdots\rangle\langle n_1 n_2 \cdots| = \hat{1}$$

---

[11] It is evident the transformation preserves $[A_{\mathbf{k}\lambda}, A_{\mathbf{k}'\lambda'}] = [A_{\mathbf{k}\lambda}^{\dagger}, A_{\mathbf{k}'\lambda'}^{\dagger}] = 0$.

Hence the normalization condition on $|\underline{0}\rangle$ can be written as

$$\sum_{n_1 n_2 \cdots} |\langle n_1 n_2 \cdots |\underline{0}\rangle|^2 = \langle \underline{0}|\underline{0}\rangle = 1$$

Substitution of the above amplitudes, and identification of the exponential for each mode, gives

$$\exp\left\{\frac{e^2 c^2}{\varepsilon_0} \sum_{\mathbf{k},\lambda} \frac{|j(\mathbf{k}\lambda)|^2}{2\hbar\omega_k^3}\right\} |\langle 0|\underline{0}\rangle|^2 = 1$$

Therefore, the overlap of the new and the old vacuua is given by

$$|\langle 0|\underline{0}\rangle|^2 = \exp\left\{-\frac{e^2 c^2}{\varepsilon_0} \sum_{\mathbf{k},\lambda} \frac{|j(\mathbf{k}\lambda)|^2}{2\hbar\omega_k^3}\right\}$$

(e) Let us compute the number of photons $n_{\mathbf{k}\lambda}$ in the new ground state

$$n_{\mathbf{k}\lambda} = \langle \underline{0}|a_{\mathbf{k}\lambda}^\dagger a_{\mathbf{k}\lambda}|\underline{0}\rangle$$

Expressed in terms of the new operators, this is immediately evaluated as

$$n_{\mathbf{k}\lambda} = \langle \underline{0}| \left[A_{\mathbf{k}\lambda}^\dagger + \left(\frac{c^2}{2\hbar\omega_k^3 \varepsilon_0}\right)^{1/2} e j(\mathbf{k}\lambda) \right] \left[A_{\mathbf{k}\lambda} + \left(\frac{c^2}{2\hbar\omega_k^3 \varepsilon_0}\right)^{1/2} e j^\dagger(\mathbf{k}\lambda) \right] |\underline{0}\rangle$$

$$= \frac{e^2 c^2}{\varepsilon_0} \frac{|j(\mathbf{k}\lambda)|^2}{2\hbar\omega_k^3}$$

The total number of photons in the new ground state is[12]

$$N = \sum_{\mathbf{k},\lambda} n_{\mathbf{k}\lambda} = \frac{e^2 c^2}{\varepsilon_0} \sum_{\mathbf{k},\lambda} \frac{|j(\mathbf{k}\lambda)|^2}{2\hbar\omega_k^3}$$

Hence the result in (d) can be written as

$$|\langle 0|\underline{0}\rangle|^2 = e^{-N}$$

(f) As $\Omega \to \infty$, the sum over modes can be converted to an integral in the familiar manner

$$\sum_{\mathbf{k}} \to \frac{\Omega}{(2\pi)^3} \int d^3k \qquad ; \Omega \to \infty$$

---

[12] Note that $\hat{H}$ and $\hat{N}$ do not commute in this problem, and there is a non-vanishing mean-square-deviation in the photon number.

In this limit, N takes the form

$$N = \frac{e^2 c^2}{\varepsilon_0} \frac{\Omega}{(2\pi)^3} \sum_\lambda \int d^3k \, \frac{|j(\mathbf{k}\lambda)|^2}{2\hbar\omega_k^3}$$

$$= \frac{e^2}{\hbar c \varepsilon_0} \frac{\Omega}{2(2\pi)^3} \sum_\lambda \int \frac{dk}{k} \int d\Omega_k \, |j(\mathbf{k}\lambda)|^2$$

where we have used $\omega_k = kc$. If $|j(\mathbf{0}\lambda)|^2 \neq 0$, then the resulting integral $\int dk/k$ diverges as $k \to 0$.[13] Hence we conclude that there are an *infinite number* of long-wavelength photons in the true ground state.[14] In contrast, the *energy* carried by these long-wavelength photons exhibits no such divergence

$$E = \sum_{\mathbf{k}\lambda} \hbar\omega_k n_{\mathbf{k}\lambda}$$

$$= \frac{e^2}{\varepsilon_0} \frac{\Omega}{2(2\pi)^3} \sum_\lambda \int dk \int d\Omega_k \, |j(\mathbf{k}\lambda)|^2$$

Problem 11.3 Verify the boson matrix element in Eq. (11.3).

Solution to Problem 11.3

The boson interaction is given in Eq. (11.1)

$$\hat{V} = \frac{G}{2\Omega} \sum_{\mathbf{k}_1} \cdots \sum_{\mathbf{k}_4} \delta_{\mathbf{k}_1+\mathbf{k}_2, \mathbf{k}_3+\mathbf{k}_4} \, a_{\mathbf{k}_3}^\dagger a_{\mathbf{k}_4}^\dagger a_{\mathbf{k}_2} a_{\mathbf{k}_1}$$

The initial and final states under consideration are (compare Fig. 8.5 in the text)

$$|i\rangle = |\mathbf{k}, -\mathbf{k}\rangle \qquad ; \; |f\rangle = |\mathbf{k}', -\mathbf{k}'\rangle$$

The contributing operators in $\hat{V}$ are then

$$\hat{V} \doteq \frac{G}{2\Omega} \left(a_{\mathbf{k}'}^\dagger a_{-\mathbf{k}'}^\dagger + a_{-\mathbf{k}'}^\dagger a_{\mathbf{k}'}^\dagger \right) (a_{\mathbf{k}} a_{-\mathbf{k}} + a_{-\mathbf{k}} a_{\mathbf{k}})$$

$$= \frac{4G}{2\Omega} a_{\mathbf{k}'}^\dagger a_{-\mathbf{k}'}^\dagger a_{\mathbf{k}} a_{-\mathbf{k}}$$

---

[13] This is the *infrared* divergence of QED.

[14] Note also that $N \to \infty$ as the volume $\Omega \to \infty$ (it is the photon density N/Ω that is then meaningful), and hence the overlap $|\langle 0|\underline{0}\rangle|^2 = e^{-N}$ *vanishes* in the infinite volume limit!

The matrix element is readily evaluated as

$$\langle \mathbf{k}', -\mathbf{k}' | a_{\mathbf{k}'}^\dagger a_{-\mathbf{k}'}^\dagger a_{\mathbf{k}} a_{-\mathbf{k}} | \mathbf{k}, -\mathbf{k} \rangle = \langle 0 | a_{-\mathbf{k}'} a_{\mathbf{k}'} a_{\mathbf{k}'}^\dagger a_{-\mathbf{k}'}^\dagger a_{\mathbf{k}} a_{-\mathbf{k}} a_{\mathbf{k}}^\dagger a_{-\mathbf{k}}^\dagger | 0 \rangle$$
$$= \langle 0 | a_{\mathbf{k}'} a_{\mathbf{k}'}^\dagger a_{-\mathbf{k}'} a_{-\mathbf{k}'}^\dagger a_{\mathbf{k}} a_{\mathbf{k}}^\dagger a_{-\mathbf{k}} a_{-\mathbf{k}}^\dagger | 0 \rangle$$
$$= 1$$

This produces the stated answer[15]

$$\langle \mathbf{k}', -\mathbf{k}' | \hat{V} | \mathbf{k}, -\mathbf{k} \rangle = \frac{4G}{2\Omega}$$

Problem 11.4 Consider the scattering problem of two identical spin-zero bosons in the C-M system. The wave function must be symmetric under particle interchange, or under $\mathbf{r} \to -\mathbf{r}$ where $\mathbf{r}$ is the relative coordinate.

(a) Assume the prepared initial state has the form[16] $\psi_{\mathbf{k}}(\mathbf{r}) = e^{i\mathbf{k}\cdot\mathbf{r}} + e^{-i\mathbf{k}\cdot\mathbf{r}}$. Show this wave function has vanishing flux. Interpret the first term as representing an incident beam moving with relative momentum $\hbar\mathbf{k}$ and the second term as representing a beam moving with $-\hbar\mathbf{k}$. Argue that the appropriate incident flux can be calculated from the first term just as if the particles were distinguishable;

(b) In the scattering region, one must employ the full wave function $\psi_{\mathbf{k}}(\mathbf{r})$. Argue from Prob. 4.10 that the first term will give rise to a scattered wave $f(k, \theta)e^{ikr}/r$ while the second term gives rise to a similar scattered wave, but with $\mathbf{r} \to -\mathbf{r}$, or $f(k, \pi - \theta)e^{ikr}/r$;

(c) Hence conclude that the scattering amplitude for this problem is $f = f(k, \theta) + f(k, \pi - \theta)$, where $f(k, \theta)$ is calculated as if the particles were distinguishable.

Solution to Problem 11.4

(a) We assume that the initial state is described by the wave function

$$\psi_{\mathbf{k}}(\mathbf{r}) = e^{i\mathbf{k}\cdot\mathbf{r}} + e^{-i\mathbf{k}\cdot\mathbf{r}} = \psi_{\mathbf{k}}^\star(\mathbf{r})$$

---

[15]Readers can convince themselves that, although there is then only a single term in $\hat{V}$, since $a_0 a_0 a_0^\dagger a_0^\dagger | 0 \rangle = 2|0\rangle$, this result continues to hold if $\mathbf{k}$ (and/or $\mathbf{k}'$) vanishes.

[16]Since the scattering cross section is the ratio of fluxes, the overall normalization is irrelevant.

The probability current is then given by

$$\mathbf{S} = \frac{\hbar}{2im} \left\{ \psi_{\mathbf{k}}^*(\mathbf{r}) \nabla \psi_{\mathbf{k}}(\mathbf{r}) - [\nabla \psi_{\mathbf{k}}(\mathbf{r})]^* \psi_{\mathbf{k}}(\mathbf{r}) \right\}$$
$$= \frac{\hbar}{2im} \left\{ \psi_{\mathbf{k}}(\mathbf{r}) \nabla \psi_{\mathbf{k}}(\mathbf{r}) - [\nabla \psi_{\mathbf{k}}(\mathbf{r})] \psi_{\mathbf{k}}(\mathbf{r}) \right\} = 0$$

where the last line has been obtained exploiting the reality of $\psi_{\mathbf{k}}(\mathbf{r})$. We conclude that the total flux associated with $\psi_{\mathbf{k}}(\mathbf{r})$ vanishes, as anticipated.

We now write the wave function as

$$\psi_{\mathbf{k}}(\mathbf{r}) = \psi_{+\mathbf{k}}(\mathbf{r}) + \psi_{-\mathbf{k}}(\mathbf{r})$$

where $\psi_{\pm\mathbf{k}}(\mathbf{r}) \equiv e^{\pm i \mathbf{k} \cdot \mathbf{r}}$ describe a beam moving with momentum $\pm \hbar \mathbf{k}$. The total probability current can then be expressed as

$$\mathbf{S} = \mathbf{S}_+ + \mathbf{S}_- + \mathbf{S}_\pm$$

where the first two terms are the probability current associated with the wave functions $\psi_{+\mathbf{k}}(\mathbf{r})$ and $\psi_{-\mathbf{k}}(\mathbf{r})$, while the last term is the interference between waves with $+\hbar\mathbf{k}$ and $-\hbar\mathbf{k}$, which vanishes, $\mathbf{S}_\pm = 0$.

As a result, the incident flux can be calculated from the first term in the wave function, as if the particles were distinguishable.

(b) As we are dealing with two identical spin-zero bosons, their wave function must be symmetric under particle interchange, or, in the C-M system, under the change $\mathbf{r} \to -\mathbf{r}$.[17] As discussed in part (b) of problem 4.10, the first term in the wave function gives rise to a scattered wave $f(k, \theta) e^{ikr}/r$; we may obtain the scattered wave generated by the second term in the wave function by requiring the symmetry of the total scattered wave function under the change $\mathbf{r} \to -\mathbf{r}$. Under these conditions we have a second scattered wave $f(k, \pi - \theta) e^{ikr}/r$.

(c) The scattered wave generated by the full wave function is thus

$$\psi_{\mathbf{k}}(\mathbf{r}) \to f(k, \theta) \frac{e^{ikr}}{r} + f(k, \pi - \theta) \frac{e^{ikr}}{r} \qquad ; |\mathbf{r}| \to \infty$$

We conclude that the scattering amplitude for the two identical spin-zero bosons is now

$$f = f(k, \theta) + f(k, \pi - \theta)$$

where $f(k, \theta)$ is calculated as if the particles were distinguishable.

---

[17]The scattering solution takes the form $\psi_i^{(+)}(\mathbf{r}) = \psi_{+\mathbf{k}}^{(+)}(\mathbf{r}) + \psi_{-\mathbf{k}}^{(+)}(\mathbf{r})$.

Problem 11.5 Show the required integral in the calculation of the depletion of the ground state for the weakly interacting Bose system is

$$\int_0^\infty x^2\,dx\,\frac{1}{2}\left[\frac{(x^2/2+1)}{[(x^2/2+1)^2-1]^{1/2}}-1\right]=\frac{2}{3}$$

Solution to Problem 11.5

Introduce a cut-off L, and a new variable $y \equiv x^2/4$, with $dy = x\,dx/2$

$$I(L) \equiv \int_0^L x^2\,dx\,\frac{1}{2}\left[\frac{(x^2/2+1)}{[(x^2/2+1)^2-1]^{1/2}}-1\right]$$

$$= \int_0^L x^2\,dx\,\frac{1}{2}\frac{(x^2/2+1)}{(x^4/4+x^2)^{1/2}}-\frac{L^3}{6}$$

$$= \int_0^{L^2/4} dy\,\frac{2y+1}{\sqrt{y+1}}-\frac{L^3}{6}$$

The first integral gives

$$\int_0^{L^2/4} dy\,\frac{2y+1}{\sqrt{y+1}} = \int_0^{L^2/4} dy\left[2\sqrt{y+1}-\frac{1}{\sqrt{y+1}}\right]$$

$$= \left[\frac{4}{3}(y+1)^{3/2}-2(y+1)^{1/2}\right]_0^{L^2/4}$$

$$= \frac{L^3}{6}+\frac{2}{3}+O\left(\frac{1}{L}\right)$$

Thus the desired integral is

$$I(L) = \frac{2}{3}+O\left(\frac{1}{L}\right)$$

Now let the cut-off $L \to \infty$

$$I(\infty) = \int_0^\infty x^2\,dx\,\frac{1}{2}\left[\frac{(x^2/2+1)}{[(x^2/2+1)^2-1]^{1/2}}-1\right]=\frac{2}{3}$$

Problem 11.6 (a) Show that if the two-body potential is of the form $V(|\mathbf{x}-\mathbf{y}|)$, then $\langle \mathbf{k}_1\mathbf{k}_2|V|\mathbf{k}_3\mathbf{k}_4\rangle \propto \delta_{\mathbf{k}_1+\mathbf{k}_2,\mathbf{k}_3+\mathbf{k}_4}$ in Eqs. (11.1) and (11.57);

(b) Use a change of dummy summation variables and the symmetry properties of the potential to show that in Eq. (11.65)

$$\sum_{\mathbf{k}} \sum_{\mathbf{k'}} [\langle \mathbf{kk'}|V|\mathbf{kk'}\rangle - \langle \mathbf{kk'}|V|\mathbf{k'k}\rangle]\, v_{k'}^2 :a_{\mathbf{k}\downarrow}^\dagger a_{\mathbf{k}\downarrow}: =$$

$$\sum_{\mathbf{k}} \sum_{\mathbf{k'}} [\langle \mathbf{kk'}|V|\mathbf{kk'}\rangle - \langle \mathbf{kk'}|V|\mathbf{k'k}\rangle]\, v_{k'}^2 :a_{-\mathbf{k}\downarrow}^\dagger a_{-\mathbf{k}\downarrow}:$$

Solution to Problem 11.6

(a) The matrix element of the potential is given in Eq. (11.57) as

$$\langle \mathbf{k_1 k_2}|V|\mathbf{k_3 k_4}\rangle = \frac{1}{\Omega^2}\int d^3x \int d^3y\, e^{-i\mathbf{k_1}\cdot\mathbf{x}} e^{-i\mathbf{k_2}\cdot\mathbf{y}}\, V(|\mathbf{x}-\mathbf{y}|)\, e^{i\mathbf{k_3}\cdot\mathbf{x}} e^{i\mathbf{k_4}\cdot\mathbf{y}}$$

Change variables to C-M and relative coordinates

$$\mathbf{R} = \frac{1}{2}(\mathbf{x}+\mathbf{y}) \qquad ;\, \mathbf{r} = \mathbf{x}-\mathbf{y}$$

The jacobian for this transformation is unity.[18] Thus

$$\langle \mathbf{k_1 k_2}|V|\mathbf{k_3 k_4}\rangle = \frac{1}{\Omega^2}\int_{\text{Box}} d^3R\, e^{i(\mathbf{k_3}+\mathbf{k_4}-\mathbf{k_1}-\mathbf{k_2})\cdot\mathbf{R}} \times$$
$$\int d^3r\, V(r)\, e^{i(\mathbf{k_3}-\mathbf{k_1}+\mathbf{k_2}-\mathbf{k_4})\cdot\mathbf{r}/2}$$

The integral over the C-M coordinate then gives

$$\frac{1}{\Omega}\int_{\text{Box}} d^3R\, e^{i(\mathbf{k_3}+\mathbf{k_4}-\mathbf{k_1}-\mathbf{k_2})\cdot\mathbf{R}} = \delta_{\mathbf{k_1}+\mathbf{k_2},\, \mathbf{k_3}+\mathbf{k_4}}$$

Therefore

$$\langle \mathbf{k_1 k_2}|V|\mathbf{k_3 k_4}\rangle = \delta_{\mathbf{k_1}+\mathbf{k_2},\, \mathbf{k_3}+\mathbf{k_4}}\, \frac{1}{\Omega}\int d^3r\, V(r)\, e^{i\mathbf{q}\cdot\mathbf{r}} \qquad ;\, \mathbf{q} \equiv \mathbf{k_3} - \mathbf{k_1}$$

(b) Change dummy summation variables $(\mathbf{k}, \mathbf{k'}) \rightarrow (-\mathbf{k}, -\mathbf{k'})$. Then

$$\sum_{\mathbf{k}} \sum_{\mathbf{k'}} [\langle \mathbf{kk'}|V|\mathbf{kk'}\rangle - \langle \mathbf{kk'}|V|\mathbf{k'k}\rangle]\, v_{k'}^2 :a_{\mathbf{k}\downarrow}^\dagger a_{\mathbf{k}\downarrow}: =$$

$$\sum_{\mathbf{k}} \sum_{\mathbf{k'}} [\langle -\mathbf{k}, -\mathbf{k'}|V|-\mathbf{k}, -\mathbf{k'}\rangle - \langle -\mathbf{k}, -\mathbf{k'}|V|-\mathbf{k'}, -\mathbf{k}\rangle]\, v_{k'}^2 :a_{-\mathbf{k}\downarrow}^\dagger a_{-\mathbf{k}\downarrow}:$$

---

[18]See, for example, the solution to Prob. A.1 in [Amore and Walecka (2014)].

Now use Eqs. (11.57)

$$\langle \mathbf{k}_1\mathbf{k}_2|V|\mathbf{k}_3\mathbf{k}_4 \rangle = \langle \mathbf{k}_2\mathbf{k}_1|V|\mathbf{k}_4\mathbf{k}_3 \rangle = \langle -\mathbf{k}_3, -\mathbf{k}_4|V| -\mathbf{k}_1, -\mathbf{k}_2 \rangle$$

This reduces the r.h.s. of the above to

$$\text{r.h.s.} = \sum_{\mathbf{k}} \sum_{\mathbf{k}'} \left[\langle \mathbf{k}\mathbf{k}'|V|\mathbf{k}\mathbf{k}' \rangle - \langle \mathbf{k}\mathbf{k}'|V|\mathbf{k}'\mathbf{k} \rangle \right] v_{k'}^2 : a^\dagger_{-\mathbf{k}\downarrow} a_{-\mathbf{k}\downarrow} :$$

which is the stated result.

Problem 11.7 Verify Eqs. (11.69).

Solution to Problem 11.7

Consider the thermodynamic potential at zero temperature given in Eq. (11.56)

$$\hat{K} = \hat{H} - \mu \hat{N}$$
$$= \sum_{\mathbf{k},\lambda} a^\dagger_{\mathbf{k}\lambda} a_{\mathbf{k}\lambda} \left(\epsilon_k^0 - \mu \right) - \frac{1}{2} \sum_{\mathbf{k}_1,\lambda_1} \cdots \sum_{\mathbf{k}_4,\lambda_4} \langle \mathbf{k}_1\lambda_1\mathbf{k}_2\lambda_2|V|\mathbf{k}_3\lambda_3\mathbf{k}_4\lambda_4 \rangle \times$$
$$a^\dagger_{\mathbf{k}_1\lambda_1} a^\dagger_{\mathbf{k}_2\lambda_2} a_{\mathbf{k}_4\lambda_4} a_{\mathbf{k}_3\lambda_3}$$

We will now perform the Bogoliubov-Valatin canonical transformation in Eqs. (11.49)

$$\alpha_{\mathbf{k}} \equiv u_k a_{\mathbf{k}\uparrow} - v_k a^\dagger_{-\mathbf{k}\downarrow} \qquad ; \ u_k^2 + v_k^2 = 1$$
$$\beta_{-\mathbf{k}} \equiv u_k a_{-\mathbf{k}\downarrow} + v_k a^\dagger_{\mathbf{k}\uparrow}$$

where u_k and v_k are real, depend only on $|\mathbf{k}|$, and $u_k^2 + v_k^2 = 1$. The inverse transformation is given in Eqs. (11.50)–(11.51)

$$a_{\mathbf{k}\uparrow} = u_k \alpha_{\mathbf{k}} + v_k \beta^\dagger_{-\mathbf{k}} \qquad ; \ a^\dagger_{\mathbf{k}\uparrow} = u_k \alpha^\dagger_{\mathbf{k}} + v_k \beta_{-\mathbf{k}}$$
$$a_{-\mathbf{k}\downarrow} = u_k \beta_{-\mathbf{k}} - v_k \alpha^\dagger_{\mathbf{k}} \qquad ; \ a^\dagger_{-\mathbf{k}\downarrow} = u_k \beta^\dagger_{-\mathbf{k}} - v_k \alpha_{\mathbf{k}}$$

The goal is to express the thermodynamic potential in terms of the new operators. The expression for the one-body term is worked out explicitly in Eq. (11.63), so we will focus on the two-body term in $\hat{K}$, reported in Eq. (11.64), which reads

$$-\frac{1}{2} \sum_{\mathbf{k}_1} \cdots \sum_{\mathbf{k}_4} \langle \mathbf{k}_1\mathbf{k}_2|V|\mathbf{k}_3\mathbf{k}_4 \rangle \left[a^\dagger_{\mathbf{k}_1\uparrow} a^\dagger_{\mathbf{k}_2\uparrow} a_{\mathbf{k}_4\uparrow} a_{\mathbf{k}_3\uparrow} + \right.$$
$$\left. a^\dagger_{\mathbf{k}_1\downarrow} a^\dagger_{\mathbf{k}_2\downarrow} a_{\mathbf{k}_4\downarrow} a_{\mathbf{k}_3\downarrow} + a^\dagger_{\mathbf{k}_1\uparrow} a^\dagger_{\mathbf{k}_2\downarrow} a_{\mathbf{k}_4\downarrow} a_{\mathbf{k}_3\uparrow} + a^\dagger_{\mathbf{k}_1\downarrow} a^\dagger_{\mathbf{k}_2\uparrow} a_{\mathbf{k}_4\uparrow} a_{\mathbf{k}_3\downarrow} \right] \equiv \hat{V}_a + \hat{V}_b$$

Here $\hat{V}_a$ corresponds to the first two terms and $\hat{V}_b$ to the last two terms.[19]

We now apply Wick's theorem to the operators in $\hat{V}_a$, following section 11.2.2.3. For example,

$$a^\dagger_{\mathbf{k}_1\uparrow}a^\dagger_{\mathbf{k}_2\uparrow}a_{\mathbf{k}_4\uparrow}a_{\mathbf{k}_3\uparrow} = (A) + (B) + (C)$$

where (A), (B), and (C) contain the normal-ordered contributions with zero, one, and two contractions respectively. Explicitly, (A) reads

$$(A) = \;:a^\dagger_{\mathbf{k}_1\uparrow}a^\dagger_{\mathbf{k}_2\uparrow}a_{\mathbf{k}_4\uparrow}a_{\mathbf{k}_3\uparrow}:$$

Similarly, we have[20]

$$(B) = \;:a^{\dagger\boldsymbol{\cdot}}_{\mathbf{k}_1\uparrow}a^\dagger_{\mathbf{k}_2\uparrow}a^{\boldsymbol{\cdot}}_{\mathbf{k}_4\uparrow}a_{\mathbf{k}_3\uparrow}: + :a^{\dagger\boldsymbol{\cdot}}_{\mathbf{k}_1\uparrow}a^\dagger_{\mathbf{k}_2\uparrow}a_{\mathbf{k}_4\uparrow}a^{\boldsymbol{\cdot}}_{\mathbf{k}_3\uparrow}: +$$
$$:a^\dagger_{\mathbf{k}_1\uparrow}a^{\dagger\boldsymbol{\cdot}}_{\mathbf{k}_2\uparrow}a^{\boldsymbol{\cdot}}_{\mathbf{k}_4\uparrow}a_{\mathbf{k}_3\uparrow}: + :a^\dagger_{\mathbf{k}_1\uparrow}a^{\dagger\boldsymbol{\cdot}}_{\mathbf{k}_2\uparrow}a_{\mathbf{k}_4\uparrow}a^{\boldsymbol{\cdot}}_{\mathbf{k}_3\uparrow}:$$
$$= -\delta_{\mathbf{k}_1\mathbf{k}_4}v^2_{k_1}:a^\dagger_{\mathbf{k}_2\uparrow}a_{\mathbf{k}_3\uparrow}: + \delta_{\mathbf{k}_1\mathbf{k}_3}v^2_{k_1}:a^\dagger_{\mathbf{k}_2\uparrow}a_{\mathbf{k}_4\uparrow}:$$
$$+\delta_{\mathbf{k}_2\mathbf{k}_4}v^2_{k_2}:a^\dagger_{\mathbf{k}_1\uparrow}a_{\mathbf{k}_3\uparrow}: - \delta_{\mathbf{k}_2\mathbf{k}_3}v^2_{k_2}:a^\dagger_{\mathbf{k}_1\uparrow}a_{\mathbf{k}_4\uparrow}:$$

and

$$(C) = \;:a^{\dagger\boldsymbol{\cdot}}_{\mathbf{k}_1\uparrow}a^{\dagger\boldsymbol{\cdot\cdot}}_{\mathbf{k}_2\uparrow}a^{\boldsymbol{\cdot}}_{\mathbf{k}_4\uparrow}a^{\boldsymbol{\cdot\cdot}}_{\mathbf{k}_3\uparrow}: + :a^{\dagger\boldsymbol{\cdot}}_{\mathbf{k}_1\uparrow}a^{\dagger\boldsymbol{\cdot\cdot}}_{\mathbf{k}_2\uparrow}a^{\boldsymbol{\cdot\cdot}}_{\mathbf{k}_4\uparrow}a^{\boldsymbol{\cdot}}_{\mathbf{k}_3\uparrow}:$$
$$= -\delta_{\mathbf{k}_1\mathbf{k}_4}\delta_{\mathbf{k}_2\mathbf{k}_3}v^2_{k_1}v^2_{k_2} + \delta_{\mathbf{k}_1\mathbf{k}_3}\delta_{\mathbf{k}_2\mathbf{k}_4}v^2_{k_1}v^2_{k_2}$$

An analogous result holds for the second term in $\hat{V}_a$ involving $a^\dagger_{\mathbf{k}_1\downarrow}a^\dagger_{\mathbf{k}_2\downarrow}a_{\mathbf{k}_4\downarrow}a_{\mathbf{k}_3\downarrow}$. We may now substitute these results inside the expression for $\hat{V}_a$, remembering that the translational invariance provides a further Kronecker delta-function $\delta_{\mathbf{k}_1+\mathbf{k}_2,\,\mathbf{k}_3+\mathbf{k}_4}$. The term $\hat{V}_a$ then reads

$$\hat{V}_a = \;:\hat{V}_a: \;- \frac{1}{2}\sum_{\mathbf{k}}\sum_{\mathbf{k}'}\left\{\langle\mathbf{k}\mathbf{k}'|V|\mathbf{k}\mathbf{k}'\rangle - \langle\mathbf{k}\mathbf{k}'|V|\mathbf{k}'\mathbf{k}\rangle\right\} \times$$
$$\left\{\left[v^2_k v^2_{k'} + 2v^2_{k'}\;:a^\dagger_{\mathbf{k}\uparrow}a_{\mathbf{k}\uparrow}:\right] + [\uparrow\leftrightharpoons\downarrow]\right\}$$

This is precisely Eq. (11.65). Notice that in obtaining this result, we have used the properties of the matrix element in Eqs. (11.57).

We now come to the calculation of $\hat{V}_b$. Since the two terms in $\hat{V}_b$ are identical, we just need to work on the first one involving

$$a^\dagger_{\mathbf{k}_1\uparrow}a^\dagger_{\mathbf{k}_2\downarrow}a_{\mathbf{k}_4\downarrow}a_{\mathbf{k}_3\uparrow} = (A') + (B') + (C')$$

---

[19]Notice that a (negative) spin-independent interaction is assumed, and therefore the matrix elements of the potential are independent of the spin, and the sums on the spin provide only 4 of the $2^4 = 16$ possible terms.

[20]The non-zero contractions are given in Eqs. (11.62).

where, as before, (A'), (B'), and (C') contain zero, one, and two contractions respectively.

Explicitly (A') reads

$$(A') = \, : a^\dagger_{k_1\uparrow} a^\dagger_{k_2\downarrow} a_{k_4\downarrow} a_{k_3\uparrow} :$$

Similarly

$$(B') = \, : a^{\dagger\,\centerdot}_{k_1\uparrow} a^{\dagger\,\centerdot}_{k_2\downarrow} a_{k_4\downarrow} a_{k_3\uparrow} : \; + \; : a^{\dagger\,\centerdot}_{k_1\uparrow} a^\dagger_{k_2\downarrow} a_{k_4\downarrow} a^{\centerdot}_{k_3\uparrow} : \; +$$

$$: a^\dagger_{k_1\uparrow} a^{\dagger\,\centerdot}_{k_2\downarrow} a^{\centerdot}_{k_4\downarrow} a_{k_3\uparrow} : \; + \; : a^\dagger_{k_1\uparrow} a^\dagger_{k_2\downarrow} a^{\centerdot}_{k_4\downarrow} a^{\centerdot}_{k_3\uparrow} :$$

$$= \delta_{k_1,\,-k_2} u_{k_1} v_{k_1} : a_{k_4\downarrow} a_{k_3\uparrow} : \; + \; \delta_{k_1 k_3} v^2_{k_1} : a^\dagger_{k_2\downarrow} a_{k_4\downarrow} : \; +$$

$$\delta_{k_2 k_4} v^2_{k_2} : a^\dagger_{k_1\uparrow} a_{k_3\uparrow} : \; + \; \delta_{k_4,\,-k_3} u_{k_3} v_{k_3} : a^\dagger_{k_1\uparrow} a^\dagger_{k_2\downarrow} :$$

and

$$(C') = \, : a^{\dagger\,\centerdot}_{k_1\uparrow} a^{\dagger\,\centerdot\centerdot}_{k_2\downarrow} a^{\centerdot\centerdot}_{k_4\downarrow} a^{\centerdot}_{k_3\uparrow} : \; + \; : a^{\dagger\,\centerdot}_{k_1\uparrow} a^{\dagger\,\centerdot\centerdot}_{k_2\downarrow} a^{\centerdot\centerdot}_{k_4\downarrow} a^{\centerdot}_{k_3\uparrow} :$$

$$= \delta_{k_1,\,-k_2} \delta_{k_4,\,-k_3} u_{k_1} v_{k_1} u_{k_3} v_{k_3} + \delta_{k_1 k_3} \delta_{k_2 k_4} v^2_{k_1} v^2_{k_2}$$

After substituting these results in the expression for $\hat{V}_b$, using Eqs. (11.57), and suitably re-labeling the integration variables, we obtain

$$\hat{V}_b = \, : \hat{V}_b : \; - \sum_k \sum_{k'} \langle k, -k' | V | k, -k' \rangle v^2_{k'} \left[v^2_k + : a^\dagger_{k\uparrow} a_{k\uparrow} : + : a^\dagger_{k\downarrow} a_{k\downarrow} : \right] -$$

$$\sum_k \sum_{k'} \langle k, -k | V | k', -k' \rangle u_{k'} v_{k'} \left[u_k v_k + : a^\dagger_{k\uparrow} a^\dagger_{-k\downarrow} : + : a_{-k\downarrow} a_{k\uparrow} : \right]$$

which is precisely Eq. (11.66).

Now notice that the two-body term in $\hat{K}$ still needs to be expressed in terms of the operators α_k and β_k and their conjugates. The operators in $\hat{K}$ contain the combinations $: a^\dagger_{k\uparrow} a_{k\uparrow} + a^\dagger_{-k\downarrow} a_{-k\downarrow} :$ and $: a^\dagger_{k\uparrow} a^\dagger_{-k\downarrow} + a_{-k\downarrow} a_{k\uparrow} :$,[21] whose expressions in terms of the new operators are given in Eqs. (11.63) and (11.67), respectively. The substitution of these relations into $\hat{K}$, using the symmetry properties of the potential, then yields Eqs. (11.69).

Problem 11.8 Obtain Eq. (11.1) from Eqs. (2.113)–(2.114). What assumptions have you made?

---

[21]Recall Prob. 11.6(b).

Solution to Problem 11.8

Use plane-wave single-particle wave functions in a big box with periodic boundary conditions

$$\phi_{\mathbf{k}}(\mathbf{x}) = \frac{1}{\sqrt{\Omega}} e^{i\mathbf{k}\cdot\mathbf{x}} \qquad ; \text{p.b.c.}$$

Now substitute the field expansion in Eq. (2.114) into the potential term in Eq. (2.113), and make use of the result obtained in Prob. 11.6(a)

$$\langle \mathbf{k}_1 \mathbf{k}_2 | V | \mathbf{k}_3 \mathbf{k}_4 \rangle = \delta_{\mathbf{k}_1+\mathbf{k}_2,\,\mathbf{k}_3+\mathbf{k}_4} \frac{1}{\Omega} \int d^3r\, V(r)\, e^{i\mathbf{q}\cdot\mathbf{r}} \qquad ; \mathbf{q} \equiv \mathbf{k}_3 - \mathbf{k}_1$$

$$= \delta_{\mathbf{k}_1+\mathbf{k}_2,\,\mathbf{k}_3+\mathbf{k}_4} \frac{1}{\Omega} \tilde{V}(\mathbf{q})$$

As long as the Fourier transform of the potential $\tilde{V}(\mathbf{q})$ does not vary much over the range of momentum transfers of relevance to the problem, one can replace it by a constant and hence obtain Eq. (11.1). A potential of the form $V(r) \propto \delta^{(3)}(r)$ will obviously accomplish this for us.

Problem 11.9 (a) Given the state $(1/\sqrt{2})(\alpha_{\mathbf{k}}^{\dagger} + \alpha_{-\mathbf{k}}^{\dagger})|0\rangle$ in a Fermi system, compute the expectation values of $\hat{K}_0$ and $\hat{\mathbf{P}}$;
(b) Interpret $\alpha_{\mathbf{k}}^{\dagger}|0\rangle$ in terms of particles and holes.

Solution to Problem 11.9

(a) We work with the state

$$|\Psi_{\mathbf{k}}\rangle \equiv \frac{1}{\sqrt{2}}(\alpha_{\mathbf{k}}^{\dagger} + \alpha_{-\mathbf{k}}^{\dagger})\,|0\rangle$$

The diagonalized part of the zero-temperature thermodynamic potential is given in Eq. (11.78), and it reads

$$\hat{K}_0 = U + \sum_{\mathbf{k}'} E_{\mathbf{k}'} \left(\alpha_{\mathbf{k}'}^{\dagger} \alpha_{\mathbf{k}'} + \beta_{\mathbf{k}'}^{\dagger} \beta_{\mathbf{k}'} \right)$$

where U is c-number, and from Eq. (11.77), $E_k = [\Delta_k^2 + (\varepsilon_k - \mu)^2]^{1/2}$.

The expectation value $\langle \Psi_{\mathbf{k}} | \hat{K}_0 | \Psi_{\mathbf{k}} \rangle$ can be obtained in terms of the following expectation values

$$\eta_{\mathbf{k}}^{(1)} \equiv \langle \Psi_{\mathbf{k}} | 1 | \Psi_{\mathbf{k}} \rangle = 1$$

and

$$\begin{aligned}
\eta^{(2)}_{\mathbf{k},\mathbf{k}'} &\equiv \langle \Psi_{\mathbf{k}} | \alpha^\dagger_{\mathbf{k}'} \alpha_{\mathbf{k}'} + \beta^\dagger_{\mathbf{k}'} \beta_{\mathbf{k}'} | \Psi_{\mathbf{k}} \rangle \\
&= \langle \Psi_{\mathbf{k}} | \alpha^\dagger_{\mathbf{k}'} \alpha_{\mathbf{k}'} | \Psi_{\mathbf{k}} \rangle \\
&= \frac{1}{2} \left[\langle \underline{0} | \alpha_{\mathbf{k}} \alpha^\dagger_{\mathbf{k}'} \alpha_{\mathbf{k}'} \alpha^\dagger_{\mathbf{k}} | \underline{0} \rangle + \langle \underline{0} | \alpha_{-\mathbf{k}} \alpha^\dagger_{\mathbf{k}'} \alpha_{\mathbf{k}'} \alpha^\dagger_{-\mathbf{k}} | \underline{0} \rangle \right] \\
&= \frac{1}{2} \left[\delta_{\mathbf{k}\mathbf{k}'} + \delta_{-\mathbf{k},\,\mathbf{k}'} \right]
\end{aligned}$$

Thus we have

$$\langle \Psi_{\mathbf{k}} | \hat{K}_0 | \Psi_{\mathbf{k}} \rangle = U + E_k$$

Since U is the expectation value of $\hat{K}_0$ in its ground state $|\underline{0}\rangle$, this relation provides the change of the expectation value in the state $|\Psi_{\mathbf{k}}\rangle$

$$\langle \Psi_{\mathbf{k}} | \hat{K}_0 | \Psi_{\mathbf{k}} \rangle - \langle \underline{0} | \hat{K}_0 | \underline{0} \rangle = E_k$$

We now come to the calculation of the expectation value in the state $|\Psi_{\mathbf{k}}\rangle$ of the momentum operator $\hat{\mathbf{P}}$, which is given in Eq. (11.90)

$$\hat{\mathbf{P}} = \sum_{\mathbf{k}'} \hbar \mathbf{k}' \left(\alpha^\dagger_{\mathbf{k}'} \alpha_{\mathbf{k}'} + \beta^\dagger_{\mathbf{k}'} \beta_{\mathbf{k}'} \right)$$

This quantity is given in terms of the $\eta^{(2)}_{\mathbf{k},\mathbf{k}'}$ calculated above, and one has

$$\begin{aligned}
\langle \Psi_{\mathbf{k}} | \hat{\mathbf{P}} | \Psi_{\mathbf{k}} \rangle &= \sum_{\mathbf{k}'} \hbar \mathbf{k}' \, \eta^{(2)}_{\mathbf{k},\mathbf{k}'} \\
&= \frac{1}{2} (\hbar \mathbf{k} - \hbar \mathbf{k}) = 0
\end{aligned}$$

   (b) The paired superconducting state $|\underline{0}\rangle$ is the ground state of $\hat{K}_0$, and it is characterized by an occupation number distribution as represented in Fig. 11.3 in the text. Therefore, when $\alpha^\dagger_{\mathbf{k}}$ acts on this state

$$\alpha^\dagger_{\mathbf{k}} |\underline{0}\rangle = \left(u_k a^\dagger_{\mathbf{k}\uparrow} - v_k a_{-\mathbf{k}\downarrow} \right) |\underline{0}\rangle$$

it produces a new state which is a linear superposition of two states: one state where a fermion with momentum $\hbar\mathbf{k}$ and spin $\uparrow$ has been created in $|\underline{0}\rangle$ (a particle), and a second state where a fermion with momentum $-\hbar\mathbf{k}$ and spin $\downarrow$ has been destroyed in $|\underline{0}\rangle$ (a hole).[22] The state $|\Psi_{\mathbf{k}}\rangle$ then consists of *unpaired* particles and holes, with a thermodynamic potential increased by $\Delta K_0 = E_k$ and a vanishing momentum expectation value, $\mathbf{P} = 0$.

---

[22]Since $\alpha_{\mathbf{k}}|\underline{0}\rangle = 0$, the corresponding hole and particle amplitudes are paired.

Appendix A

Multipole Analysis of the Radiation Field

Problem A.1 The spherical harmonics satisfy $Y_{lm}^*(\theta, \phi) = (-1)^m Y_{l,-m}$ (θ, ϕ), and the spherical basis vector satisfy $\mathbf{e}_{\mathbf{k}\lambda}^\dagger = (-1)^\lambda \mathbf{e}_{\mathbf{k},-\lambda}$. Use these relations to prove Eq. (A.25).

Solution to Problem A.1

The vector spherical harmonics are defined in Eq. (A.8)

$$\mathcal{Y}_{LJ}^M(\Omega_x) \equiv \sum_{M', q'} \langle LM'1q' | L1JM \rangle Y_{LM'}(\Omega_x) \mathbf{e}_{1q'}$$

Take the complex conjugate

$$\mathcal{Y}_{LJ}^M(\Omega_x)^\dagger \equiv \sum_{M', q'} \langle LM'1q' | L1JM \rangle Y_{LM'}^*(\Omega_x) \mathbf{e}_{1q'}^\dagger$$

where the C-G coefficients are real. Now use the behavior of the spherical harmonics and spherical basis vectors given in the statement of the problem

$$\mathcal{Y}_{LJ}^M(\Omega_x)^\dagger \equiv \sum_{M', q'} (-1)^{M'+q'} \langle LM'1q' | L1JM \rangle Y_{L,-M'}(\Omega_x) \mathbf{e}_{1,-q'}$$

The C-G coefficients behave as follows under reflection of the m-values[1]

$$\langle LM'1q' | L1JM \rangle = (-1)^{L+1-J} \langle L, -M'1, -q' | L1J, -M \rangle$$

They also vanish unless $M' + q' = M$. A change of dummy summation variables $(M', q') \to (-M', -q')$ then leads to Eq. (A.25)

$$\mathcal{Y}_{JJ}^M(\Omega_x)^\dagger = (-1)^{1+M} \mathcal{Y}_{JJ}^{-M}(\Omega_x)$$

---

[1]See Prob. 3.7(a). Note Eq. (A.25) has $L = J$.

Problem A.2 Start from the transition amplitude for the photon coming off in an arbitrary direction with respect to the target quantization axis in Eq. (A.36), and reproduce Eq. (A.33) for the total transition rate.

Solution to Problem A.2

Equation (A.36) for the transition amplitude for the photon coming off in an arbitrary direction with respect to the target quantization axis reads

$$\langle \Psi_{J_f M_f}| \int d^3 x\, e^{-i\mathbf{k}\cdot\mathbf{x}}\, \mathbf{e}_{\mathbf{k}\lambda}^\dagger \cdot \hat{\mathbf{J}}(\mathbf{x})\, |\Psi_{J_i M_i}\rangle = -\sum_{J \geq 1}\sum_M \sqrt{2\pi(2J+1)}\,(-i)^J \times$$
$$\langle J_f M_f|\lambda\hat{T}_{JM}^{\mathrm{mag}}(k) + \hat{T}_{JM}^{\mathrm{el}}(k)|J_i M_i\rangle\, \mathcal{D}_{M,-\lambda}^J(-\phi_k, -\theta_k, \phi_k)$$

When integrated over photon directions for a given photon polarization λ in the transition rate, the orthonormality of the rotation matrices can be invoked[2]

$$\int d\Omega_k \mathcal{D}_{M,-\lambda}^J(-\phi_k, -\theta_k, \phi_k)^\star \mathcal{D}_{M',-\lambda}^{J'}(-\phi_k, -\theta_k, \phi_k) = \frac{4\pi}{2J+1}\delta_{JJ'}\delta_{MM'}$$

The Wigner-Eckart theorem gives

$$\langle J_f M_f|\lambda\hat{T}_{JM}^{\mathrm{mag}}(k) + \hat{T}_{JM}^{\mathrm{el}}(k)|J_i M_i\rangle =$$
$$(-1)^{J_f - M_f} \begin{pmatrix} J_f & J & J_i \\ -M_f & M & M_i \end{pmatrix} \langle J_f||\lambda\hat{T}_J^{\mathrm{mag}}(k) + \hat{T}_J^{\mathrm{el}}(k)||J_i\rangle$$

The sums over target orientations in the transition rate can then be evaluated with the aid of the orthonormality of the 3-j symbols in Eq. (A.31)

$$\sum_{M_i}\sum_{M_f} \begin{pmatrix} J_f & J & J_i \\ -M_f & M & M_i \end{pmatrix} \begin{pmatrix} J_f & J' & J_i \\ -M_f & M' & M_i \end{pmatrix} = \frac{1}{2J+1}\delta_{MM'}\delta_{JJ'}$$

A subsequent sum over M yields $1/(2J+1)\sum_M = 1$.

These results then allow one to evaluate the total transition rate in

---

[2]See [Edmonds (1974)].

Eq. (A.24)

$$\mathcal{R}_{fi} = \frac{\alpha \omega_k}{2\pi} \frac{1}{2J_i+1} \sum_{\lambda=\pm 1} \sum_{M_i} \sum_{M_f} \int d\Omega_k \times$$

$$\left| \langle \Psi_{J_f M_f} | \int d^3x \, e^{-i\mathbf{k}\cdot\mathbf{x}} \, \mathbf{e}_{\mathbf{k}\lambda}^{\dagger} \cdot \hat{J}(\mathbf{x}) | \Psi_{J_i M_i} \rangle \right|^2$$

$$= 4\pi\alpha\omega_k \frac{1}{2J_i+1} \sum_{\lambda=\pm 1} \sum_{J\geq 1} |\langle J_f || \lambda \hat{T}_J^{\mathrm{mag}}(k) + \hat{T}_J^{\mathrm{el}}(k) || J_i \rangle|^2$$

The result is the transition rate given in Eq. (A.33)[3]

$$\mathcal{R}_{fi} = 8\pi\alpha\,\omega_k \frac{1}{2J_i+1} \sum_{J\geq 1} \left(\left| \langle J_f || \hat{T}_J^{\mathrm{mag}}(k) || J_i \rangle \right|^2 + \left| \langle J_f || \hat{T}_J^{\mathrm{el}}(k) || J_i \rangle \right|^2 \right)$$

Problem A.3 (a) Verify the commutation relations in Eq. (A.21) for the creation and destruction operators for circularly polarized photons; (b) Verify Eqs. (A.22).

Solution to Problem A.3

(a) The photon creation and destruction operators and spherical basis vectors for circularly polarized photons are defined in Eqs. (A.20)

$$a_{\mathbf{k},\pm 1}^{\dagger} \equiv \mp \frac{1}{\sqrt{2}} (a_{\mathbf{k}1}^{\dagger} \pm i a_{\mathbf{k}2}^{\dagger}) \qquad ; \; \mathbf{e}_{\mathbf{k},\pm 1} = \mp \frac{1}{\sqrt{2}} (\mathbf{e}_{\mathbf{k}1} \pm i \mathbf{e}_{\mathbf{k}2})$$

$$a_{\mathbf{k},\pm 1} \equiv \mp \frac{1}{\sqrt{2}} (a_{\mathbf{k}1} \mp i a_{\mathbf{k}2}) \qquad ; \; \mathbf{e}_{\mathbf{k},\pm 1}^{\dagger} = \mp \frac{1}{\sqrt{2}} (\mathbf{e}_{\mathbf{k}1} \mp i \mathbf{e}_{\mathbf{k}2})$$

Let us calculate a couple of the relevant new commutators using the original commutation relations

$$[a_{\mathbf{k}s}, a_{\mathbf{k}'s'}^{\dagger}] = \delta_{\mathbf{k}\mathbf{k}'} \delta_{ss'}$$

---

[3]Note that the corresponding photon polarization asymmetry is given by

$$\frac{\mathcal{R}_{fi}(\lambda=1) - \mathcal{R}_{fi}(\lambda=-1)}{\mathcal{R}_{fi}(\lambda=1) + \mathcal{R}_{fi}(\lambda=-1)} = \frac{\sum_{J\geq 1} 2\mathrm{Re}\,\langle J_f || \hat{T}_J^{\mathrm{mag}}(k) || J_i \rangle^{\star} \langle J_f || \hat{T}_J^{\mathrm{el}}(k) || J_i \rangle}{\sum_{J\geq 1} \left(\left| \langle J_f || \hat{T}_J^{\mathrm{mag}}(k) || J_i \rangle \right|^2 + \left| \langle J_f || \hat{T}_J^{\mathrm{el}}(k) || J_i \rangle \right|^2 \right)}$$

Since the electric and magnetic multipoles have opposite parity, this vanishes if parity is conserved in the target system; here we leave the hats on the operators in the reduced matrix elements.

For example

$$[a_{k,+1}, a^\dagger_{k,+1}] = \frac{1}{2}\left([a_{k1}, a^\dagger_{k1}] + [a_{k2}, a^\dagger_{k2}]\right) = 1$$

$$[a_{k,+1}, a_{k,+1}] = [a^\dagger_{k,+1}, a^\dagger_{k,+1}] = 0$$

In a similar fashion

$$[a_{k,+1}, a^\dagger_{k,-1}] = -\frac{1}{2}\left([a_{k1}, a^\dagger_{k1}] - [a_{k2}, a^\dagger_{k2}]\right) = 0$$

$$[a_{k,+1}, a_{k,-1}] = [a^\dagger_{k,+1}, a^\dagger_{k,-1}] = 0$$

These new commutation relations are summarized in Eqs. (A.21)

$$[a_{k\lambda}, a^\dagger_{k'\lambda'}] = \delta_{kk'}\delta_{\lambda\lambda'} \qquad\qquad ; (\lambda, \lambda') = \pm1$$

$$[a_{k\lambda}, a_{k'\lambda'}] = [a^\dagger_{k\lambda}, a^\dagger_{k'\lambda'}] = 0$$

(b) Let us evaluate the first sum in part (b)

$$\sum_{\lambda=\pm1} a_{k\lambda}\,\mathbf{e}_{k\lambda} = a_{k,+1}\mathbf{e}_{k,+1} + a_{k,-1}\mathbf{e}_{k,-1}$$

$$= \frac{1}{2}\left(a_{k1}\mathbf{e}_{k1} + ia_{k1}\mathbf{e}_{k2} - ia_{k2}\mathbf{e}_{k1} + a_{k2}\mathbf{e}_{k2}\right) +$$

$$\frac{1}{2}\left(a_{k1}\mathbf{e}_{k1} - ia_{k1}\mathbf{e}_{k2} + ia_{k2}\mathbf{e}_{k1} + a_{k2}\mathbf{e}_{k2}\right)$$

$$= a_{k1}\mathbf{e}_{k1} + a_{k2}\mathbf{e}_{k2}$$

This yields the first result

$$\sum_{\lambda=\pm1} a_{k\lambda}\,\mathbf{e}_{k\lambda} = \sum_{s=1}^{2} a_{ks}\mathbf{e}_{ks}$$

The second result is just the adjoint of the first one[4]

$$\sum_{\lambda=\pm1} a^\dagger_{k\lambda}\,\mathbf{e}^\dagger_{k\lambda} = \sum_{s=1}^{2} a^\dagger_{ks}\,\mathbf{e}_{ks}$$

---

[4]Recall the basis vector $\mathbf{e}_{ks}$ is real.

Appendix B

Functions of a Complex Variable

Problem B.1 (a) Use Cauchy's theorem to demonstrate that a contour integral of an analytic function $f(z)$ around a closed curve C can be deformed in any fashion through a region of analyticity;

(b) Consider the contour integral between two points of an analytic function $f(z)$ in a simply-connected region of analyticity. Demonstrate from Cauchy's theorem that the integral is independent of the path.

Solution to Problem B.1

(a) Consider an analytic function $f(z)$ in a simply-connected region R, and let C_1 be a closed curve in R running in the counter-clockwise direction. Cauchy's theorem says that

$$\oint_{C_1} f(z)dz = 0$$

Let C_2 be a similar closed curve in R with an element in common with C_1 (see Fig. B.1).

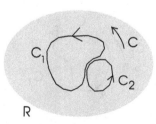

Fig. B.1 Contours C_1 and C_2 with a common element in a simply-connected region R, around which an analytic function $f(z)$ is integrated. The contour C is the union of C_1 and C_2 with the common element excluded.

Cauchy's theorem again says

$$\oint_{C_2} f(z)dz = 0$$

Add this to the first relation

$$\oint_{C_1} f(z)dz = \oint_{C_1} f(z)dz + \oint_{C_2} f(z)dz$$

The integrals run in opposite directions along the common element, and hence *cancel* along the common element. Thus

$$\oint_{C_1} f(z)dz = \oint_{C} f(z)dz$$

where the contour C is the union of C_1 and C_2 with the common element excluded.

This result demonstrates that a contour integral of an analytic function $f(z)$ around a closed curve C can be deformed in any fashion through a region of analyticity.

(b) Cauchy's theorem states that the integral of an analytic function around a closed contour C in a simply-connected region vanishes

$$\oint_{C} f(z)dz = 0$$

Take two points (A, B) on the closed contour C. Let C_1 be that part of the contour running from A to B, and C_2 be the remainder of the contour, but running in the *opposite* direction (see Fig. B.2).

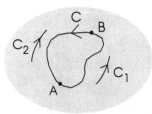

Fig. B.2   Two points (A, B) on the closed contour C, with C_1 that part of the contour running from A to B, and C_2 the remainder of the contour, but running in the *opposite* direction.

Then

$$\oint_C f(z)dz = \int_{C_1} f(z)dz - \int_{C_2} f(z)dz = 0$$

Hence

$$\int_{C_1} f(z)dz = \int_{C_2} f(z)dz$$

Since C_1 and C_2 are now just two arbitrary curves connecting the points (A, B), we conclude that that the integral is independent of the path.

Problem B.2 Consider the integral

$$I \equiv \int_{-\infty}^{\infty} \frac{dx}{1+x^2} = 2 \int_0^{\infty} \frac{dx}{1+x^2}$$

(a) Convert this to the integral $\oint_C dz/(1+z^2)$ around a closed contour C by adding the contribution of a large semi-circle in either half-plane, and showing that this additional contribution is negligible in the limit that the radius of the semi-circle $R \to \infty$;

(b) Locate the singularity of the integrand within that contour, and make a Laurent expansion about that singularity;

(c) Use the theory of residues to show that $I = \pi$.

Solution to Problem B.2

(a) Let us add the contribution from a large semi-circle of radius R in the upper half-plane as in Fig. 4.1 in the text, and take the limit $R \to \infty$ of the contour integral

$$I(R) = \oint_C \frac{dz}{1+z^2}$$

On the large semi-circle, write

$$z = Re^{i\phi}$$
$$dz = iRd\phi\, e^{i\phi}$$

Hence the contribution to I(R) from the semi-circle is

$$\int_0^{\pi} \frac{iRd\phi\, e^{i\phi}}{1 + R^2 e^{2i\phi}}$$

As $R \to \infty$, the 1 in the denominator is negligible. Therefore

$$\int_0^\pi \frac{iRd\phi e^{i\phi}}{1 + R^2 e^{2i\phi}} \approx i \int_0^\pi d\phi \frac{1}{Re^{i\phi}} \qquad ; R \to \infty$$

This clearly vanishes as $R \to \infty$. Hence

$$\text{Lim}_{R \to \infty} I(R) = \int_{-\infty}^\infty \frac{dx}{1 + x^2} = I$$

(b) One can write

$$\frac{1}{1 + z^2} = \frac{1}{z + i} \frac{1}{z - i}$$

There is a simple pole inside the contour C at $z = i$, and a Laurent expansion gives

$$\frac{1}{1 + z^2} = \frac{1}{z - i} \frac{1}{2i + z - i}$$

$$= \frac{1}{z - i} \frac{1}{2i} \left[1 - \left(\frac{z - i}{2i} \right) + \left(\frac{z - i}{2i} \right)^2 + \cdots \right]$$

The residue at the pole is $1/2i$.

(c) The method of residues then gives for the contour integral

$$I(R) = 2\pi i \, \text{Res} \left[\frac{1}{1 + z^2} \right]_{z=i}$$

$$= \pi$$

This result holds as long as the pole lies inside of the contour C.

Problem B.3 Use the theory of residues to verify the derivation of Eq. (4.107) starting from Eq. (4.100).

Solution to Problem B.3

The analysis in the text reduces the scattering Green's function to the form in Eq. (4.104)

$$G_0(\mathbf{x} - \mathbf{y}) = \frac{2m}{\hbar^2} \frac{4\pi}{(2\pi)^3} \frac{1}{2ir} \int_{-\infty}^\infty t dt \, e^{itr} \frac{1}{t^2 - k^2 - i\varepsilon} \qquad ; \mathbf{r} \equiv \mathbf{x} - \mathbf{y}$$

Call the integral I

$$I \equiv \int_{-\infty}^\infty t dt \, e^{itr} \frac{1}{t^2 - k^2 - i\varepsilon}$$

This is evaluated just as in the previous problem. Add a contribution from a semi-circle of radius R in the upper-half plane as in Fig. 4.1 in the text, and consider the contour integral

$$I(R) = \oint_C dz \frac{ze^{izr}}{z^2 - k^2 - i\varepsilon}$$

It was shown in Prob. 4.4 that the contribution from the large semi-circle vanishes as $R \to \infty$ for any non-zero r. Hence

$$\mathrm{Lim}_{R\to\infty} I(R) = I$$

The pole at $z_0 = k + i\varepsilon$ now lies within the contour.[1] In the vicinity of the pole, write

$$\frac{1}{z^2 - k^2 - i\varepsilon} = \frac{1}{z^2 - z_0^2} = \frac{1}{z - z_0} \frac{1}{z + z_0} \qquad ; z_0 = k + i\varepsilon$$

where we re-define $(2k)\varepsilon \to \varepsilon$. Now make a Laurent expansion around z_0

$$\frac{1}{z^2 - z_0^2} = \frac{1}{z - z_0} \frac{1}{2z_0} \left[1 - \left(\frac{z - z_0}{2z_0} \right) + \cdots \right]$$

$$ze^{izr} = z_0 e^{iz_0 r} \left[1 + \left(\frac{z - z_0}{z_0} \right) \right] [1 + ir(z - z_0) + \cdots]$$

Hence the residue of the integrand at the pole at z_0 is $z_0 e^{iz_0 r}/2z_0$. In the limit $\varepsilon \to 0$, this is $e^{ikr}/2$. Therefore

$$I(R) = 2\pi i \, \mathrm{Res} \left[\frac{ze^{izr}}{z^2 - k^2 - i\varepsilon} \right]_{z=k+i\varepsilon}$$

$$= i\pi e^{ikr}$$

This yields Eq. (4.107)

$$G_0(\mathbf{x} - \mathbf{y}) = \frac{2m}{\hbar^2} \frac{e^{ikr}}{4\pi r} \qquad ; \mathbf{r} \equiv \mathbf{x} - \mathbf{y}$$

Problem B.4 Consider the following integral along the contour C illustrated in Fig. B.3. Assume that $f(z)$ is well-behaved at x_0.

$$\int_C dz \frac{f(z)}{z - x_0}$$

---

[1] Here $k > 0$. Recall that ε is a positive infinitesimal, and in the end $\varepsilon \to 0$.

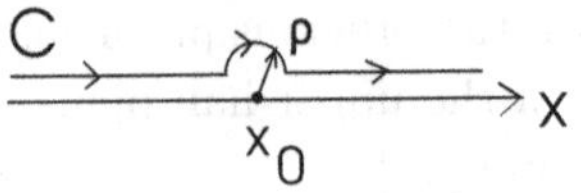

Fig. B.3　Contour C in Prob. B.4. Here ρ is the radius of the small semi-circle in the upper-half plane centered on x_0.

(a) Break the integral up into three contributions: one up to $x_0 - \rho$; one from $x_0 + \rho$; and one around the small semi-circle in the upper-half plane with radius ρ centered on x_0 . Define the Cauchy principal value by

$$\mathcal{P} \int dx \, \frac{f(x)}{x - x_0} \equiv \mathrm{Lim}\,_{\rho \to 0} \left(\int^{x_0 - \rho} + \int_{x_0 + \rho} \right) dx \, \frac{f(x)}{x - x_0}$$

$$; \text{ Cauchy principal value}$$

(b) Explicitly evaluate the integral on the small semi-circle in polar coordinates and show that as $\rho \to 0$

$$\int_C dz \, \frac{f(z)}{z - x_0} = \mathcal{P} \int dx \, \frac{f(x)}{x - x_0} - i\pi f(x_0)$$

(c) Hence justify the symbolic replacement

$$\frac{1}{x - x_0 + i\eta} = \mathcal{P}\frac{1}{x - x_0} - i\pi\delta(x - x_0)$$

Solution to Problem B.4

(a) Break the integral along the contour C in Fig. B.3 into three contributions, as noted in the problem. As the radius ρ of the small semi-circle goes to zero, the first and third contributions give the Cauchy principal value defined by

$$\mathcal{P} \int dx \, \frac{f(x)}{x - x_0} \equiv \mathrm{Lim}\,_{\rho \to 0} \left(\int^{x_0 - \rho} + \int_{x_0 + \rho} \right) dx \, \frac{f(x)}{x - x_0}$$

(b) On the small semi-circle, write

$$z - x_0 = \rho e^{i\phi}$$
$$dz = i\rho d\phi \, e^{i\phi}$$

The contribution from the small semi-circle is then given by

$$\int_{s\text{-}c} dz \, \frac{f(z)}{z - x_0} = i \int_\pi^0 d\phi \, f(x_0 + \rho e^{i\phi})$$

In the limit $\rho \to 0$, this becomes[2]

$$\int_{s-c} dz \, \frac{f(z)}{z - x_0} = if(x_0) \int_{\pi}^{0} d\phi \qquad ; \rho \to 0$$
$$= -i\pi f(x_0)$$

Thus, in the limit $\rho \to 0$, we have

$$\int_C dz \, \frac{f(z)}{z - x_0} = \mathcal{P} \int dx \, \frac{f(x)}{x - x_0} - i\pi f(x_0) \qquad ; \rho \to 0$$

(c) This result justifies the symbolic replacement

$$\frac{1}{x - x_0 + i\eta} = \mathcal{P} \frac{1}{x - x_0} - i\pi\delta(x - x_0)$$

Here the role of the infinitesimal $i\eta$ is to indicate that the singularity at $x = x_0 - i\eta$ lies just slightly below a contour along the real axis in Fig. B.3.

Problem B.5 Consider the function $f(x) = 1/\sqrt{x}$ for $x > 0$. Show the analytic continuation of this function to the cut-plane, where the cut runs along the negative real axis, is $f(z) = 1/\sqrt{z}$. Evaluate this function along the positive imaginary axis and show $f(z) = 1/\sqrt{|z|e^{i\pi/2}}$.

Solution to Problem B.5

The function $f(z) = 1/\sqrt{z}$ is analytic and single-valued in the cut z-plane, with a cut running along the negative real axis from the origin to $-\infty$ (see Fig. B.4).

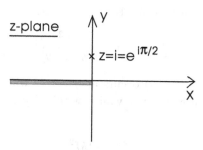

Fig. B.4   Cut z-plane in Prob. B.5. Here $z = x + iy$, and the point $z = i = e^{i\pi/2}$ is indicated.

---

[2]This is " half-a-pole".

The function $f(z)$ agrees with the function $1/\sqrt{x}$ along the entire positive real axis. Thus $f(z)$ provides the unique analytic continuation of $1/\sqrt{x}$ to the entire cut z-plane.

When z lies along the positive imaginary axis where $z = |z|e^{i\pi/2}$ (see Fig. B.4), then

$$f\left(|z|e^{i\pi/2}\right) = \frac{1}{\sqrt{|z|e^{i\pi/2}}}$$

Problem B.6 Determine the analytic properties of $F_{\mathrm{Mag}}^{(2)}(q^2)$ in Eq. (9.79).

Solution to Problem B.6

The magnetic form factor of the electron $F_{\mathrm{Mag}}^{(2)}(q^2)$ is given in QED by Eq. (9.79)

$$F_{\mathrm{Mag}}^{(2)}(q^2) = \frac{\alpha}{2\pi}\left[2M^2\int_0^1 dx\int_0^x dy\,\frac{x(1-x)}{M^2x^2 + q^2y(x-y)}\right]$$

As long as the denominator of the integral over the parameters (x, y) does not vanish within the domain of integration, this expression can be differentiated any number of times with respect to q^2 to give a well-defined analytic function.

The condition that the denominator vanish is

$$\frac{q^2}{M^2} = -\frac{x^2}{y(x-y)}$$

These values of q^2 lie along the negative q^2 axis, and the maximum value of q^2/M^2 for which this vanishes is

$$\frac{q^2}{M^2} = -\min\left[\frac{x^2}{y(x-y)}\right]$$

The largest value of $y(x-y)$ occurs at $y = x/2$ where $y(x-y) = x^2/4$. Hence the denominator will not vanish within the domain of integration provided

$$q^2 > -4M^2$$

Thus $F_{\mathrm{Mag}}^{(2)}(q^2)$ is an analytic function of q^2, provided we stay way from the negative real axis where $-\infty < q^2 < -4M^2$. This can be accomplished by cutting the q^2-plane along the negative q^2 axis from $-4M^2$ to $-\infty$ (see Fig. B.5).

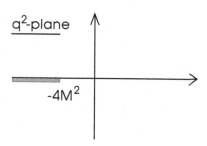

Fig. B.5   The function $F_{\text{Mag}}^{(2)}(q^2)$ in Prob. B.6 is analytic in the cut q^2-plane.

Appendix C

Electromagnetic Field

Problem C.1 Show that the action computed from the lagrangian density in Eq. (C.8) is gauge invariant. (*Hint*: make use of partial integration. Note that the addition of a constant to the action changes nothing.)

Solution to Problem C.1

The lagrangian density in Eq. (C.8) is[1]

$$\mathcal{L}\left(\frac{\partial A_\lambda}{\partial x_\mu}, A_\lambda\right) = -\frac{\varepsilon_0}{4} F_{\mu\nu} F_{\mu\nu} + ec\, j_\mu A_\mu$$

Under a gauge transformation, the vector potential changes according to Eq. (C.7)

$$A_\mu \to A_\mu + \frac{\partial \Lambda}{\partial x_\mu}$$

The field tensor $F_{\mu\nu}$ is *gauge invariant*, and hence under the gauge transformation the lagrangian density is changed to

$$\mathcal{L} \to \mathcal{L} + ec\, j_\mu \frac{\partial \Lambda}{\partial x_\mu}$$

The action is given by Eq. (5.57)

$$S = \frac{1}{c} \int_{t_1}^{t_2} \int_\Omega d^4 x\, \mathcal{L} \qquad ; \ d^4 x \equiv d^3 x (cdt)$$

The change in the action under the gauge transformation is therefore

$$\delta S = e \int_{t_1}^{t_2} \int_\Omega d^4 x\, j_\mu \frac{\partial \Lambda}{\partial x_\mu}$$

---

[1] Recall that we are working in SI units.

Now carry out a partial integration to arrive at

$$\delta S = -e \int_{t_1}^{t_2} \int_{\Omega} d^4x \, \Lambda \frac{\partial j_\mu}{\partial x_\mu} + \text{boundary terms}$$

The current is conserved

$$\frac{\partial j_\mu}{\partial x_\mu} = 0$$

Therefore

$$\delta S = \text{boundary terms}$$

Periodic boundary conditions, or a localized disturbance, eliminate any contributions from the spatial boundaries, and the times surfaces contribute at most an additional constant term to the action. As stated in the problem, the addition of a constant to the action changes nothing. Hence the action is effectively *gauge invariant*.

Problem C.2 Show the energy-flux Poynting vector implied by Eq. (C.23) reproduces the result quoted in Vol. I

$$\mathbf{S} = \frac{1}{\mu_0} \mathbf{E} \times \mathbf{B}$$

Solution to Problem C.2

Equation (C.23) gives the total momentum in the field

$$\mathbf{P} = \varepsilon_0 \int_{\Omega} d^3x \, \mathbf{E} \times \mathbf{B}$$

The momentum *density* is therefore

$$\boldsymbol{\mathcal{P}} = \varepsilon_0 \mathbf{E} \times \mathbf{B} \qquad ; \text{momentum density}$$

The momentum *flux* is the momentum density times the velocity of light $\boldsymbol{\mathcal{P}}c$. The ratio of the *energy flux* to the *momentum flux* is $\mathbf{S}/\boldsymbol{\mathcal{P}}c = c$. Therefore the energy-flux Poynting vector is

$$\mathbf{S} = \boldsymbol{\mathcal{P}}c^2$$
$$= \frac{1}{\mu_0} \mathbf{E} \times \mathbf{B} \qquad ; \varepsilon_0 \mu_0 = \frac{1}{c^2}$$

This is the result quoted in Vol. I.

Problem C.3 Insert the interaction-picture representation of $\hat{A}(\mathbf{x}, t)$ and derive the expression for the total momentum contained in the free electromagnetic field in Eq. (C.36).

Solution to Problem C.3

We start from Eq. (C.35) for the quantized vector potential in the interaction picture

$$\hat{\mathbf{A}}(\mathbf{x}, t) = \sum_{\mathbf{k}} \sum_{s=1}^{2} \left(\frac{\hbar}{2\omega_k \varepsilon_0 \Omega} \right)^{1/2} \left[a_{\mathbf{k}s} \mathbf{e}_{\mathbf{k}s} e^{i(\mathbf{k}\cdot\mathbf{x}-\omega_k t)} + a_{\mathbf{k}s}^{\dagger} \mathbf{e}_{\mathbf{k}s} e^{-i(\mathbf{k}\cdot\mathbf{x}-\omega_k t)} \right]$$

The electric and magnetic fields are then obtained using Eqs. (C.25)

$$\hat{\mathbf{E}} = -\frac{\partial \hat{\mathbf{A}}}{\partial t} = -\sum_{\mathbf{k}} \sum_{s=1}^{2} i \left(\frac{\hbar \omega_k}{2\varepsilon_0 \Omega} \right)^{1/2} \times$$

$$\left[-a_{\mathbf{k}s} \mathbf{e}_{\mathbf{k}s} e^{i(\mathbf{k}\cdot\mathbf{x}-\omega_k t)} + a_{\mathbf{k}s}^{\dagger} \mathbf{e}_{\mathbf{k}s} e^{-i(\mathbf{k}\cdot\mathbf{x}-\omega_k t)} \right]$$

$$\hat{\mathbf{B}} = \nabla \times \hat{\mathbf{A}} = \sum_{\mathbf{k}} \sum_{s=1}^{2} i \left(\frac{\hbar}{2\omega_k \varepsilon_0 \Omega} \right)^{1/2} \times$$

$$\left[a_{\mathbf{k}s}(\mathbf{k} \times \mathbf{e}_{\mathbf{k}s}) e^{i(\mathbf{k}\cdot\mathbf{x}-\omega_k t)} - a_{\mathbf{k}s}^{\dagger}(\mathbf{k} \times \mathbf{e}_{\mathbf{k}s}) e^{-i(\mathbf{k}\cdot\mathbf{x}-\omega_k t)} \right]$$

The total momentum contained in the free electromagnetic field is given in Eq. (C.23)

$$\hat{\mathbf{P}} = \varepsilon_0 \int_{\Omega} d^3x \, \hat{\mathbf{E}} \times \hat{\mathbf{B}}$$

Substitution of the above leads to

$$\hat{\mathbf{P}} = \sum_{\mathbf{k}} \sum_{s=1}^{2} \sum_{s'=1}^{2} \frac{\hbar}{2} \times$$

$$\left[-\left(a_{\mathbf{k}s} a_{-\mathbf{k},s'} e^{-2i\omega_k t} + a_{\mathbf{k}s}^{\dagger} a_{-\mathbf{k},s'}^{\dagger} e^{2i\omega_k t} \right) \mathbf{e}_{\mathbf{k}s} \times (-\mathbf{k} \times \mathbf{e}_{-\mathbf{k},s'}) \right.$$

$$\left. + \left(a_{\mathbf{k}s} a_{\mathbf{k}s'}^{\dagger} + a_{\mathbf{k}s}^{\dagger} a_{\mathbf{k}s'} \right) \mathbf{e}_{\mathbf{k}s} \times (\mathbf{k} \times \mathbf{e}_{\mathbf{k}s'}) \right]$$

Here we have performed the integration over space in a big box with p.b.c.

$$\frac{1}{\Omega} \int_{\Omega} d^3x \, e^{i(\mathbf{k}-\mathbf{k}')\cdot\mathbf{x}} = \delta_{\mathbf{k}\mathbf{k}'}$$

and used the resulting Kronecker delta-functions to eliminate one of the sums over momenta.

Now use

$$\mathbf{e}_{\mathbf{k}s} \times (\mathbf{k} \times \mathbf{e}_{\mathbf{k}s'}) = \mathbf{k}(\mathbf{e}_{\mathbf{k}s} \cdot \mathbf{e}_{\mathbf{k}s'}) = \mathbf{k}\delta_{ss'}$$
$$\mathbf{e}_{\mathbf{k}s} \times (\mathbf{k} \times \mathbf{e}_{-\mathbf{k},s'}) = \mathbf{k}(\mathbf{e}_{\mathbf{k}s} \cdot \mathbf{e}_{-\mathbf{k},s'})$$

The first contribution in the expression for $\mathbf{P}$ vanishes because the summand changes sign under the change of summation variables $(\mathbf{k}, s) \leftrightarrows (-\mathbf{k}, s')$. Therefore, the total momentum of the free field takes the form

$$\hat{\mathbf{P}} = \sum_{\mathbf{k}} \sum_{s=1}^{2} \hbar \mathbf{k} \frac{1}{2} \left(a_{\mathbf{k}s} a_{\mathbf{k}s}^{\dagger} + a_{\mathbf{k}s}^{\dagger} a_{\mathbf{k}s} \right)$$

$$= \sum_{\mathbf{k}} \sum_{s=1}^{2} \hbar \mathbf{k} \left(a_{\mathbf{k}s}^{\dagger} a_{\mathbf{k}s} + \frac{1}{2} \right)$$

where we have used the commutator of the creation and destruction operators in Eq. (C.33) to obtain the second line. After realizing that $\sum_{\mathbf{k}} \sum_{s=1}^{2} \hbar \mathbf{k} = 0$ (total momentum of the vacuum), the total momentum in the field takes the form of Eq. (C.36)

$$\hat{\mathbf{P}} = \sum_{\mathbf{k}} \sum_{s=1}^{2} \hbar \mathbf{k} \, \hat{N}_{\mathbf{k}s}$$

Problem C.4 Insert the interaction-picture representation of $\hat{A}(\mathbf{x}, t)$ and discuss the processes described by the interaction $\hat{H}_I(t)$ in Eq. (C.71).

Solution to Problem C.4

Consider the interaction hamiltonian in Eq. (C.71)

$$\hat{H}_I(t) = -e \int_{\Omega} d^3x \, \mathbf{j}^{\text{ext}}(\mathbf{x}, t) \cdot \hat{\mathbf{A}}(\mathbf{x}, t)$$

where $\mathbf{j}^{\text{ext}}(\mathbf{x}, t)$ is an external current, and the quantized vector potential is

$$\hat{\mathbf{A}}(\mathbf{x}, t) = \sum_{\mathbf{k}} \sum_{s=1}^{2} \left(\frac{\hbar}{2\omega_k \varepsilon_0 \Omega} \right)^{1/2} \left[a_{\mathbf{k}s} \mathbf{e}_{\mathbf{k}s} e^{i(\mathbf{k}\cdot\mathbf{x} - \omega_k t)} + a_{\mathbf{k}s}^{\dagger} \mathbf{e}_{\mathbf{k}s} e^{-i(\mathbf{k}\cdot\mathbf{x} - \omega_k t)} \right]$$

Since $\mathbf{j}^{\text{ext}}$ is a classical field, $\hat{H}_I$ can act upon quantum states containing n photons changing the number of photons by one unit. When the destruction operator $a_{\mathbf{k}s}$ acts on a state, it destroys a photon with momentum $\hbar\mathbf{k}$ and polarization s (absorption), whereas when the creation operator $a_{\mathbf{k}s}^{\dagger}$

acts on a state, it creates a photon with momentum $\hbar\mathbf{k}$ and polarization s (emission).

Problem C.5 It is a theorem that any vector field can be uniquely separated into a part with zero divergence and a part with zero curl. That is, $\mathbf{v}(\mathbf{x}) = \mathbf{v}^T(\mathbf{x}) + \mathbf{v}^L(\mathbf{x})$ where $\nabla \cdot \mathbf{v}^T(\mathbf{x}) = 0$ and $\nabla \times \mathbf{v}^L(\mathbf{x}) = 0$.
(a) Provide a derivation of Eq. (C.46);
(b) Use that result to prove the theorem;
(c) Show $\int_\Omega d^3x\, \mathbf{v}^T(\mathbf{x}) \cdot \mathbf{v}^L(\mathbf{x}) = 0$.

Solution to Problem C.5

(a) An arbitrary vector field has a Fourier series representation

$$\mathbf{v}(\mathbf{x}) = \frac{1}{\sqrt{\Omega}} \sum_{\mathbf{k}} \mathbf{v}(\mathbf{k})\, e^{i\mathbf{k}\cdot\mathbf{x}}$$

The vector coefficients $\mathbf{v}(\mathbf{k})$ can be expanded in a complete, orthonormal set of unit vectors $\mathbf{e}_{\mathbf{k}s}$, where $\mathbf{e}_{\mathbf{k}1}$ and $\mathbf{e}_{\mathbf{k}2}$ are two unit vectors orthogonal to $\mathbf{k}$ (see Fig. C.1 in the text), and $\mathbf{e}_{\mathbf{k}3} = \mathbf{k}/k$

$$\mathbf{v}(\mathbf{k}) = \sum_{s=1}^{3} v(\mathbf{k}, s)\, \mathbf{e}_{\mathbf{k}s}$$

Substitution into the above gives Eq. (C.46)[2]

$$\mathbf{v}(\mathbf{x}) = \frac{1}{\sqrt{\Omega}} \sum_{\mathbf{k}} \sum_{s=1}^{3} v_{\mathbf{k}s}\mathbf{e}_{\mathbf{k}s}\, e^{i\mathbf{k}\cdot\mathbf{x}}$$

(b) The vector field is then separated, quite generally, into transverse and longitudinal parts

$$\mathbf{v}(\mathbf{x}) = \mathbf{v}^T(\mathbf{x}) + \mathbf{v}^L(\mathbf{x})$$

where

$$\mathbf{v}^T(\mathbf{x}) = \frac{1}{\sqrt{\Omega}} \sum_{\mathbf{k}} \sum_{s=1}^{2} v(\mathbf{k}, s)\, \mathbf{e}_{\mathbf{k}s}\, e^{i\mathbf{k}\cdot\mathbf{x}} \qquad ;\ \nabla \cdot \mathbf{v}^T(\mathbf{x}) = 0$$

$$\mathbf{v}^L(\mathbf{x}) = \frac{1}{\sqrt{\Omega}} \sum_{\mathbf{k}} v(\mathbf{k}, 3)\, \mathbf{e}_{\mathbf{k}3}\, e^{i\mathbf{k}\cdot\mathbf{x}} \qquad ;\ \nabla \times \mathbf{v}^L(\mathbf{x}) = 0$$

---

[2] Here $v(\mathbf{k}, s) \equiv v_{\mathbf{k}s}$.

(c) One can use the orthonormality of the plane waves

$$\frac{1}{\Omega}\int_{\Omega} d^3x\, e^{i(\mathbf{k}-\mathbf{k'})\cdot\mathbf{x}} = \delta_{\mathbf{kk'}}$$

to compute

$$\int_{\Omega} d^3x\, \mathbf{v}^L(\mathbf{x})\cdot\mathbf{v}^T(\mathbf{x}) = \sum_{\mathbf{k}}\sum_{s=1}^{2} v(\mathbf{k},s)v(-\mathbf{k},3)\,\mathbf{e}_{\mathbf{k}s}\cdot\mathbf{e}_{-\mathbf{k},3}$$

Since the transverse and longitudinal unit vectors are orthogonal, this vanishes. Hence

$$\int_{\Omega} d^3x\, \mathbf{v}^L(\mathbf{x})\cdot\mathbf{v}^T(\mathbf{x}) = 0$$

Problem C.6 Consider a neutral massive vector meson field V_μ with inverse Compton wavelength $m = m_0 c/\hbar$ and lagrangian density[3]

$$\mathcal{L} = -\frac{1}{4}V_{\mu\nu}V_{\mu\nu} - \frac{1}{2}m^2 V_\mu V_\mu \qquad ; \; V_{\mu\nu} = \frac{\partial V_\nu}{\partial x_\mu} - \frac{\partial V_\mu}{\partial x_\nu}$$

(a) Show the field equations are

$$\frac{\partial}{\partial x_\nu}V_{\mu\nu} + m^2 V_\mu = 0 \qquad ; \; (\mu,\nu) = 1,2,3,4$$

(b) Note that with $m^2 \neq 0$, the field equation for $\mu = 4$ can be solved for V_4, and the resulting expression used to eliminate V_4 as an independent dynamical variable.[4] Show that the momentum density and V_4 are then given by

$$\Pi_j = \frac{1}{ic}V_{j4} \qquad ; \; V_4 = \frac{ic}{m^2}\nabla\cdot\mathbf{\Pi}$$

(c) use the canonical procedure and show the hamiltonian is given by

$$H = \frac{1}{2}\int_{\Omega} d^3x\left[c^2\mathbf{\Pi}^2 + (\nabla\times\mathbf{V})^2 + m^2\mathbf{V}^2 + \frac{c^2}{m^2}(\nabla\cdot\mathbf{\Pi})^2\right]$$

---

[3]The analogy is to QED in H-L units.
[4]This is important in Prob. C.7, since here $\Pi_4 \equiv 0$. Although Probs. C.6–C.7 are algebraically challenging, they are very interesting.

Solution to Problem C.6

(a) We first need to compute the following derivative

$$\frac{\partial \mathcal{L}}{\partial(\partial V_\mu/\partial x_\nu)} = -\frac{1}{2}\frac{\partial}{\partial(\partial V_\mu/\partial x_\nu)}\left[\frac{\partial V_\mu}{\partial x_\nu}\frac{\partial V_\mu}{\partial x_\nu} - \frac{\partial V_\mu}{\partial x_\nu}\frac{\partial V_\nu}{\partial x_\mu}\right]$$

$$= \frac{\partial V_\nu}{\partial x_\mu} - \frac{\partial V_\mu}{\partial x_\nu} = V_{\mu\nu}$$

Lagrange's equations then gives the field equations

$$\frac{\partial}{\partial x_\nu}V_{\mu\nu} + m^2 V_\mu = 0$$

Since $V_{\mu\nu}$ is antisymmetric, the four-divergence of this relation gives

$$m^2\frac{\partial V_\mu}{\partial x_\mu} = 0$$

If $m^2 \neq 0$, the four-divergence of the field V_μ thus vanishes

$$\frac{\partial V_\mu}{\partial x_\mu} = 0 \qquad ; m^2 \neq 0$$

It follows that

$$\frac{\partial V_{\mu\nu}}{\partial x_\nu} = \frac{\partial}{\partial x_\nu}\left[\frac{\partial V_\nu}{\partial x_\mu} - \frac{\partial V_\mu}{\partial x_\nu}\right] = -\Box V_\mu$$

The field equations can therefore be re-written as

$$(\Box - m^2)V_\mu = 0$$

Each component of the field satisfies the Klein-Gordon equation.

(b) The canonical momentum density is given as

$$\Pi_\mu = \frac{\partial \mathcal{L}}{\partial(\partial V_\mu/\partial t)} = \frac{1}{ic}\frac{\partial \mathcal{L}}{\partial(\partial V_\mu/\partial x_4)} = \frac{1}{ic}V_{\mu 4}$$

where we have used the result from part (a). Thus

$$\Pi_j = \frac{1}{ic}V_{j4} \qquad ; \Pi_4 = 0$$

Since $\Pi_4 = 0$, the field V_4 is not an independent dynamical variable. In fact, from the field equations in part (a)

$$V_4 = -\frac{1}{m^2}\frac{\partial}{\partial x_\nu}V_{4\nu} = -\frac{1}{m^2}\frac{\partial}{\partial x_j}(-ic\Pi_j)$$

$$V_4 = \frac{ic}{m^2}\boldsymbol{\nabla}\cdot\boldsymbol{\Pi}$$

Note it is essential here that the mass be non-zero.

(c) Since $\Pi_4 = 0$, The hamiltonian density is given by

$$\mathcal{H} = \Pi_j \frac{\partial V_j}{\partial t} - \mathcal{L} = ic\Pi_j \frac{\partial V_j}{\partial x_4} - \mathcal{L}$$

$$= ic\Pi_j \left[\frac{\partial V_4}{\partial x_j} - V_{j4} \right] - \mathcal{L}$$

We are interested in the hamiltonian H, and we can perform partial integrations on the hamiltonian density, since the boundary conditions will get rid of any boundary terms. Thus

$$\mathcal{H} \doteq -ic \frac{\partial \Pi_j}{\partial x_j} V_4 - ic\Pi_j V_{j4} - \mathcal{L} = \frac{c^2}{m^2} (\boldsymbol{\nabla} \cdot \boldsymbol{\Pi})^2 + c^2 \boldsymbol{\Pi}^2 - \mathcal{L}$$

Here the symbol $\doteq$ implies that a partial integration has been performed. With the insertion of $\mathcal{L}$, this gives

$$\mathcal{H} \doteq \frac{c^2}{m^2} (\boldsymbol{\nabla} \cdot \boldsymbol{\Pi})^2 + c^2 \boldsymbol{\Pi}^2 + \frac{1}{4} V_{ij} V_{ij} + \frac{1}{2} V_{j4} V_{j4} + \frac{m^2}{2} \mathbf{V}^2 + \frac{m^2}{2} V_4^2$$

$$= \frac{c^2}{m^2} (\boldsymbol{\nabla} \cdot \boldsymbol{\Pi})^2 + c^2 \boldsymbol{\Pi}^2 + \frac{1}{4} V_{ij} V_{ij} - \frac{c^2}{2} \boldsymbol{\Pi}^2 + \frac{m^2}{2} \mathbf{V}^2 - \frac{c^2}{2m^2} (\boldsymbol{\nabla} \cdot \boldsymbol{\Pi})^2$$

Now use

$$(\boldsymbol{\nabla} \times \mathbf{V})^2 = \epsilon_{ijk} \epsilon_{ilm} \frac{\partial V_j}{\partial x_k} \frac{\partial V_l}{\partial x_m} = (\delta_{jl} \delta_{km} - \delta_{jm} \delta_{kl}) \frac{\partial V_j}{\partial x_k} \frac{\partial V_l}{\partial x_m}$$

$$= \frac{\partial V_j}{\partial x_k} \frac{\partial V_j}{\partial x_k} - \frac{\partial V_j}{\partial x_k} \frac{\partial V_k}{\partial x_j} = \frac{1}{2} V_{jk} V_{jk}$$

where V_{jk} is the previous field tensor.

Thus, the hamiltonian for this system is given by

$$H = \frac{1}{2} \int_\Omega d^3x \left[c^2 \boldsymbol{\Pi}^2 + (\boldsymbol{\nabla} \times \mathbf{V})^2 + m^2 \mathbf{V}^2 + \frac{c^2}{m^2} (\boldsymbol{\nabla} \cdot \boldsymbol{\Pi})^2 \right]$$

The dynamical variables are now $H(\boldsymbol{\Pi}, \mathbf{V})$.

Problem C.7 (a) Make use of the result in Prob. C.5(c), and show that the hamiltonian in Prob. C.6(c) separates into transverse and longitudinal contributions;

(b) Show that the following expansions of the field operators put the problem into normal modes

$$\mathbf{V}_T(\mathbf{x},t) = \sum_{\mathbf{k}}\sum_{s=1}^{2}\left(\frac{\hbar c^2}{2\omega_k\Omega}\right)^{1/2}\left[c_{\mathbf{k}s}\mathbf{e}_{\mathbf{k}s}e^{i(\mathbf{k}\cdot\mathbf{x}-\omega_k t)} + c_{\mathbf{k}s}^{*}\mathbf{e}_{\mathbf{k}s}e^{-i(\mathbf{k}\cdot\mathbf{x}-\omega_k t)}\right]$$

$$\mathbf{\Pi}_T(\mathbf{x},t) = \frac{1}{i}\sum_{\mathbf{k}}\sum_{s=1}^{2}\left(\frac{\hbar\omega_k}{2c^2\Omega}\right)^{1/2}\left[c_{\mathbf{k}s}\mathbf{e}_{\mathbf{k}s}e^{i(\mathbf{k}\cdot\mathbf{x}-\omega_k t)} - c_{\mathbf{k}s}^{*}\mathbf{e}_{\mathbf{k}s}e^{-i(\mathbf{k}\cdot\mathbf{x}-\omega_k t)}\right]$$

$$\mathbf{V}_L(\mathbf{x},t) = \sum_{\mathbf{k}}\left(\frac{\hbar\omega_k}{2m^2\Omega}\right)^{1/2}\left[c_{\mathbf{k}3}\mathbf{e}_{\mathbf{k}3}e^{i(\mathbf{k}\cdot\mathbf{x}-\omega_k t)} + c_{\mathbf{k}3}^{*}\mathbf{e}_{\mathbf{k}3}e^{-i(\mathbf{k}\cdot\mathbf{x}-\omega_k t)}\right]$$

$$\mathbf{\Pi}_L(\mathbf{x},t) = \frac{1}{i}\sum_{\mathbf{k}}\left(\frac{\hbar m^2}{2\omega_k\Omega}\right)^{1/2}\left[c_{\mathbf{k}3}\mathbf{e}_{\mathbf{k}3}e^{i(\mathbf{k}\cdot\mathbf{x}-\omega_k t)} - c_{\mathbf{k}3}^{*}\mathbf{e}_{\mathbf{k}3}e^{-i(\mathbf{k}\cdot\mathbf{x}-\omega_k t)}\right]$$

(c) Now identify the normal-mode amplitudes $(c_{\mathbf{k}s}^{*}, c_{\mathbf{k}'s'})$ with the creation and destruction operators $(c_{\mathbf{k}s}^{\dagger}, c_{\mathbf{k}'s'})$. Show that imposing $[c_{\mathbf{k}s}, c_{\mathbf{k}'s'}^{\dagger}] = \delta_{\mathbf{k}'\mathbf{k}}\delta_{s's}$ produces the proper interaction-picture canonical commutation relations for the field operators $\hat{\mathbf{V}}(\mathbf{x},t)$ and $\hat{\mathbf{\Pi}}(\mathbf{x},t)$.

Solution to Problem C.7

(a) From Prob. C.5, we can write

$$\mathbf{\Pi} = \mathbf{\Pi}_T + \mathbf{\Pi}_L \qquad ; \ \boldsymbol{\nabla}\cdot\mathbf{\Pi}_T = \boldsymbol{\nabla}\times\mathbf{\Pi}_L = 0$$
$$\mathbf{V} = \mathbf{V}_T + \mathbf{V}_L \qquad ; \ \boldsymbol{\nabla}\cdot\mathbf{V}_T = \boldsymbol{\nabla}\times\mathbf{V}_L = 0$$
$$\int_{\Omega} d^3x\,\mathbf{\Pi}_T\cdot\mathbf{\Pi}_L = \int_{\Omega} d^3x\,\mathbf{V}_T\cdot\mathbf{V}_L = 0$$

The hamiltonian for the massive vector field in Prob. C.6 then separates into two distinct contributions

$$H = H_T + H_L$$
$$H_T = \frac{1}{2}\int_{\Omega} d^3x\,\left[c^2\mathbf{\Pi}_T^2 + m^2\mathbf{V}_T^2 + (\boldsymbol{\nabla}\times\mathbf{V}_T)^2\right]$$
$$H_L = \frac{1}{2}\int_{\Omega} d^3x\,\left[c^2\mathbf{\Pi}_L^2 + m^2\mathbf{V}_L^2 + \frac{c^2}{m^2}(\boldsymbol{\nabla}\cdot\mathbf{\Pi}_L)^2\right]$$

(b) Now insert the following normal-mode expansions, which satisfy the field equations and the above conditions

$$\mathbf{V}_T(\mathbf{x},t) = \sum_{\mathbf{k}}\sum_{s=1}^{2}\left(\frac{\hbar c^2}{2\omega_k\Omega}\right)^{1/2}\left[c_{\mathbf{k}s}\mathbf{e}_{\mathbf{k}s}e^{i(\mathbf{k}\cdot\mathbf{x}-\omega_k t)} + c^{\star}_{\mathbf{k}s}\mathbf{e}_{\mathbf{k}s}e^{-i(\mathbf{k}\cdot\mathbf{x}-\omega_k t)}\right]$$

$$\mathbf{\Pi}_T(\mathbf{x},t) = \frac{1}{i}\sum_{\mathbf{k}}\sum_{s=1}^{2}\left(\frac{\hbar\omega_k}{2c^2\Omega}\right)^{1/2}\left[c_{\mathbf{k}s}\mathbf{e}_{\mathbf{k}s}e^{i(\mathbf{k}\cdot\mathbf{x}-\omega_k t)} - c^{\star}_{\mathbf{k}s}\mathbf{e}_{\mathbf{k}s}e^{-i(\mathbf{k}\cdot\mathbf{x}-\omega_k t)}\right]$$

$$\mathbf{V}_L(\mathbf{x},t) = \sum_{\mathbf{k}}\left(\frac{\hbar\omega_k}{2m^2\Omega}\right)^{1/2}\left[c_{\mathbf{k}3}\mathbf{e}_{\mathbf{k}3}e^{i(\mathbf{k}\cdot\mathbf{x}-\omega_k t)} + c^{\star}_{\mathbf{k}3}\mathbf{e}_{\mathbf{k}3}e^{-i(\mathbf{k}\cdot\mathbf{x}-\omega_k t)}\right]$$

$$\mathbf{\Pi}_L(\mathbf{x},t) = \frac{1}{i}\sum_{\mathbf{k}}\left(\frac{\hbar m^2}{2\omega_k\Omega}\right)^{1/2}\left[c_{\mathbf{k}3}\mathbf{e}_{\mathbf{k}3}e^{i(\mathbf{k}\cdot\mathbf{x}-\omega_k t)} - c^{\star}_{\mathbf{k}3}\mathbf{e}_{\mathbf{k}3}e^{-i(\mathbf{k}\cdot\mathbf{x}-\omega_k t)}\right]$$

The longitudinal hamiltonian becomes

$$H_L = \frac{1}{4\Omega}\int_{\Omega}d^3x\sum_{\mathbf{k}}\sum_{\mathbf{k}'}\frac{\hbar}{\sqrt{\omega_k\omega_{k'}}}\times$$

$$\left\{-c^2 m^2\left(\mathbf{e}_{\mathbf{k}3}\cdot\mathbf{e}_{\mathbf{k}'3}\right)\left[c_{\mathbf{k}3}e^{i(\mathbf{k}\cdot\mathbf{x}-\omega_k t)} - c^{\star}_{\mathbf{k}3}e^{-i(\mathbf{k}\cdot\mathbf{x}-\omega_k t)}\right]\times\right.$$

$$\left[c_{\mathbf{k}'3}e^{i(\mathbf{k}'\cdot\mathbf{x}-\omega_{k'} t)} - c^{\star}_{\mathbf{k}'3}e^{-i(\mathbf{k}'\cdot\mathbf{x}-\omega_{k'} t)}\right]$$

$$+\omega_k\omega_{k'}\left(\mathbf{e}_{\mathbf{k}3}\cdot\mathbf{e}_{\mathbf{k}'3}\right)\left[c_{\mathbf{k}3}e^{i(\mathbf{k}\cdot\mathbf{x}-\omega_k t)} + c^{\star}_{\mathbf{k}3}e^{-i(\mathbf{k}\cdot\mathbf{x}-\omega_k t)}\right]\times$$

$$\left[c_{\mathbf{k}'3}e^{i(\mathbf{k}'\cdot\mathbf{x}-\omega_{k'} t)} + c^{\star}_{\mathbf{k}'3}e^{-i(\mathbf{k}'\cdot\mathbf{x}-\omega_{k'} t)}\right]$$

$$+c^2\left(\mathbf{k}\cdot\mathbf{e}_{\mathbf{k}3}\right)\left(\mathbf{k}'\cdot\mathbf{e}_{\mathbf{k}'3}\right)\left[c_{\mathbf{k}3}e^{i(\mathbf{k}\cdot\mathbf{x}-\omega_k t)} + c^{\star}_{\mathbf{k}3}e^{-i(\mathbf{k}\cdot\mathbf{x}-\omega_k t)}\right]\times$$

$$\left.\left[c_{\mathbf{k}'3}e^{i(\mathbf{k}'\cdot\mathbf{x}-\omega_{k'} t)} + c^{\star}_{\mathbf{k}'3}e^{-i(\mathbf{k}'\cdot\mathbf{x}-\omega_{k'} t)}\right]\right\}$$

In a big box with p.b.c., the spatial integral gives

$$\frac{1}{\Omega}\int_{\Omega}d^3x\,e^{i(\mathbf{k}-\mathbf{k}')\cdot\mathbf{x}} = \delta_{\mathbf{k}\mathbf{k}'}$$

The hamiltonian H_L then takes the form[5]

$$H_L = \frac{1}{4}\sum_{\mathbf{k}}\frac{\hbar}{\omega_k}\times$$

$$\left\{c^2 m^2\left[c_{\mathbf{k}3}c^{\star}_{\mathbf{k}3} + c^{\star}_{\mathbf{k}3}c_{\mathbf{k}3} + \left(c_{\mathbf{k}3}c_{-\mathbf{k},3}e^{-2i\omega_k t} + c^{\star}_{\mathbf{k}3}c^{\star}_{-\mathbf{k},3}e^{2i\omega_k t}\right)\right] + \right.$$

$$\omega_k^2\left[c_{\mathbf{k}3}c^{\star}_{\mathbf{k}3} + c^{\star}_{\mathbf{k}3}c_{\mathbf{k}3} - \left(c_{\mathbf{k}3}c_{-\mathbf{k},3}e^{-2i\omega_k t} + c^{\star}_{\mathbf{k}3}c^{\star}_{-\mathbf{k},3}e^{2i\omega_k t}\right)\right] + $$

$$\left.c^2\mathbf{k}^2\left[c_{\mathbf{k}3}c^{\star}_{\mathbf{k}3} + c^{\star}_{\mathbf{k}3}c_{\mathbf{k}3} + \left(c_{\mathbf{k}3}c_{-\mathbf{k},3}e^{-2i\omega_k t} + c^{\star}_{\mathbf{k}3}c^{\star}_{-\mathbf{k},3}e^{2i\omega_k t}\right)\right]\right\}$$

---

[5]Recall $\mathbf{e}_{\mathbf{k}3} = \mathbf{k}/|\mathbf{k}|$.

Now use the dispersion relation

$$\omega_k = c\sqrt{\mathbf{k}^2 + m^2}$$

The longitudinal hamiltonian is then expressed in normal modes as

$$H_L = \sum_{\mathbf{k}} \hbar\omega_k \frac{1}{2} \left(c_{\mathbf{k}3} c_{\mathbf{k}3}^\star + c_{\mathbf{k}3}^\star c_{\mathbf{k}3} \right)$$

The transverse hamiltonian is similarly calculated as

$$H_T = \frac{1}{4\Omega} \int_\Omega d^3x \sum_{\mathbf{k}} \sum_{\mathbf{k}'} \sum_{s=1}^{2} \sum_{s'=1}^{2} \frac{\hbar}{\sqrt{\omega_k \omega_{k'}}} \times$$

$$\Big\{ m^2 c^2 \left(\mathbf{e}_{\mathbf{k}s} \cdot \mathbf{e}_{\mathbf{k}'s'} \right) \left[c_{\mathbf{k}s} e^{i(\mathbf{k}\cdot\mathbf{x} - \omega_k t)} + c_{\mathbf{k}s}^\star e^{-i(\mathbf{k}\cdot\mathbf{x} - \omega_k t)} \right] \times$$

$$\left[c_{\mathbf{k}'s'} e^{i(\mathbf{k}'\cdot\mathbf{x} - \omega_{k'}t)} + c_{\mathbf{k}'s'}^\star e^{-i(\mathbf{k}'\cdot\mathbf{x} - \omega_{k'}t)} \right]$$

$$-\omega_k \omega_{k'} \left(\mathbf{e}_{\mathbf{k}s} \cdot \mathbf{e}_{\mathbf{k}'s'} \right) \left[c_{\mathbf{k}s} e^{i(\mathbf{k}\cdot\mathbf{x} - \omega_k t)} - c_{\mathbf{k}s}^\star e^{-i(\mathbf{k}\cdot\mathbf{x} - \omega_k t)} \right] \times$$

$$\left[c_{\mathbf{k}'s'} e^{i(\mathbf{k}'\cdot\mathbf{x} - \omega_{k'}t)} - c_{\mathbf{k}'s'}^\star e^{-i(\mathbf{k}'\cdot\mathbf{x} - \omega_{k'}t)} \right]$$

$$-c^2 \left(\mathbf{k} \times \mathbf{e}_{\mathbf{k}s} \right) \cdot \left(\mathbf{k}' \times \mathbf{e}_{\mathbf{k}'s'} \right) \left[c_{\mathbf{k}s} e^{i(\mathbf{k}\cdot\mathbf{x} - \omega_k t)} - c_{\mathbf{k}s}^\star e^{-i(\mathbf{k}\cdot\mathbf{x} - \omega_k t)} \right] \times$$

$$\left[c_{\mathbf{k}'s'} e^{i(\mathbf{k}'\cdot\mathbf{x} - \omega_{k'}t)} - c_{\mathbf{k}'s'}^\star e^{-i(\mathbf{k}'\cdot\mathbf{x} - \omega_{k'}t)} \right] \Big\}$$

The spatial integration then gives[6]

$$H_T = \frac{1}{4} \sum_{\mathbf{k}} \sum_{s=1}^{2} \sum_{s'=1}^{2} \frac{\hbar c^2}{\omega_k} \times$$

$$\Big\{ m^2 \left[\left(c_{\mathbf{k}s} c_{\mathbf{k}s}^\star + c_{\mathbf{k}s}^\star c_{\mathbf{k}s} \right) \delta_{ss'} + \left(c_{\mathbf{k}s} c_{-\mathbf{k},s'} e^{-2i\omega_k t} + c_{\mathbf{k}s}^\star c_{-\mathbf{k},s'}^\star e^{2i\omega_k t} \right) \mathbf{e}_{\mathbf{k}s} \cdot \mathbf{e}_{-\mathbf{k},s'} \right]$$

$$+ \frac{\omega_k^2}{c^2} \left[\left(c_{\mathbf{k}s} c_{\mathbf{k}s}^\star + c_{\mathbf{k}s}^\star c_{\mathbf{k}s} \right) \delta_{ss'} - \left(c_{\mathbf{k}s} c_{-\mathbf{k},s'} e^{-2i\omega_k t} + c_{\mathbf{k}s}^\star c_{-\mathbf{k},s'}^\star e^{2i\omega_k t} \right) \mathbf{e}_{\mathbf{k}s} \cdot \mathbf{e}_{-\mathbf{k},s'} \right]$$

$$+ \mathbf{k}^2 \left[\left(c_{\mathbf{k}s} c_{\mathbf{k}s}^\star + c_{\mathbf{k}s}^\star c_{\mathbf{k}s} \right) \delta_{ss'} + \left(c_{\mathbf{k}s} c_{-\mathbf{k},s'} e^{-2i\omega_k t} + c_{\mathbf{k}s}^\star c_{-\mathbf{k},s'}^\star e^{2i\omega_k t} \right) \mathbf{e}_{\mathbf{k}s} \cdot \mathbf{e}_{-\mathbf{k},s'} \right] \Big\}$$

Here we have used

$$(\mathbf{k} \times \mathbf{e}_{\mathbf{k}s}) \cdot (\mathbf{k} \times \mathbf{e}_{\mathbf{k}s'}) = \mathbf{k}^2 \delta_{ss'}$$

$$(\mathbf{k} \times \mathbf{e}_{\mathbf{k}s}) \cdot (-\mathbf{k} \times \mathbf{e}_{-\mathbf{k},s'}) = -\mathbf{k}^2 \left(\mathbf{e}_{\mathbf{k}s} \cdot \mathbf{e}_{-\mathbf{k},s'} \right)$$

---

[6] Recall $\mathbf{e}_{\mathbf{k}s} \cdot \mathbf{e}_{\mathbf{k}s'} = \delta_{ss'}$.

With the use of the previous dispersion relation, this gives

$$H_T = \sum_{\mathbf{k}} \sum_{s=1}^{2} \hbar\omega_k \frac{1}{2} \left(c_{\mathbf{k}s} c_{\mathbf{k}s}^{\star} + c_{\mathbf{k}s}^{\star} c_{\mathbf{k}s} \right)$$

The transverse hamiltonian is now also expressed in normal modes.

(c) To quantize this system, we replace the Fourier coefficients $(c_{\mathbf{k}s}^{\star}, c_{\mathbf{k}s})$ by creation and destruction operators $(c_{\mathbf{k}s}^{\dagger}, c_{\mathbf{k}s})$, and impose the canonical commutation relations on them

$$[c_{\mathbf{k}s}, c_{\mathbf{k}'s'}^{\dagger}] = \delta_{ss'} \delta_{\mathbf{k}\mathbf{k}'}$$
$$[c_{\mathbf{k}s}^{\dagger}, c_{\mathbf{k}'s'}^{\dagger}] = [c_{\mathbf{k}s}, c_{\mathbf{k}'s'}] = 0$$

Let us then compute[7]

$$[\hat{V}_i(\mathbf{x}, t), \hat{\Pi}_j(\mathbf{x}', t')]_{t=t'} = [\hat{V}_i^T(\mathbf{x}, t), \hat{\Pi}_j^T(\mathbf{x}', t')]_{t=t'} + [\hat{V}_i^L(\mathbf{x}, t), \hat{\Pi}_j^L(\mathbf{x}', t')]_{t=t'}$$

$$= \frac{i\hbar}{2\Omega} \sum_{\mathbf{k}} \sum_{s=1}^{3} (\mathbf{e}_{\mathbf{k}s})_i (\mathbf{e}_{\mathbf{k}s})_j \left[e^{i\mathbf{k}\cdot(\mathbf{x}-\mathbf{x}')} + e^{-i\mathbf{k}\cdot(\mathbf{x}-\mathbf{x}')} \right]$$

$$= i\hbar \delta_{ij} \delta^{(3)}(\mathbf{x}-\mathbf{x}')$$

Furthermore

$$[\hat{V}_i(\mathbf{x}, t), \hat{V}_j(\mathbf{x}', t')]_{t=t'} = [\hat{\Pi}_i(\mathbf{x}, t), \hat{\Pi}_j(\mathbf{x}', t')]_{t=t'} = 0$$

Hence, with the above normal-mode expansions, the canonical commutation relations for the creation and destruction operators imply the proper canonical commutation relations for the fields.

---

[7]Note that [compare Eqs. (C.43)]

$$[\hat{V}_i^T(\mathbf{x}, t), \hat{\Pi}_j^T(\mathbf{x}', t')]_{t=t'} = i\hbar \delta_{ij}^T(\mathbf{x}-\mathbf{x}')$$

Appendix D

Irreducible Representations of SU(n)

Problem D.1 Show that Eq. (D.34) produces the singlet state [1] for any SU(n).

Solution to Problem D.1

The singlet state for SU(n) is defined in terms of the fundamental basis and its adjoint as

$$|n[1]\rangle \equiv |i_p\rangle \, |\bar{i}_p\rangle \qquad ; \; \text{SU(n) singlet}$$

Here the repeated index is summed from 1 to n. It follows from Eqs. (D.1), (D.3), and (D.21) that this direct-product state transforms according to

$$e^{i\omega^a \hat{G}^a} |i_p\rangle |\bar{i}_p\rangle = r_{i'_p i_p}(\omega)\, r^{\star}_{i''_p i_p}(\omega)\, |i'_p\rangle |\bar{i}''_p\rangle$$

The $n \times n$ matrix $\underline{r}(\omega)$ is unitary, and hence

$$r_{i'_p i_p}(\omega)\, r^{\star}_{i''_p i_p}(\omega) = \delta_{i'_p i''_p}$$

Therefore the state $|n[1]\rangle$ is invariant for any SU(n)

$$e^{i\omega^a \hat{G}^a} |i_p\rangle |\bar{i}_p\rangle = |i'_p\rangle |\bar{i}'_p\rangle = |i_p\rangle |\bar{i}_p\rangle$$

Problem D.2 Consider an octet of scalar mesons interacting with a triplet of Dirac particles in the Sakata model. To find the irreducible representations of SU(3) available to this system, one must evaluate $[8] \otimes [3]$. This is done with the aid of the Young tableaux rules as shown in Fig. D.2 in the text. Compute the dimension of each of the resulting representations, and show that $[8] \otimes [3] = [3] \oplus [\bar{6}] \oplus [15]$.

Solution to Problem D.2

The Young tableaux needed to carry out $[8] \otimes [3]$ in SU(3) are shown in Fig. D.2 in the text. To compute the dimension of the representations on the r.h.s. in that figure, we must first construct the required analog of Table D.2 in the text

Table D.1

$[\lambda]$	l	3	$[\lambda]$	l	3	$[\lambda]$	l	3
3	5	2	2	4	2	2	4	2
1	2	1	2	3	1	1	2	1
0	0	0	0	0	0	1	1	0

Then forming the ratio of decreasing differences according to the rule in Eq. (D.23), one has

$$^{(3)}N_{[3,1,0]} = \frac{5 \cdot 3 \cdot 2}{2 \cdot 1 \cdot 1} = 15$$

$$^{(3)}N_{[2,2,0]} = \frac{4 \cdot 3 \cdot 1}{2 \cdot 1 \cdot 1} = 6$$

$$^{(3)}N_{[2,1,1]} = \frac{3 \cdot 2 \cdot 1}{2 \cdot 1 \cdot 1} = 3$$

Here the partition in the Young tableaux is denoted with $[\lambda_1, \lambda_2, \lambda_3]$.

Hence, in summary, $[8] \otimes [3] = [15] \oplus [\bar{6}] \oplus [3]$.[1]

Problem D.3 Consider two nucleons outside of a light closed-shell nucleus and assume SU(4) symmetry with an inert core.

(a) Show that $[4] \otimes [4] = [6] \oplus [10]$, and hence determine the irreducible representations available to this system;

(b) Argue from the symmetry of the representations that the spin and isospin content $[(2S + 1) \otimes (2T + 1)]$ of these representations is $[6] = [(3) \otimes (1)] \oplus [(1) \otimes (3)]$ and $[10] = [(3) \otimes (3)] \oplus [(1) \otimes (1)]$;

(c) Given that the overall wave function of the pair must be antisymmetric, and that the attractive nuclear force favors a symmetric spatial state, which supermultiplet would you expect to lie lower in energy?

(d) The nuclei ($^6_4$Be, $^6_3$Li, $^6_2$He) effectively consist of two nucleons, each in the $1p$-state and coupled to total orbital angular momentum $L = 0$, moving about an inert $^4_2$He core. Compare your answer in (c) with the experimental spectra of these nuclei.[2]

---

[1]As with $[3]$, we label it $[\bar{6}]$ since it corresponds to two *holes*.

[2]Recall the discussion of "pairing" in Vol. I; for the data, see [National Nuclear Data Center (2009)].

Solution to Problem D.3

(a) The Young tableaux for reducing the direct product $[4] \otimes [4]$ in $SU(4)$ to a direct sum of irreducible representations are shown in Fig. D.1 (compare Fig. D.4 in the text).

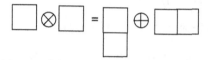

Fig. D.1   Young tableaux for $[4] \otimes [4] = [6] \oplus [10]$ in $SU(4)$.

The dimensions of the resulting irreducible representations are computed from Eq. (D.23) by first constructing the requisite analog of Table D.2 in the text

Table D.2

$[\lambda]$	l	4	$[\lambda]$	l	4
1	4	3	2	5	3
1	3	2	0	2	2
0	1	1	0	1	1
0	0	0	0	0	0

and then computing the ratio of decreasing differences as in Eqs. (D.28)

$$^{(4)}N_{[1,1,0,0]} = \frac{4 \cdot 3 \cdot 1 \cdot 3 \cdot 2 \cdot 1}{3 \cdot 2 \cdot 1 \cdot 2 \cdot 1 \cdot 1} = 6$$

$$^{(4)}N_{[2,0,0,0]} = \frac{5 \cdot 4 \cdot 3 \cdot 2 \cdot 1 \cdot 1}{3 \cdot 2 \cdot 1 \cdot 2 \cdot 1 \cdot 1} = 10$$

Here the partition in the Young tableaux is denoted with $[\lambda_1, \lambda_2, \lambda_3, \lambda_4]$.

(b) The Young operator in Eq. (D.12) associated with the tableaux, which produces the irreducible representations, symmetrizes along the rows, and then antisymmetrizes along the columns. Hence the representation corresponding to the first tableau on the r.h.s. in Fig. D.1 is antisymmetric in the spin-isospin indices, while that corresponding to the second tableau is symmetric in these indices. This observation allows us to immediately identify the the spin and isospin content of these representations. The triplet spin states with $S = 1$ for two nucleons are given by

$$\psi^{[3]}(1,2) = \psi_{1/2}(1)\psi_{1/2}(2) \qquad\qquad ; M_S = 1$$
$$= \frac{1}{\sqrt{2}}\left[\psi_{1/2}(1)\psi_{-1/2}(2) + \psi_{-1/2}(1)\psi_{1/2}(2)\right] \qquad ; M_S = 0$$
$$= \psi_{-1/2}(1)\psi_{-1/2}(2) \qquad\qquad ; M_S = -1$$

The singlet spin state with $S = 0$ is given by

$$\psi^{[1]}(1,2) = \frac{1}{\sqrt{2}}\left[\psi_{1/2}(1)\psi_{-1/2}(2) - \psi_{-1/2}(1)\psi_{1/2}(1)\right] \qquad ; M_S = 0$$

The former are symmetric in particle state labels, while the latter is anti-symmetric. An exactly analogous result holds for isospin. Hence to get a spin-isospin wave function that is antisymmetric in the exchange of state labels, one must merely take the product of the symmetric spin states with the antisymmetric isospin state, and *vice versa*. To get a totally symmetric spin-isospin wave function, one combines the symmetric spin states with the symmetric isospin states and the antisymmetric spin state with the antisymmetric isospin state. We can therefore immediately identify the spin and isospin content $[(2S + 1) \otimes (2T + 1)]$ of the previous SU(4) representations

$$[6] = [(3) \otimes (1)] \oplus [(1) \otimes (3)]$$
$$[10] = [(3) \otimes (3)] \oplus [(1) \otimes (1)]$$

(c) The spin-isospin wave function must be combined with the spatial wave function, which for two nucleons in the p-shell takes the form

$$\psi_{(1p)^2}(1,2) = \sum_{m_1,m_2} \langle 1m_1 1m_2 | 11LM_L\rangle \psi_{1m_1}(1)\psi_{1m_2}(2)$$

Under the exchange of spatial coordinates, this wave function behaves as

$$\psi_{(1p)^2}(2,1) = (-1)^{1+1-L}\psi_{(1p)^2}(1,2)$$

The maximum overlap with the attractive nuclear potential will be obtained in the symmetric case with $L = 0$. To get a totally antisymmetric wave function, one must then combine this with the [6]-dimensional SU(4) supermultiplet above. *Hence the* [6] *will be the lowest-lying supermultiplet.*

(d) The experimental low-lying spectrum for the $A = 6$ system is shown in Fig. D.2. All the states of the [6]-dimensional SU(4) supermultiplet are clearly identified, and the $^6_3\text{Li}_3$ ground state indeed has $(T = 0, S = 1)$.

MeV

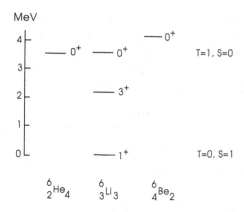

Fig. D.2   Level spectrum for the $A = 6$ nuclear system (atomic masses). $^6_2\text{He}_4$ decays by β^- emission, while $^6_4\text{Be}_2$ is particle unstable [National Nuclear Data Center (2009)].

The splitting of this ground state from the $(T = 1, S = 0)$ iso-multiplet provides a measure of the validity of $SU(4)$ symmetry in this system.[3]

Problem D.4 Consider the Fermi-Yang model of mesons as $(\bar{B}B)$ bound states in the Sakata model.

(a) Use the Young tableaux to show that $[\bar{3}] \otimes [3] = [1] \oplus [8]$ (note Prob. D.1). Make a weight diagram for the octet where Y is plotted as the ordinate and T_3 as abscissa;

(b) Compare with the observed low-lying 0^- and 1^- mesons;[4]

(c) Explain why this is the same set of multiplets, with the same quantum numbers, that one obtains in the quark model.

Solution to Problem D.4

(a) The Young tableaux for $[\bar{3}] \otimes [3]$ in $SU(3)$ are shown in Fig. D.3.

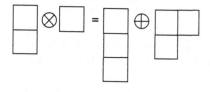

Fig. D.3   Young tableaux for $[\bar{3}] \otimes [3] = [1] \oplus [8]$ in $SU(3)$.

---

[3]Within this framework, the 3^+ excited state in $^6_3\text{Li}_3$ can be identified as the lowest member with $(T = 0, S = 1)$ of a [6]-dimensional supermultiplet built on the symmetric $(1p)^2$ configuration with $L = 2$. Readers should search for the other members!

[4]See [Particle Data Group (2009)]; note Fig. 7.2 and Table 7.4 in Vol. I.

The first representation on the r.h.s. is the singlet in SU(3), and, as shown in section D.6.2, the dimension of the second is [8]. Hence $[\bar{3}] \otimes [3] = [1] \oplus [8]$.

In the Fermi-Yang model of mesons as $(\bar{B}B)$ bound states in the Sakata model, the additive meson quantum numbers can be read off from those of the baryons and antibaryons shown in Table D.3.[5]

Table D.3

particle	B	S	T	T_3	$Y = B + S$	$Q = T_3 + Y/2$
p	1	0	1/2	1/2	1	1
n	1	0	1/2	-1/2	1	0
Λ	1	-1	0	0	0	0
$\bar{p}$	-1	0	1/2	-1/2	-1	-1
$\bar{n}$	-1	0	1/2	1/2	-1	0
$\bar{\Lambda}$	-1	1	0	0	0	0

The corresponding weight diagram for the meson octet is shown in Fig. D.3, where $Y = B + S$ is the ordinate and T_3 the abscissa.

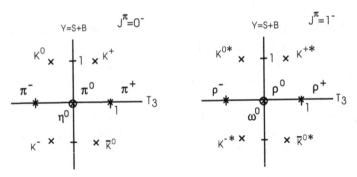

Fig. D.4   Weight diagram for the octet in the Fermi-Yang model of mesons as $(\bar{B}B)$ bound states in the Sakata model. The ordinate is $Y = B + S$ and the abscissa is T_3. The observed low-lying pseudoscalar mesons with $J^\pi = 0^-$ and vector mesons with $J^\pi = 1^-$ are also indicated [Particle Data Group (2009)].

(b) The observed low-lying pseudoscalar mesons with $J^\pi = 0^-$ and vector mesons with $J^\pi = 1^-$ have also been placed on Fig. D.3. Note that every site is occupied.

(c) The baryon quantum number B *cancels* for the $(\bar{B}B)$ bound states. Hence, exactly the same meson quantum numbers are obtained for the $(\bar{q}q)$ bound states in the quark model, where the only difference is the baryon number $B = 1/3$ of the quarks.

---

[5]For example, the state $|\bar{p}n\rangle$ has $T_3 = -1$, $Y = 0$, and $Q = -1$.

Appendix E

Lorentz Transformations in Quantum Field Theory

Problem E.1 (a) Use the expressions for $(P_\mu, M_{\mu\nu})$ in Eqs. (E.44), and derive the equations for the generators of the inhomogeneous Lorentz group in Eqs. (E.45);

(b) Show that replacing $M_{\mu\nu} \to M_{\mu\nu} + \hbar\sigma_{\mu\nu}/2$ leaves the commutation relations in part (a) unaltered;

(c) Show that γ_μ now satisfies the same commutation relation with $M_{\mu\nu}$ as does P_μ.

Solution to Problem E.1

(a) Equations (E.44) read

$$P_\mu \equiv \frac{\hbar}{i} \frac{\partial}{\partial x_\mu}$$

$$M_{\mu\nu} \equiv \frac{\hbar}{i} \left(x_\mu \frac{\partial}{\partial x_\nu} - x_\nu \frac{\partial}{\partial x_\mu} \right)$$

It is clear that the components P_μ commute with each other. Consider

$$
\begin{aligned}
\frac{i}{\hbar}[M_{\mu\nu}, P_\lambda] &= \frac{\hbar}{i} \left[x_\mu \frac{\partial}{\partial x_\nu} - x_\nu \frac{\partial}{\partial x_\mu}, \frac{\partial}{\partial x_\lambda} \right] \\
&= \frac{\hbar}{i} \left(-\delta_{\mu\lambda} \frac{\partial}{\partial x_\nu} + \delta_{\nu\lambda} \frac{\partial}{\partial x_\mu} \right) \\
&= \delta_{\nu\lambda} P_\mu - \delta_{\mu\lambda} P_\nu
\end{aligned}
$$

Now work out the last commutator

$$\frac{i}{\hbar}[M_{\mu\nu}, M_{\rho\sigma}] = \frac{\hbar}{i} \left[x_\mu \frac{\partial}{\partial x_\nu} - x_\nu \frac{\partial}{\partial x_\mu}, x_\rho \frac{\partial}{\partial x_\sigma} - x_\sigma \frac{\partial}{\partial x_\rho} \right]$$

This yields

$$\frac{i}{\hbar}[M_{\mu\nu}, M_{\rho\sigma}] = \frac{\hbar}{i}\left(\delta_{\nu\rho}x_\mu\frac{\partial}{\partial x_\sigma} - \delta_{\mu\sigma}x_\rho\frac{\partial}{\partial x_\nu} - \delta_{\nu\sigma}x_\mu\frac{\partial}{\partial x_\rho} + \delta_{\mu\rho}x_\sigma\frac{\partial}{\partial x_\nu} - \right.$$
$$\left. \delta_{\mu\rho}x_\nu\frac{\partial}{\partial x_\sigma} + \delta_{\nu\sigma}x_\rho\frac{\partial}{\partial x_\mu} + \delta_{\mu\sigma}x_\nu\frac{\partial}{\partial x_\rho} - \delta_{\nu\rho}x_\sigma\frac{\partial}{\partial x_\mu}\right)$$
$$= \delta_{\mu\sigma}M_{\nu\rho} + \delta_{\nu\rho}M_{\mu\sigma} - \delta_{\mu\rho}M_{\nu\sigma} - \delta_{\nu\sigma}M_{\mu\rho}$$

Thus we have established Eqs. (E.45).

(b) Now replace

$$M_{\mu\nu} \rightarrow M_{\mu\nu} + \frac{\hbar}{2}\sigma_{\mu\nu} \equiv M_{\mu\nu} + \tilde{M}_{\mu\nu}$$

where $\tilde{M}_{\mu\nu} = \hbar\sigma_{\mu\nu}/2$ and $\sigma_{\mu\nu} = [\gamma_\mu, \gamma_\nu]/2i$. Clearly the commutation relations involving P_μ are unaffected. We just have to examine

$$\frac{i}{\hbar}[\tilde{M}_{\mu\nu}, \tilde{M}_{\rho\sigma}] = \frac{i\hbar}{4}[\sigma_{\mu\nu}, \sigma_{\rho\sigma}] = \frac{\hbar}{16i}[\gamma_\mu\gamma_\nu - \gamma_\nu\gamma_\mu, \gamma_\rho\gamma_\sigma - \gamma_\sigma\gamma_\rho]$$

Now use

$$[\gamma_\mu\gamma_\nu, \gamma_\rho\gamma_\sigma] = \gamma_\mu\gamma_\nu\gamma_\rho\gamma_\sigma - \gamma_\rho\gamma_\sigma\gamma_\mu\gamma_\nu$$
$$= \gamma_\mu\gamma_\rho\gamma_\sigma\gamma_\nu + 2\delta_{\nu\rho}\gamma_\mu\gamma_\sigma - 2\delta_{\nu\sigma}\gamma_\mu\gamma_\rho - \gamma_\rho\gamma_\sigma\gamma_\mu\gamma_\nu$$
$$= 2\delta_{\mu\rho}\gamma_\sigma\gamma_\nu - 2\delta_{\mu\sigma}\gamma_\rho\gamma_\nu + 2\delta_{\nu\rho}\gamma_\mu\gamma_\sigma - 2\delta_{\nu\sigma}\gamma_\mu\gamma_\rho$$

Thus

$$\frac{i}{\hbar}[\tilde{M}_{\mu\nu}, \tilde{M}_{\rho\sigma}] = \frac{\hbar}{8i}\left(\delta_{\mu\rho}\gamma_\sigma\gamma_\nu - \delta_{\mu\sigma}\gamma_\rho\gamma_\nu + \delta_{\nu\rho}\gamma_\mu\gamma_\sigma - \delta_{\nu\sigma}\gamma_\mu\gamma_\rho - \right.$$
$$\delta_{\nu\rho}\gamma_\sigma\gamma_\mu + \delta_{\nu\sigma}\gamma_\rho\gamma_\mu - \delta_{\mu\rho}\gamma_\nu\gamma_\sigma + \delta_{\mu\sigma}\gamma_\nu\gamma_\rho -$$
$$\delta_{\mu\sigma}\gamma_\rho\gamma_\nu + \delta_{\mu\rho}\gamma_\sigma\gamma_\nu - \delta_{\nu\sigma}\gamma_\mu\gamma_\rho + \delta_{\nu\rho}\gamma_\mu\gamma_\sigma +$$
$$\left.\delta_{\nu\sigma}\gamma_\rho\gamma_\mu - \delta_{\nu\rho}\gamma_\sigma\gamma_\mu + \delta_{\mu\sigma}\gamma_\nu\gamma_\rho - \delta_{\mu\rho}\gamma_\nu\gamma_\sigma\right)$$

A combination of terms, and re-identification of $\sigma_{\mu\nu}$, then gives

$$\frac{i}{\hbar}[\tilde{M}_{\mu\nu}, \tilde{M}_{\rho\sigma}] = \delta_{\mu\sigma}\tilde{M}_{\nu\rho} + \delta_{\nu\rho}\tilde{M}_{\mu\sigma} - \delta_{\mu\rho}\tilde{M}_{\nu\sigma} - \delta_{\nu\sigma}\tilde{M}_{\mu\rho}$$

Hence Eqs. (E.45) continue to hold in the presence of the additive contribution $\tilde{M}_{\mu\nu} = \hbar\sigma_{\mu\nu}/2$.

(c) Consider

$$\frac{i}{\hbar}[\tilde{M}_{\mu\nu}, \gamma_\lambda] = \frac{1}{4}[\gamma_\mu\gamma_\nu - \gamma_\nu\gamma_\mu, \gamma_\lambda]$$

$$= \frac{1}{4}(\gamma_\mu\gamma_\nu\gamma_\lambda - \gamma_\lambda\gamma_\mu\gamma_\nu - \gamma_\nu\gamma_\mu\gamma_\lambda + \gamma_\lambda\gamma_\nu\gamma_\mu)$$

$$= \delta_{\nu\lambda}\gamma_\mu - \delta_{\mu\lambda}\gamma_\nu$$

This demonstrates that γ_λ satisfies the same commutation relation with $M_{\mu\nu} + \tilde{M}_{\mu\nu}$ as does P_λ.

Problem E.2 Start from Eqs. (E.55), go to infinitesimals as in the text, and derive Eqs. (E.54).

Solution to Problem E.2

For the Dirac field, the arguments following from the infinitesimal form of the translation operator are unchanged from the spin-zero case in Eqs. (E.24)–(E.28). The result following from an infinitesimal rotation in four-dimensions is summarized in the spin-zero case in Eqs. (E.34) and (E.41)

$$\hat{U}(L)\hat{\phi}(x)\hat{U}(L)^{-1} = \hat{\phi}(x') \qquad ; \; x'_\mu = a_{\mu\nu}(-v)x_\nu$$

$$\hat{U}(L) = \exp\left\{-\frac{i}{\hbar}\frac{\Omega}{2}\alpha_{\mu\nu}\hat{M}_{\mu\nu}\right\}$$

$$\frac{i}{\hbar}[\hat{M}_{\mu\nu}, \hat{\phi}(x)] = -\left(x_\mu\frac{\partial}{\partial x_\nu} - x_\nu\frac{\partial}{\partial x_\mu}\right)\hat{\phi}(x)$$

Evidently a factor of $-(\Omega/2)\alpha_{\mu\nu}$ has been cancelled from both sides in arriving at the final infinitesimal result, where the rotation angle $\Omega = \epsilon \to 0$.

The corresponding result in the Dirac case is that given in Eqs. (E.55)

$$\hat{U}(L)\hat{\psi}(x_\mu)\hat{U}(L)^{-1} = \mathcal{S}(\Omega)\,\hat{\psi}(x'_\mu)$$

$$\mathcal{S}(\Omega) \equiv \exp\left\{\frac{i}{4}\Omega\alpha_{\mu\nu}\sigma_{\mu\nu}\right\}$$

An expansion for infinitesimal rotation, and cancellation of the same factor of $-(\Omega/2)\alpha_{\mu\nu}$ then produces the second of Eqs. (E.54)

$$\frac{i}{\hbar}[\hat{P}_\mu, \hat{\psi}(x)] = -\frac{\partial}{\partial x_\mu}\hat{\psi}(x)$$

$$\frac{i}{\hbar}[\hat{M}_{\mu\nu}, \hat{\psi}(x)] = -\left(x_\mu\frac{\partial}{\partial x_\nu} - x_\nu\frac{\partial}{\partial x_\mu} + \frac{i}{2}\sigma_{\mu\nu}\right)\hat{\psi}(x)$$

Problem E.3 A two-component, spin-1/2, fermion field transforms in the following fashion under a real spatial rotation [compare Eq. (6.49)]

$$e^{-i\boldsymbol{\omega}\cdot\hat{\mathbf{J}}}\,\hat{\psi}(\mathbf{x})\,e^{i\boldsymbol{\omega}\cdot\hat{\mathbf{J}}} = \exp\left\{\frac{i}{2}\boldsymbol{\omega}\cdot\boldsymbol{\sigma}\right\}\hat{\psi}(\mathbf{x}') \qquad ; \; x_i' = a_{ij}(-\boldsymbol{\omega})\,x_j$$

where $\boldsymbol{\sigma}$ are the Pauli matrices, and the underlining of the column vectors and matrices is suppressed. Specialize the Lorentz transformation of the Dirac field in Eq. (E.55) to a real rotation, and show that both the upper and lower components individually transform in this fashion.[1]

Solution to Problem E.3

For a real rotation, Eqs. (E.43) and (E.55) read

$$\hat{U}(R)\hat{\psi}(x_\mu)\hat{U}(R)^{-1} = S(\omega)\,\hat{\psi}(x_\mu')$$

$$\hat{U}(R) = \exp\left\{-\frac{i\omega}{2\hbar}\alpha_{ij}\hat{M}_{ij}\right\}$$

$$S(\omega) = \exp\left\{\frac{i\omega}{4}\alpha_{ij}\sigma_{ij}\right\}$$

From Eq. (E.31), one has

$$\frac{1}{2\hbar}\alpha_{ij}\hat{M}_{ij} = (\mathbf{n}_1 \times \mathbf{n}_2)\cdot\hat{\mathbf{J}} = \mathbf{n}\cdot\hat{\mathbf{J}}$$

In an exactly analogous calculation, one has from Eq. (E.51)

$$\frac{1}{2}\alpha_{ij}\sigma_{ij} = (\mathbf{n}_1 \times \mathbf{n}_2)\cdot\boldsymbol{\Sigma} = \mathbf{n}\cdot\boldsymbol{\Sigma}$$

where the 4×4 matrix $\boldsymbol{\Sigma}$ is given in 2×2 form by

$$\boldsymbol{\Sigma} = \begin{pmatrix} \boldsymbol{\sigma} & 0 \\ 0 & \boldsymbol{\sigma} \end{pmatrix}$$

It follows from the above that for a real rotation, the Lorentz transformation of the Dirac field specializes to

$$e^{-i\boldsymbol{\omega}\cdot\hat{\mathbf{J}}}\,\hat{\psi}(\mathbf{x})\,e^{i\boldsymbol{\omega}\cdot\hat{\mathbf{J}}} = \exp\left\{\frac{i}{2}\boldsymbol{\omega}\cdot\boldsymbol{\Sigma}\right\}\hat{\psi}(\mathbf{x}') \qquad ; \; \boldsymbol{\omega} = \omega\mathbf{n}$$

Since the matrix $\boldsymbol{\Sigma}$ is block diagonal in 2×2 form, the pair of upper and lower components of the Dirac spinor are un-mixed by the rotation,

---

[1] Recall that $\boldsymbol{\Sigma} = \begin{pmatrix} \boldsymbol{\sigma} & 0 \\ 0 & \boldsymbol{\sigma} \end{pmatrix}$.

and hence both the upper and lower components individually transform in the stated fashion

$$e^{-i\boldsymbol{\omega}\cdot\hat{\mathbf{J}}}\,\hat{\psi}(\mathbf{x})\,e^{i\boldsymbol{\omega}\cdot\hat{\mathbf{J}}} = \exp\left\{\frac{i}{2}\boldsymbol{\omega}\cdot\boldsymbol{\sigma}\right\}\hat{\psi}(\mathbf{x}') \qquad ; \; x_i' = a_{ij}(-\omega)\,x_j$$

Problem E.4 The Dirac spinor transformation matrix in Eq. (E.55) is

$$S(\Omega) \equiv \exp\left\{\frac{i}{4}\Omega\alpha_{\mu\nu}\sigma_{\mu\nu}\right\} \qquad ; \; \sigma_{\mu\nu} = \frac{1}{2i}[\gamma_\mu, \gamma_\nu]$$

$$; \; \alpha_{\mu\nu} = (n_1)_\mu(n_2)_\nu - (n_2)_\mu(n_1)_\nu$$

(a) Show $\gamma_4 S(\Omega)^\dagger \gamma_4 = S(\Omega)^{-1}$;

(b) It is a general result that

$$S(\Omega)^{-1}\gamma_\mu S(\Omega) = a_{\mu\nu}(-v)\gamma_\nu \qquad ; \; \tan i\Omega = -iv/c$$

Establish this relation for a Lorentz transformation in the z-direction where the unit vectors characterizing the plane of rotation are $n_1 = (0,0,1,0)$ and $n_2 = (0,0,0,i)$. Use the expansion in repeated commutators, and the commutation relations in Prob. E.1(c), to show

$$S(\Omega)^{-1}\gamma_3 S(\Omega) = \gamma_3 \cos i\Omega + \gamma_4 \sin i\Omega$$
$$S(\Omega)^{-1}\gamma_4 S(\Omega) = -\gamma_3 \sin i\Omega + \gamma_4 \cos i\Omega$$

Hence reproduce the Lorentz transformation matrix of Eq. (E.2).

Solution to Problem E.4

(a) Since $\sigma_{\mu\nu} = [\gamma_\mu, \gamma_\nu]/2i$ is hermitian, and $\alpha_{\mu\nu}$ has an imaginary fourth component, one has

$$\alpha_{\mu\nu}^\star \sigma_{\mu\nu}^\dagger \gamma_4 = \alpha_{\mu\nu}^\star \sigma_{\mu\nu} \gamma_4 = \gamma_4 \alpha_{\mu\nu}\sigma_{\mu\nu}$$

Hence

$$S(\Omega)^\dagger\gamma_4 = \exp\left\{-\frac{i}{4}\Omega\,\alpha_{\mu\nu}^\star \sigma_{\mu\nu}^\dagger\right\}\gamma_4 = \gamma_4 \exp\left\{-\frac{i}{4}\Omega\,\alpha_{\mu\nu}\sigma_{\mu\nu}\right\} = \gamma_4 S(\Omega)^{-1}$$

(b) for a Lorentz transformation in the z-direction where the unit vectors characterizing the plane of rotation are $n_1 = (0,0,1,0)$ and $n_2 = (0,0,0,i)$, one has

$$\alpha_{\mu\nu} = (n_1)_\mu(n_2)_\nu - (n_2)_\mu(n_1)_\nu = i[\delta_{\mu3}\delta_{\nu4} - \delta_{\mu4}\delta_{\nu3}]$$

Therefore, in this case,

$$\alpha_{\mu\nu}\sigma_{\mu\nu} = 2\gamma_3\gamma_4$$

$$S(\Omega) = \exp\left\{\frac{i}{2}\Omega\,\gamma_3\gamma_4\right\}$$

We can now compute

$$\exp\left\{-\frac{i}{2}\Omega\,\gamma_3\gamma_4\right\}\gamma_3\exp\left\{\frac{i}{2}\Omega\,\gamma_3\gamma_4\right\} = \gamma_3 + \left(\frac{-i\Omega}{2}\right)[\gamma_3\gamma_4, \gamma_3] +$$

$$\left(\frac{-i\Omega}{2}\right)^2\frac{1}{2!}[\gamma_3\gamma_4, [\gamma_3\gamma_4, \gamma_3]] + \left(\frac{-i\Omega}{2}\right)^3\frac{1}{3!}[\gamma_3\gamma_4, [\gamma_3\gamma_4, [\gamma_3\gamma_4, \gamma_3]]] + \cdots$$

Evaluation of the commutators gives

$$\exp\left\{-\frac{i}{2}\Omega\,\gamma_3\gamma_4\right\}\gamma_3\exp\left\{\frac{i}{2}\Omega\,\gamma_3\gamma_4\right\} = \gamma_3 + (i\Omega)\gamma_4 - \frac{(i\Omega)^2}{2!}\gamma_3$$

$$-\frac{(i\Omega)^3}{3!}\gamma_4 + \cdots$$

Therefore

$$S(\Omega)^{-1}\gamma_3 S(\Omega) = \gamma_3\cos i\Omega + \gamma_4\sin i\Omega$$

In a similar fashion

$$\exp\left\{-\frac{i}{2}\Omega\,\gamma_3\gamma_4\right\}\gamma_4\exp\left\{\frac{i}{2}\Omega\,\gamma_3\gamma_4\right\} = \gamma_4 + \left(\frac{-i\Omega}{2}\right)[\gamma_3\gamma_4, \gamma_4] +$$

$$\left(\frac{-i\Omega}{2}\right)^2\frac{1}{2!}[\gamma_3\gamma_4, [\gamma_3\gamma_4, \gamma_4]] + \left(\frac{-i\Omega}{2}\right)^3\frac{1}{3!}[\gamma_3\gamma_4, [\gamma_3\gamma_4, [\gamma_3\gamma_4, \gamma_4]]] + \cdots$$

Evaluation of the commutators now gives

$$\exp\left\{-\frac{i}{2}\Omega\,\gamma_3\gamma_4\right\}\gamma_4\exp\left\{\frac{i}{2}\Omega\,\gamma_3\gamma_4\right\} = \gamma_4 - (i\Omega)\gamma_3 - \frac{(i\Omega)^2}{2!}\gamma_4$$

$$+\frac{(i\Omega)^3}{3!}\gamma_3 + \cdots$$

Thus, in this case,

$$S(\Omega)^{-1}\gamma_4 S(\Omega) = -\gamma_3\sin i\Omega + \gamma_4\cos i\Omega$$

These are the stated results. Notice that

$$S(\Omega)^{-1}\gamma_1 S(\Omega) = \gamma_1 \qquad ; \ S(\Omega)^{-1}\gamma_2 S(\Omega) = \gamma_2$$

The Lorentz transformation matrix in Eq. (E.2) is

$$\underline{a}(-v) = \begin{bmatrix} 1 & 0 & 0 & 0 \\ 0 & 1 & 0 & 0 \\ 0 & 0 & \dfrac{1}{\sqrt{1-v^2/c^2}} & \dfrac{-iv/c}{\sqrt{1-v^2/c^2}} \\ 0 & 0 & \dfrac{iv/c}{\sqrt{1-v^2/c^2}} & \dfrac{1}{\sqrt{1-v^2/c^2}} \end{bmatrix}$$

Hence if we identify

$$\tan i\Omega = -i\frac{v}{c}$$

we obtain[2]

$$\mathcal{S}(\Omega)^{-1}\gamma_\mu \mathcal{S}(\Omega) = a_{\mu\nu}(-v)\gamma_\nu$$

Problem E.5 Use the results in Prob. E.4 to show that:

(a) The bilinear combination $\hat{\bar{\psi}}(x)\hat{\psi}(x)$ transforms as a scalar under homogeneous Lorentz transformations;

(b) The Dirac current $i\hat{\bar{\psi}}(x)\gamma_\mu\hat{\psi}(x)$ transforms as a four-vector [see Eq. (E.56)].

Solution to Problem E.5

(a) The behavior of the Dirac field under a Lorentz transformation is given in Eq. (E.55)

$$\hat{U}\hat{\psi}(x_\mu)\hat{U}^{-1} = \mathcal{S}(\Omega)\,\hat{\psi}(x'_\mu) \qquad ; \; x'_\mu = a_{\mu\nu}(-v)x_\nu$$

and $\mathcal{S}(\Omega)$ is given in Prob. E.4, and its properties are analyzed there.

The bilinear combination $\hat{\bar{\psi}}(x)\hat{\psi}(x)$ then transforms as follows

$$\begin{aligned} \hat{U}\hat{\bar{\psi}}(x)\hat{\psi}(x)\hat{U}^{-1} &= \hat{\psi}^\dagger(x')\mathcal{S}^\dagger(\Omega)\gamma_4\mathcal{S}(\Omega)\hat{\psi}(x') \\ &= \hat{\bar{\psi}}(x')\mathcal{S}(\Omega)^{-1}\mathcal{S}(\Omega)\hat{\psi}(x') \\ &= \hat{\bar{\psi}}(x')\hat{\psi}(x') \end{aligned}$$

This is a Lorentz scalar.

---

[2]Note that if the sign of v is reversed so that $\tan i\Omega = iv/c$, one obtains

$$\mathcal{S}(\Omega)^{-1}\gamma_\mu\mathcal{S}(\Omega) = a_{\mu\nu}(v)\gamma_\nu = a_{\nu\mu}(-v)\gamma_\nu$$

See the solution to Prob. 5.19.

(b) The current $i\hat{\bar{\psi}}(x)\gamma_\mu\hat{\psi}(x)$ then transforms as a four-vector

$$\hat{U}i\hat{\bar{\psi}}(x)\gamma_\mu\hat{\psi}(x)\hat{U}^{-1} = i\hat{\psi}^\dagger(x')\mathcal{S}^\dagger(\Omega)\gamma_4\gamma_\mu\mathcal{S}(\Omega)\hat{\psi}(x')$$
$$= i\hat{\bar{\psi}}(x')\mathcal{S}(\Omega)^{-1}\gamma_\mu\mathcal{S}(\Omega)\hat{\psi}(x')$$
$$= a_{\mu\nu}(-v)i\hat{\bar{\psi}}(x')\gamma_\nu\hat{\psi}(x')$$

This is Eq. (E.56).

Problem E.6 Given a set of generators $\hat{G}_\alpha$ in the abstract Hilbert space satisfying $[\hat{G}_\alpha,\ \hat{\phi}(x)] = -G_\alpha\hat{\phi}(x)$ where the G_α form a Lie algebra:
   (a) Prove the relation $[[\hat{G}_\alpha,\ \hat{G}_\beta],\ \hat{\phi}(x)] = -[G_\alpha,\ G_\beta]\hat{\phi}(x)$;
   (b) Hence verify the statement made before Eqs. (E.44).

Solution to Problem E.6

(a) Consider

$$[[\hat{G}_\alpha,\ \hat{G}_\beta],\ \hat{\phi}(x)] = \left(\hat{G}_\alpha\hat{G}_\beta - \hat{G}_\beta\hat{G}_\alpha\right)\hat{\phi}(x) - \hat{\phi}(x)\left(\hat{G}_\alpha\hat{G}_\beta - \hat{G}_\beta\hat{G}_\alpha\right)$$

Use the following identities

$$\hat{G}_\alpha\hat{G}_\beta\hat{\phi}(x) \equiv \hat{G}_\alpha[\hat{G}_\beta,\ \hat{\phi}(x)] + [\hat{G}_\alpha,\ \hat{\phi}(x)]\hat{G}_\beta + \hat{\phi}(x)\hat{G}_\alpha\hat{G}_\beta$$
$$\hat{G}_\beta\hat{G}_\alpha\hat{\phi}(x) \equiv \hat{G}_\beta[\hat{G}_\alpha,\ \hat{\phi}(x)] + [\hat{G}_\beta,\ \hat{\phi}(x)]\hat{G}_\alpha + \hat{\phi}(x)\hat{G}_\beta\hat{G}_\alpha$$

Thus

$$[[\hat{G}_\alpha,\ \hat{G}_\beta],\ \hat{\phi}(x)] = \hat{G}_\alpha[\hat{G}_\beta,\ \hat{\phi}(x)] + [\hat{G}_\alpha,\ \hat{\phi}(x)]\hat{G}_\beta$$
$$-\hat{G}_\beta[\hat{G}_\alpha,\ \hat{\phi}(x)] - [\hat{G}_\beta,\ \hat{\phi}(x)]\hat{G}_\alpha$$

Now use the relation for the generators given in the statement of the problem

$$[\hat{G}_\alpha,\ \hat{\phi}(x)] = -G_\alpha\hat{\phi}(x) \qquad ;\ [\hat{G}_\beta,\ \hat{\phi}(x)] = -G_\beta\hat{\phi}(x)$$

Since (G_α, G_β) are no longer operators in the abstract Hilbert space, it follows that

$$[[\hat{G}_\alpha,\ \hat{G}_\beta],\ \hat{\phi}(x)] = -G_\beta\hat{G}_\alpha\hat{\phi}(x) - G_\alpha\hat{\phi}(x)\hat{G}_\beta$$
$$+G_\alpha\hat{G}_\beta\hat{\phi}(x) + G_\beta\hat{\phi}(x)\hat{G}_\alpha$$
$$= G_\alpha[\hat{G}_\beta,\ \hat{\phi}(x)] - G_\beta[\hat{G}_\alpha,\ \hat{\phi}(x)]$$

Now use the previous relations once more

$$[\,[\hat{G}_\alpha,\ \hat{G}_\beta],\ \hat{\phi}(x)] = -G_\alpha\,G_\beta\,\hat{\phi}(x) + G_\beta\,G_\alpha\,\hat{\phi}(x)$$
$$= -[G_\alpha,\ G_\beta]\hat{\phi}(x)$$

This is the desired result.

(b) Now we can just work backwards. Suppose the G_α form a Lie algebra

$$[G_\alpha,\ G_\beta] = i f^{\alpha\beta\gamma} G_\gamma$$

Then

$$[\,[\hat{G}_\alpha,\ \hat{G}_\beta],\ \hat{\phi}(x)] = -i f^{\alpha\beta\gamma} G_\gamma\,\hat{\phi}(x)$$
$$= i f^{\alpha\beta\gamma}[\hat{G}_\gamma,\ \hat{\phi}(x)]$$

Hence we can identify the corresponding operator relation

$$[\hat{G}_\alpha,\ \hat{G}_\beta] = i f^{\alpha\beta\gamma}\,\hat{G}_\gamma$$

We thus verify the statement made just before Eq. (E.44)

As with angular momentum, the commutation relations of the generators among themselves can be obtained by simply looking at the commutation relations of the covariant, first-quantized differential operators $(P_\mu, M_{\mu\nu})$.

Appendix F

Green's Functions and Other Singular Functions

Problem F.1 Show that the commutator of the interaction-picture scalar field at two distinct space-time points $[\hat{\phi}(x^{(1)}), \hat{\phi}(x^{(2)})]$ depends only on the relative space-time coordinate $x = x^{(1)} - x^{(2)}$.

Solution to Problem F.1

The massless scalar field in the interaction picture in Eq. (5.60) is

$$\hat{\phi}(\mathbf{x}, t) = \sum_{\mathbf{k}} \left(\frac{\hbar}{2\omega_k \Omega}\right)^{1/2} \left[a_{\mathbf{k}} e^{i(\mathbf{k}\cdot\mathbf{x} - \omega_k t)} + a_{\mathbf{k}}^{\dagger} e^{-i(\mathbf{k}\cdot\mathbf{x} - \omega_k t)}\right] \quad ; \; \omega_k \equiv |\mathbf{k}| c$$

The creation and destruction operators satisfy

$$[a_{\mathbf{k}}, a_{\mathbf{k}'}^{\dagger}] = \delta_{\mathbf{k}\mathbf{k}'}$$
$$[a_{\mathbf{k}}, a_{\mathbf{k}'}] = [a_{\mathbf{k}}^{\dagger}, a_{\mathbf{k}'}^{\dagger}] = 0$$

Thus the commutator of this interaction-picture scalar field at two distinct space-time points $[\hat{\phi}(x^{(1)}), \hat{\phi}(x^{(2)})]$ is given by

$$[\hat{\phi}(x^{(1)}), \hat{\phi}(x^{(2)})] = \sum_{\mathbf{k}} \frac{\hbar}{2\omega_k \Omega} \left\{e^{ik_\mu[x^{(1)} - x^{(2)}]_\mu} - e^{-ik_\mu[x^{(1)} - x^{(2)}]_\mu}\right\}$$

$$k_\mu = (\mathbf{k}, i\omega_k/c) \quad ; \; x_\mu = (\mathbf{x}, ict)$$

This is now explicitly a function of the relative space-time coordinate $x = x^{(1)} - x^{(2)}$. The same is true for the massive field with $\omega_k = c\sqrt{\mathbf{k}^2 + m^2}$.

Problem F.2 The completeness relation for the free Dirac spinors was derived in Probl. 12.7; it can be written as

$$\sum_{\lambda} \left[u_\alpha(\mathbf{k}\lambda) u_\beta^{\dagger}(\mathbf{k}\lambda) + v_\alpha(\mathbf{k}\lambda) v_\beta^{\dagger}(\mathbf{k}\lambda)\right] = \delta_{\alpha\beta}$$

The positive and negative-energy projection matrices are constructed as

$$P_\pm = \left[\frac{E_k \pm H_D}{2E_k}\right] \qquad ; \; H_D = \hbar c(\boldsymbol{\alpha} \cdot \mathbf{k} + \beta M)$$

Multiply the completeness relation by these matrices on the left, and then by $\beta = \gamma_4$ on the right, to show that

$$\sum_\lambda u_\alpha(\mathbf{k}\lambda)\bar{u}_\beta(\mathbf{k}\lambda) = \left[\frac{M - i\gamma_\mu k_\mu}{2\omega_k/c}\right]_{\alpha\beta} \qquad ; \; k_\mu = (\mathbf{k}, i\omega_k/c)$$

$$\sum_\lambda v_\alpha(-\mathbf{k}\lambda)\bar{v}_\beta(-\mathbf{k}\lambda) = \left[\frac{-M - i\gamma_\mu k_\mu}{2\omega_k/c}\right]_{\alpha\beta} \qquad \omega_k = c\sqrt{\mathbf{k}^2 + M^2}$$

Here $\gamma = i\alpha\beta$, and we have taken $\mathbf{k} \to -\mathbf{k}$ in the second result.

Solution to Problem F.2

It is important to explicitly exhibit the matrix indices and sums in this problem. The Dirac equations are

$$\hbar c \sum_\rho (\boldsymbol{\alpha} \cdot \mathbf{k} + \beta M)_{\alpha\rho} u_\rho(\mathbf{k}\lambda) = E_k u_\alpha(\mathbf{k}\lambda)$$

$$\hbar c \sum_\rho (\boldsymbol{\alpha} \cdot \mathbf{k} + \beta M)_{\alpha\rho} v_\rho(\mathbf{k}\lambda) = -E_k v_\alpha(\mathbf{k}\lambda)$$

where $E_k = \hbar c\sqrt{\mathbf{k}^2 + M^2}$, and λ is the helicity. Hence[1]

$$\sum_\rho [\hbar c(\boldsymbol{\alpha} \cdot \mathbf{k} + \beta M) - E_k]_{\alpha\rho}\, u_\rho(\mathbf{k}\lambda) = 0$$

$$\sum_\rho [\hbar c(\boldsymbol{\alpha} \cdot \mathbf{k} + \beta M) + E_k]_{\alpha\rho}\, v_\rho(\mathbf{k}\lambda) = 0$$

Also

$$\sum_\rho [\hbar c(\boldsymbol{\alpha} \cdot \mathbf{k} + \beta M) + E_k]_{\alpha\rho}\, u_\rho(\mathbf{k}\lambda) = 2E_k u_\alpha(\mathbf{k}\lambda)$$

$$\sum_\rho [\hbar c(\boldsymbol{\alpha} \cdot \mathbf{k} + \beta M) - E_k]_{\alpha\rho}\, v_\rho(\mathbf{k}\lambda) = -2E_k v_\alpha(\mathbf{k}\lambda)$$

Introduce the projection matrices

$$[P_\pm]_{\sigma\rho} = \left[\frac{E_k \pm H_D}{2E_k}\right]_{\sigma\rho} \qquad ; \; H_D = \hbar c(\boldsymbol{\alpha} \cdot \mathbf{k} + \beta M)$$

---

[1] We suppress the unit matrix.

Then the above relations take the form

$$\sum_\rho [P_+]_{\sigma\rho} u_\rho(\mathbf{k}\lambda) = u_\sigma(\mathbf{k}\lambda) \qquad ; \sum_\rho [P_+]_{\sigma\rho} v_\rho(\mathbf{k}\lambda) = 0$$

$$\sum_\rho [P_-]_{\sigma\rho} v_\rho(\mathbf{k}\lambda) = v_\sigma(\mathbf{k}\lambda) \qquad ; \sum_\rho [P_-]_{\sigma\rho} u_\rho(\mathbf{k}\lambda) = 0$$

Multiplication of the completeness relation by the projection matrix P_+ isolates the positive-energy spinors, and gives

$$\sum_\rho [P_+]_{\sigma\rho} \left[\sum_\lambda u_\rho(\mathbf{k}\lambda)\, u_\nu^\dagger(\mathbf{k}\lambda) \right] = [P_+]_{\sigma\nu}$$

or

$$\sum_\lambda u_\sigma(\mathbf{k}\lambda)\, u_\nu^\dagger(\mathbf{k}\lambda) = \left[\frac{E_k/\hbar c + \boldsymbol{\alpha} \cdot \mathbf{k} + \beta M}{2E_k/\hbar c} \right]_{\sigma\nu}$$

Perform $\sum_\nu \beta_{\nu\mu}$, and introduce

$$\boldsymbol{\gamma} = i\boldsymbol{\alpha}\beta \qquad ; \omega_k = c\sqrt{\mathbf{k}^2 + M^2} \qquad ; k_\mu = (\mathbf{k}, i\omega_k/c)$$

Then

$$\sum_\lambda u_\sigma(\mathbf{k}\lambda)\bar{u}_\mu(\mathbf{k}\lambda) = \left[\frac{M - i\gamma_\mu k_\mu}{2\omega_k/c} \right]_{\sigma\mu} \qquad ; k_\mu = (\mathbf{k}, i\omega_k/c)$$

In the same fashion, multiplication of the completeness relation by P_- gives

$$\sum_\rho [P_-]_{\sigma\rho} \left[\sum_\lambda v_\rho(\mathbf{k}\lambda)\, v_\nu^\dagger(\mathbf{k}\lambda) \right] = [P_-]_{\sigma\nu}$$

or

$$\sum_\lambda v_\sigma(\mathbf{k}\lambda)\, v_\nu^\dagger(\mathbf{k}\lambda) = \left[\frac{E_k/\hbar c - \boldsymbol{\alpha} \cdot \mathbf{k} - \beta M}{2E_k/\hbar c} \right]_{\sigma\nu}$$

Perform $\sum_\nu \beta_{\nu\mu}$, and introduce

$$\boldsymbol{\gamma} = i\boldsymbol{\alpha}\beta \qquad ; \omega_k = c\sqrt{\mathbf{k}^2 + M^2} \qquad ; k_\mu = (\mathbf{k}, i\omega_k/c)$$

Then, provided we take $\mathbf{k} \to -\mathbf{k}$,

$$\sum_\lambda v_\sigma(-\mathbf{k}\lambda)\bar{v}_\mu(-\mathbf{k}\lambda) = \left[\frac{-M - i\gamma_\mu k_\mu}{2\omega_k/c} \right]_{\sigma\mu} \qquad ; k_\mu = (\mathbf{k}, i\omega_k/c)$$

Problem F.3 Show directly that the Dirac Green's function constructed in Eq. (F.34) satisfies the defining Eq. (F.29).

Solution to Problem F.3

Equations (F.34) read

$$
G^D(x - x') = \left[\gamma_\nu \frac{\partial}{\partial x_\nu} - M \right] \frac{1}{(2\pi)^4} \int \frac{e^{ik\cdot(x-x')}}{k^2 + M^2} d^4 k
$$

$$
= \left[\gamma_\nu \frac{\partial}{\partial x_\nu} - M \right] G(x - x')
$$

where (x, x') are four-vectors. We are asked to show this satisfies Eq. (F.29)

$$
\left[\gamma_\mu \frac{\partial}{\partial x_\mu} + M \right]_{\alpha\sigma} G^D_{\sigma\beta}(x - x') = -\delta_{\alpha\beta}\, \delta^{(4)}(x - x')
$$

Make use of the same algebra as in Eq. (F.33)[2]

$$
\gamma_\nu \gamma_\mu \frac{\partial}{\partial x_\nu} \frac{\partial}{\partial x_\mu} = \frac{1}{2} \left(\gamma_\nu \gamma_\mu + \gamma_\mu \gamma_\nu \right) \frac{\partial}{\partial x_\nu} \frac{\partial}{\partial x_\mu}
$$

$$
= \frac{\partial}{\partial x_\mu} \frac{\partial}{\partial x_\mu} = \Box
$$

It follows that

$$
\left\{ \left[\gamma_\mu \frac{\partial}{\partial x_\mu} + M \right] \left[\gamma_\nu \frac{\partial}{\partial x_\nu} - M \right] \right\}_{\alpha\beta} G(x - x') = \left(\Box - M^2 \right) \delta_{\alpha\beta} G(x - x')
$$

Now use

$$
\left(\Box - M^2 \right) \delta_{\alpha\beta}\, G(x - x') = -\delta_{\alpha\beta} \int \frac{d^4 k}{(2\pi)^4} \frac{k^2 + M^2}{k^2 + M^2} e^{ik\cdot(x-x')}
$$

$$
= -\delta_{\alpha\beta} \int \frac{d^4 k}{(2\pi)^4} e^{ik\cdot(x-x')}
$$

$$
= -\delta_{\alpha\beta}\, \delta^{(4)}(x - x')
$$

where we have identified the continuum form of the Dirac delta-function. This establishes Eq. (F.29).

Problem F.4 (a) Fill in the steps in the derivation of the transverse photon propagator in Eq. (F.45);

(b) Show that one recovers the canonical commutation relations for the vector potential in the Coulomb gauge from this expression.

---

[2] We again suppress the unit matrix in this relation.

Solution to Problem F.4

(a) The vector potential in the Coulomb gauge in the interaction picture is given in Eqs. (F.2)

$$\hat{\mathbf{A}}(x) = \sum_{\mathbf{k}} \sum_{s=1}^{2} \left(\frac{\hbar}{2\omega_k \varepsilon_0 \Omega} \right)^{1/2} \left[a_{\mathbf{k}s} \mathbf{e}_{\mathbf{k}s} e^{ik \cdot x} + a^\dagger_{\mathbf{k}s} \mathbf{e}_{\mathbf{k}s} e^{-ik \cdot x} \right]$$

The derivation for the propagator follows closely that for the scalar field in section F.3.1. The differences are:

- The mass of the photon is zero, so that $\omega_k/c = |\mathbf{k}|$. We then use the notation $\Delta \to D$ for the invariant functions;
- There are polarization vectors present, and the sum over transverse polarization vectors is given in Eq. (C.42)

$$\sum_{s=1}^{2} (\mathbf{e}_{\mathbf{k}s})_i (\mathbf{e}_{\mathbf{k}s})_j = \delta_{ij} - \frac{k_i k_j}{\mathbf{k}^2}$$

- There is an additional factor of $1/\varepsilon_0$.

The analogs of Eqs. (F.43) are then

$$\langle 0|T[\hat{A}_i(x_\mu), \hat{A}_j(x'_\mu)]|0 \rangle = \frac{1}{\Omega \varepsilon_0} \sum_{\mathbf{k}} \left(\delta_{ij} - \frac{k_i k_j}{\mathbf{k}^2} \right) \frac{\hbar}{2\omega_k} e^{ik \cdot (x-x')} \qquad ; t > t'$$

$$= \frac{1}{\Omega \varepsilon_0} \sum_{\mathbf{k}} \left(\delta_{ij} - \frac{k_i k_j}{\mathbf{k}^2} \right) \frac{\hbar}{2\omega_k} e^{-ik \cdot (x-x')} \qquad ; t < t'$$

The factor in parentheses can be taken out of the sum as

$$\left(\delta_{ij} - \frac{k_i k_j}{\mathbf{k}^2} \right) \to \left(\delta_{ij} - \frac{1}{\nabla^2} \frac{\partial}{\partial x_i} \frac{\partial}{\partial x_j} \right)$$

In fact, this expression is defined through its Fourier transform. Thus

$$\langle 0|T[\hat{A}_i(x_\mu), \hat{A}_j(x'_\mu)]|0 \rangle = \frac{\hbar}{ic\varepsilon_0} \left[\delta_{ij} - \frac{1}{\nabla^2} \frac{\partial}{\partial x_i} \frac{\partial}{\partial x_j} \right] D_F(x_\mu - x'_\mu)$$

$$\equiv \frac{\hbar}{ic\varepsilon_0} D_F^T(x_\mu - x'_\mu)_{ij}$$

where

$$\frac{\hbar}{ic}D_F(x_\mu - x'_\mu) = \frac{1}{\Omega}\sum_k \frac{\hbar}{2\omega_k}e^{ik\cdot(x-x')} = \frac{\hbar}{ic}D_+(x_\mu - x'_\mu) \qquad ; t > t'$$

$$= \frac{1}{\Omega}\sum_k \frac{\hbar}{2\omega_k}e^{-ik\cdot(x-x')} = -\frac{\hbar}{ic}D_-(x_\mu - x'_\mu) \qquad ; t < t'$$

(b) Let us show how this works for the scalar field, and then the transition to the transverse radiation field is carried out as in part (a). Take the time derivative of Eq. (F.43) with respect to t' [3]

$$\langle 0|T[\hat{\dot{\phi}}(x_\mu), \hat{\phi}(x'_\mu)]|0\rangle = \frac{i\hbar}{2}\frac{1}{\Omega}\sum_k e^{ik\cdot(x-x')} \qquad ; t > t'$$

$$= -\frac{i\hbar}{2}\frac{1}{\Omega}\sum_k e^{-ik\cdot(x-x')} \qquad ; t < t'$$

Write

$$T[\hat{\dot{\phi}}(x_\mu), \hat{\phi}(x'_\mu)] = \hat{\dot{\phi}}(x_\mu)\hat{\phi}(x'_\mu) \qquad ; t > t'$$

$$= \hat{\phi}(x_\mu)\hat{\dot{\phi}}(x'_\mu) + [\hat{\dot{\phi}}(x'_\mu), \hat{\phi}(x_\mu)] \qquad ; t < t'$$

The discontinuity in the time-ordered product at $t = t'$, a c-number, is therefore[4]

$$[\hat{\dot{\phi}}(x'_\mu), \hat{\phi}(x_\mu)]_{t=t'} = -\frac{i\hbar}{\Omega}\sum_k e^{-ik\cdot(x-x')}$$

$$= \frac{\hbar}{i}\delta^{(3)}(x - x')$$

This is the canonical commutation relation for the scalar field in Eq. (F.9). The corresponding relation for the transverse radiation field is given in Eq. (C.66)

$$\left[\hat{\dot{A}}_i(x',t'), \hat{A}_j(x,t)\right]_{t=t'} = \frac{\hbar}{i\varepsilon_0}\delta^T_{ij}(x - x')$$

where δ^T_{ij} is defined in Eq. (C.43).

Problem F.5 This problem evaluates the Feynman propagator for a massive vector field; it is more difficult, but the result is valuable.

---

[3]Since the equal-time commutator of the field at two distinct spatial points vanishes, there is no additional term arising when one differentiates the time-ordered product as in Eq. (8.37).

[4]We use $\sum_k e^{ik\cdot(x-x')} = \sum_k e^{-ik\cdot(x-x')}$.

(a) Start from the interaction-picture field expansions in Prob. C.7 and show that the vacuum expectation value of the time-ordered product of fields can be written as

$$\langle 0|T[\hat{V}_\mu(x)\,\hat{V}_\nu(0)]|0\rangle = \frac{1}{\Omega}\sum_\mathbf{k}\frac{\hbar c^2}{2\omega_k}\Delta^F_{\mu\nu}(\mathbf{k})e^{i\mathbf{k}\cdot\mathbf{x}}e^{-i\omega_k t} \qquad ; t>0$$

$$= \frac{1}{\Omega}\sum_\mathbf{k}\frac{\hbar c^2}{2\omega_k}\Delta^F_{\mu\nu}(\mathbf{k})e^{-i\mathbf{k}\cdot\mathbf{x}}e^{i\omega_k t} \qquad ; t<0$$

where $\hat{V}_4(x) \equiv (ic/m^2)\boldsymbol{\nabla}\cdot\hat{\boldsymbol{\Pi}}(x)$, and

$$\Delta^F_{ij}(\mathbf{k}) = \delta_{ij} + \frac{k_i k_j}{m^2} \qquad ; (i,j)=1,2,3$$

$$\Delta^F_{j4}(\mathbf{k}) = \Delta^F_{4j}(\mathbf{k}) = \frac{ik_j\omega_k}{m^2 c} \qquad ; \Delta^F_{44}(\mathbf{k}) = -\frac{\mathbf{k}^2}{m^2}$$

(b) Show that in the limit $\Omega \to \infty$, this is the same result one gets by evaluating the following four-dimensional Fourier transform in Minkowski space using the techniques of contour integration, where the k_0-integral is defined in all cases by closing with a large semi-circle in the appropriate half-plane

$$\langle 0|T[\hat{V}_\mu(x)\,\hat{V}_\nu(0)]|0\rangle = \frac{\hbar c}{i}\Delta^F_{\mu\nu}(x;\,m^2)$$

$$\Delta^F_{\mu\nu}(x;\,m^2) = \frac{1}{(2\pi)^4}\int d^4k\,\frac{e^{ik\cdot x}}{k^2+m^2}\left(\delta_{\mu\nu} + \frac{k_\mu k_\nu}{m^2}\right) \qquad ; m\to m-i\eta$$

Solution to Problem F.5

(a) We use the normal-mode expansion of Prob. C.7, with the creation and destruction operators

$$\hat{\mathbf{V}}_T(\mathbf{x},t) = \sum_\mathbf{k}\sum_{s=1}^2\left(\frac{\hbar c^2}{2\omega_k\Omega}\right)^{1/2}\left[c_{\mathbf{k}s}\mathbf{e}_{\mathbf{k}s}e^{i(\mathbf{k}\cdot\mathbf{x}-\omega_k t)} + c^\dagger_{\mathbf{k}s}\mathbf{e}_{\mathbf{k}s}e^{-i(\mathbf{k}\cdot\mathbf{x}-\omega_k t)}\right]$$

$$\hat{\boldsymbol{\Pi}}_T(\mathbf{x},t) = \frac{1}{i}\sum_\mathbf{k}\sum_{s=1}^2\left(\frac{\hbar\omega_k}{2c^2\Omega}\right)^{1/2}\left[c_{\mathbf{k}s}\mathbf{e}_{\mathbf{k}s}e^{i(\mathbf{k}\cdot\mathbf{x}-\omega_k t)} - c^\dagger_{\mathbf{k}s}\mathbf{e}_{\mathbf{k}s}e^{-i(\mathbf{k}\cdot\mathbf{x}-\omega_k t)}\right]$$

$$\hat{\mathbf{V}}_L(\mathbf{x},t) = \sum_\mathbf{k}\left(\frac{\hbar\omega_k}{2m^2\Omega}\right)^{1/2}\left[c_{\mathbf{k}3}\mathbf{e}_{\mathbf{k}3}e^{i(\mathbf{k}\cdot\mathbf{x}-\omega_k t)} + c^\dagger_{\mathbf{k}3}\mathbf{e}_{\mathbf{k}3}e^{-i(\mathbf{k}\cdot\mathbf{x}-\omega_k t)}\right]$$

$$\hat{\boldsymbol{\Pi}}_L(\mathbf{x},t) = \frac{1}{i}\sum_\mathbf{k}\left(\frac{\hbar m^2}{2\omega_k\Omega}\right)^{1/2}\left[c_{\mathbf{k}3}\mathbf{e}_{\mathbf{k}3}e^{i(\mathbf{k}\cdot\mathbf{x}-\omega_k t)} - c^\dagger_{\mathbf{k}3}\mathbf{e}_{\mathbf{k}3}e^{-i(\mathbf{k}\cdot\mathbf{x}-\omega_k t)}\right]$$

Also

$$\hat{\mathbf{V}} = \hat{\mathbf{V}}_\mathrm{T} + \hat{\mathbf{V}}_\mathrm{L} \qquad ; \ \hat{\mathbf{\Pi}} = \hat{\mathbf{\Pi}}_\mathrm{T} + \hat{\mathbf{\Pi}}_\mathrm{L}$$

Then

$$\langle 0|T[\hat{V}_i(x)\,\hat{V}_j(0)]|0\rangle = \langle 0|T[\hat{V}_i^T(x)\,\hat{V}_j^T(0)]|0\rangle + \langle 0|T[\hat{V}_i^L(x)\,\hat{V}_j^L(0)]|0\rangle$$

$$= \frac{1}{\Omega}\sum_\mathbf{k} \frac{\hbar c^2}{2\omega_k} \Delta_{ij}^F(\mathbf{k}) e^{i\mathbf{k}\cdot\mathbf{x} - i\omega_k t} \qquad ; \ t > 0$$

$$= \frac{1}{\Omega}\sum_\mathbf{k} \frac{\hbar c^2}{2\omega_k} \Delta_{ij}^F(\mathbf{k}) e^{-i\mathbf{k}\cdot\mathbf{x} + i\omega_k t} \qquad ; \ t < 0$$

Here $\omega_k = c\sqrt{\mathbf{k}^2 + m^2}$, and

$$\Delta_{ij}^F(\mathbf{k}) = \sum_{s=1}^{2} (\mathbf{e}_{\mathbf{k}s})_i (\mathbf{e}_{\mathbf{k}s})_j + \frac{\omega_k^2}{m^2 c^2}(\mathbf{e}_{\mathbf{k}3})_i (\mathbf{e}_{\mathbf{k}3})_j$$

$$= \delta_{ij} - \frac{k_i k_j}{\mathbf{k}^2} + \frac{(m^2 + \mathbf{k}^2)}{m^2}\frac{k_i k_j}{\mathbf{k}^2}$$

$$= \delta_{ij} + \frac{k_i k_j}{m^2}$$

For the fourth component

$$\langle 0|T[\hat{V}_i(x)\,\hat{V}_4(0)]|0\rangle = \frac{ic}{m^2}\langle 0|T[\hat{V}_i(x)\,\boldsymbol{\nabla}\cdot\hat{\mathbf{\Pi}}(0)]|0\rangle$$

In this case

$$\Delta_{i4}^F(\mathbf{k}) = \Delta_{4i}^F(\mathbf{k}) = \frac{ik_i\omega_k}{m^2 c}$$

A similar calculation gives

$$\langle 0|T[\hat{V}_4(x)\,\hat{V}_4(0)]|0\rangle = -\frac{c^2}{m^4}\langle 0|T[\boldsymbol{\nabla}\cdot\hat{\mathbf{\Pi}}(x)\,\boldsymbol{\nabla}\cdot\hat{\mathbf{\Pi}}(0)]|0\rangle$$

Here

$$\Delta_{44}^F(\mathbf{k}) = -\frac{\mathbf{k}^2}{m^2}$$

(b) Consider the integral

$$\Delta_{\mu\nu}^F(x; m^2) = \frac{1}{(2\pi)^4}\int d^4k \frac{e^{ik\cdot x}}{k^2 + m^2}\left(\delta_{\mu\nu} + \frac{k_\mu k_\nu}{m^2}\right) \qquad ; \ m \to m - i\eta$$

The analysis of this expression is carried out as in section F.2. Consider the k_0-integral

$$\mathcal{I}_{\mu\nu}(\mathbf{k};t) \equiv \int_{-\infty}^{\infty} \frac{dk_0}{2\pi} \frac{e^{-ik_0 ct}}{k^2 + m^2} \left(\delta_{\mu\nu} + \frac{k_\mu k_\nu}{m^2} \right) =$$

$$- \int_{-\infty}^{\infty} \frac{dk_0}{2\pi} \frac{e^{-ik_0 ct}}{(k_0 - \omega_k/c + i\eta)(k_0 + \omega_k/c - i\eta)} \left(\delta_{\mu\nu} + \frac{k_\mu k_\nu}{m^2} \right)$$

The integrand has poles at $k_0 = \omega_k/c - i\eta$ and $k_0 = -\omega_k/c + i\eta$ (see Fig. F.3 in the text). If $t > 0$, the integral can be completed with a large semi-circle in the lower-half k_0-plane, and if $t < 0$, in the upper-half plane (see Fig. F.2 in the text). Thus, for $\mathcal{I}_{ij}(\mathbf{k};t)$

$$\mathcal{I}_{ij}(\mathbf{k};t) = i\frac{e^{-i\omega_k t}}{2\omega_k/c} \left(\delta_{ij} + \frac{k_i k_j}{m^2} \right) \qquad ; t > 0$$

$$= i\frac{e^{i\omega_k t}}{2\omega_k/c} \left(\delta_{ij} + \frac{k_i k_j}{m^2} \right) \qquad ; t < 0$$

It follows that

$$\Delta_{ij}^F(x; m^2) = i \int \frac{d^3k}{(2\pi)^3} \frac{e^{ik\cdot x}}{2\omega_k/c} \left(\delta_{ij} + \frac{k_i k_j}{m^2} \right) \qquad ; t > 0$$

$$= i \int \frac{d^3k}{(2\pi)^3} \frac{e^{-ik\cdot x}}{2\omega_k/c} \left(\delta_{ij} + \frac{k_i k_j}{m^2} \right) \qquad ; t < 0$$

where we take the remaining integration variable $\mathbf{k} \to -\mathbf{k}$ in the second line. It is evident that this reproduces the infinite-volume limit $\Omega \to \infty$ of the result obtained in part (a)

$$\langle 0|T[\hat{V}_i(x)\,\hat{V}_j(0)]|0\rangle = \frac{\hbar c}{i}\Delta_{ij}^F(x; m^2)$$

For $\mathcal{I}_{i4}(\mathbf{k};t)$, the additional factor in the integrand can be written, for $t > 0$,

$$\frac{k_i k_4}{m^2} = i\frac{k_i}{m^2} \left(k_0 - \frac{\omega_k}{c} + \frac{\omega_k}{c} \right)$$

The term $(k_0 - \omega_k/c)$ cancels the pole, and hence produces an integral that vanishes by Cauchy's theorem. Thus[5]

$$\frac{k_i k_4}{m^2} \doteq i\frac{k_i \omega_k}{m^2 c}$$

[5]Here $\doteq$ discards the vanishing contribution.

For $t < 0$, we write this as

$$\frac{k_i k_4}{m^2} = i \frac{k_i}{m^2} \left(k_0 + \frac{\omega_k}{c} - \frac{\omega_k}{c} \right)$$

The same argument then gives

$$\frac{k_i k_4}{m^2} \doteq i \frac{(-k_i)(-\omega_k)}{m^2 c}$$

Here the additional minus sign in $(-k_i)$ comes from again changing the dummy integration variable $\mathbf{k} \to -\mathbf{k}$ in the remaining integral. Hence, we again reproduce the infinite-volume limit of the result in part (a) with

$$\langle 0 | T[\hat{V}_i(x) \, \hat{V}_4(0)] | 0 \rangle = \frac{\hbar c}{i} \Delta_{i4}^F(x; \, m^2)$$

For $\mathcal{I}_{44}(\mathbf{k}; t)$, the additional factor in the integrand is

$$1 + \frac{k_4 k_4}{m^2} = 1 - \frac{1}{m^2} \left(k_0^2 - \frac{\omega_k^2}{c^2} + \frac{\omega_k^2}{c^2} \right)$$

$$\doteq 1 - \frac{\omega_k^2}{m^2 c^2} = 1 - \frac{(\mathbf{k}^2 + m^2)}{m^2}$$

$$\doteq -\frac{\mathbf{k}^2}{m^2}$$

Here the factor $k_0^2 - \omega_k^2/c^2$ removes both poles. Thus, we again reproduce the infinite-volume limit of the result in part (a) with

$$\langle 0 | T[\hat{V}_4(x) \, \hat{V}_4(0)] | 0 \rangle = \frac{\hbar c}{i} \Delta_{44}^F(x; \, m^2)$$

In *summary* we have shown that the Feynman propagator for the massive neutral vector meson is

$$\langle 0 | T[\hat{V}_\mu(x) \, \hat{V}_\nu(0)] | 0 \rangle = \frac{\hbar c}{i} \Delta_{\mu\nu}^F(x; \, m^2)$$

$$\Delta_{\mu\nu}^F(x; \, m^2) = \frac{1}{(2\pi)^4} \int d^4 k \, \frac{e^{ik \cdot x}}{k^2 + m^2} \left(\delta_{\mu\nu} + \frac{k_\mu k_\nu}{m^2} \right) \qquad ; \, m \to m - i\eta$$

where the k_0-integral is defined in all cases by closing with a large semicircle in the appropriate half-plane.

Dimensional Regularization

Problem G.1 (a) Define the vacuum polarization tensor $\Pi_{\mu\nu}(l)$ in Eq. (G.29) through dimensional regularization, which allows one to manipulate it algebraically. Prove that it satisfies the current conservation relations $l_\mu \Pi_{\mu\nu}(l) = \Pi_{\mu\nu}(l)l_\nu = 0$;[1]

(c) Hence show that it has the general form in Eq. (G.31).

Solution to Problem G.1

The expression for the vacuum polarization tensor $\Pi_{\mu\nu}(l)$ in Eq. (G.29) is

$$\Pi_{\mu\nu}(l) = -\frac{ie^2}{\hbar c \varepsilon_0} \int \frac{d^n k}{(2\pi)^4} \mathrm{Tr}\left[\frac{1}{i(\slashed{k} - \slashed{l}/2) + M}\gamma_\mu \frac{1}{i(\slashed{k} + \slashed{l}/2) + M}\gamma_\nu\right]$$

Here the integral is defined through dimensional regularization, which allows us to manipulate it algebraically.

Consider the quantity

$$\frac{1}{i\slashed{a} + M} - \frac{1}{i\slashed{b} + M}$$

Multiply the first term on the right by $1 = (i\slashed{b} + M)(i\slashed{b} + M)^{-1}$ and the second term on the left by $1 = (i\slashed{a} + M)^{-1}(i\slashed{a} + M)$. Then

$$\frac{1}{i\slashed{a} + M} - \frac{1}{i\slashed{b} + M} = \frac{1}{i\slashed{a} + M}(i\slashed{b} + M)\frac{1}{i\slashed{b} + M} - \frac{1}{i\slashed{a} + M}(i\slashed{a} + M)\frac{1}{i\slashed{b} + M}$$

$$= \frac{1}{i\slashed{a} + M}i(\slashed{b} - \slashed{a})\frac{1}{i\slashed{b} + M}$$

---

[1] *Hint:* First show that $[i\slashed{a} + M]^{-1}[i(\slashed{b} - \slashed{a})][i\slashed{b} + M]^{-1} = [i\slashed{a} + M]^{-1} - [i\slashed{b} + M]^{-1}$.

Now take $\Pi_{\mu\nu}(l)l_\nu$. Observe that the trace is invariant under cyclic permutations, and make use of this result

$$\Pi_{\mu\nu}(l)l_\nu = -\frac{e^2}{\hbar c\varepsilon_0}\int \frac{d^n k}{(2\pi)^4}\operatorname{Tr}\gamma_\mu\left[\frac{1}{i(\slashed{k}+\slashed{l}/2)+M}\,i\slashed{l}\,\frac{1}{i(\slashed{k}-\slashed{l}/2)+M}\right]$$

$$= \frac{e^2}{\hbar c\varepsilon_0}\int \frac{d^n k}{(2\pi)^4}\operatorname{Tr}\gamma_\mu\left[\frac{1}{i(\slashed{k}+\slashed{l}/2)+M}-\frac{1}{i(\slashed{k}-\slashed{l}/2)+M}\right]$$

Change integration variables to $k \to k + l/2$ in the first term, and $k \to k - l/2$ in the second. The two terms then cancel. Hence

$$\Pi_{\mu\nu}(l)l_\nu = l_\mu\Pi_{\mu\nu}(l) = 0$$

where the last equality is proven in exactly the same manner.

(b) It follows that the general form of $\Pi_{\mu\nu}(l)$ can be exhibited as in Eq. (G.31)

$$\Pi_{\mu\nu}(l) = C(l^2)(l_\mu l_\nu - l^2\delta_{\mu\nu})$$

Problem G.2 It was shown in Eq. (8.48) that the self-energy insertion for a photon in Fig. 8.3(b) in the text leads to the following S-matrix element[2]

$$[e^2\hat{S}_{02}]_{fi} = (2\pi)^4\delta^{(4)}(l'-l)\frac{ic}{\Omega}\frac{1}{\sqrt{4\omega_1\omega_2}}\varepsilon_\mu^f\Pi_{\mu\nu}(l)\varepsilon_\nu^i$$

where the photon polarization vectors are of the form $\varepsilon = (\mathbf{e}_{1s},0)$ and satisfy $l\cdot\varepsilon_f = l\cdot\varepsilon_i = 0$. Insert the general form of the polarization tensor in Eq. (G.31), use the result in Eqs. (G.30)–(G.33), and show this matrix element *vanishes* for a real photon. Hence conclude that there is no mass renormalization for the photon to this order.

Solution to Problem G.2

As demonstrated in the previous problem, the general form of the polarization tensor $\Pi_{\mu\nu}(l)$ follows from current conservation, and it is exhibited in Eq. (G.31)

$$\Pi_{\mu\nu}(l) = C(l^2)(l_\mu l_\nu - l^2\delta_{\mu\nu})$$

---

[2]Here $k \to l$ and $\mu \leftrightarrow \nu$.

If this is inserted in $[e^2\hat{S}_{02}]_{fi}$ as given in the statement of the problem, and one uses $l \cdot \varepsilon_f = l \cdot \varepsilon_i = 0$, then

$$[e^2\hat{S}_{02}]_{fi} = (2\pi)^4 \delta^{(4)}(l' - l) \frac{ic}{\Omega} \frac{1}{\sqrt{4\omega_1\omega_2}} \left[-l^2 C(l^2)\right] \varepsilon_f \cdot \varepsilon_i$$

$$= (2\pi)^4 \delta^{(4)}(l' - l) \frac{ic}{\Omega} \frac{1}{\sqrt{4\omega_1\omega_2}} \left[\frac{1}{n-1}\Pi_{\mu\mu}(l^2)\right] \varepsilon_f \cdot \varepsilon_i$$

where Eq. (G.32) has been used in obtaining the second line. Equation (G.33) states that

$$\Pi_{\mu\mu}(0) = 0$$

Hence, the vacuum polarization insertion illustrated in Fig. 8.3(b) in the text vanishes for a real photon with $l^2 = 0$

$$[e^2\hat{S}_{02}]_{fi} = 0 \qquad \text{; real photon}$$

Thus a real photon remains massless in QED, at least to this order.

Problem G.3 Start from the Clifford algebra in n-dimensions in Eq. (G.25), and show that $\text{Tr}\,\gamma_\mu = 0$. (*Hint*: make use of $\Gamma_2 \equiv \gamma_\mu\gamma_\lambda$ with $\lambda \neq \mu$.)

Solution to Problem G.3

Assume $\lambda \neq \mu$, and consider

$$\gamma_\lambda\gamma_\mu\gamma_\lambda + \gamma_\lambda\gamma_\lambda\gamma_\mu = 0 \qquad \text{; } \lambda \neq \mu \text{, no sum on } \lambda$$

Now take the trace, making use of the properties in Eqs. (8.72). It follows that

$$\text{Tr}\,\gamma_\lambda\gamma_\mu\gamma_\lambda + \text{Tr}\,\gamma_\lambda\gamma_\lambda\gamma_\mu = 2\,\text{Tr}\,\gamma_\mu = 0$$

Appendix H

Path Integrals and the Electromagnetic Field

Problem H.1 Show that the generating functional for the electromagnetic field is given by

$$\tilde{W}_0(J) = \exp\left\{ \frac{i}{2\hbar c} \int d^4x \int d^4y \, J_\mu(x) D_{\mu\nu}(x-y) J_\nu(y) \right\} \qquad ; \text{E-M field}$$

Remember appendix H is in H-L units.

Solution to Problem H.1

We start by writing down the lagrangian density for the free electromagnetic field in H-L units

$$\mathcal{L}_0(A_\mu) = -\frac{1}{4} F_{\mu\nu} F_{\mu\nu}$$

$$F_{\mu\nu} = \frac{\partial A_\nu}{\partial x_\mu} - \frac{\partial A_\mu}{\partial x_\nu}$$

The interaction of the E-M field with an external current is described by the interaction term

$$\mathcal{L}_{\text{int}} = J_\mu A_\mu$$

and the generating functional for QED is given in Eq. (H.22)[1]

$$\tilde{W}_0[J] = \int\int\int \mathcal{D}(\bar{g})\mathcal{D}(g)\mathcal{D}(A_\mu) \times$$

$$\exp\left\{ \frac{i}{\hbar c} \int d^4x \left[\mathcal{L}_0 + J_\mu A_\mu - \frac{1}{2\xi}\left(\frac{\partial A_\mu}{\partial x_\mu}\right)^2 - \frac{\partial \bar{g}}{\partial x_\mu}\frac{\partial g}{\partial x_\mu} \right] \right\} / (\cdots)_{J=0}$$

---

[1] Recall that the "ghost" contribution factors and cancels here.

With the use of Eqs. (H.23)–(H.24), we can write

$$\int d^4x \left[\mathcal{L}_0 - \frac{1}{2\xi} \left(\frac{\partial A_\mu}{\partial x_\mu} \right)^2 \right] = \frac{1}{2} \int d^4x \int d^4y \, A_\mu(x) K_{\mu\nu}(x,y) A_\nu(y)$$

where

$$K_{\mu\nu}(x,y) = \left[\Box_x \delta_{\mu\nu} - \left(1 - \frac{1}{\xi} \right) \frac{\partial}{\partial x_\mu} \frac{\partial}{\partial x_\nu} \right] \delta^{(4)}(x-y)$$

If we now consider the numerator of $\tilde{W}_0(J)$ we may perform a change of variable in the path integral

$$A_\mu(x) \to A_\mu(x) + \int d^4z \, D_{\mu\rho}(x-z) J_\rho(z)$$

Evidently

$$\int \mathcal{D}(A_\mu) \to \int \mathcal{D}(A_\mu)$$

Now make use of Eq. (H.25)[2]

$$\int d^4y \, K_{\mu\nu}(x,y) D_{\nu\rho}(y-z) = -\delta^{(4)}(x-z) \delta_{\mu\rho}$$

This allows us to write, in the path integral,

$$\int d^4x \left[\mathcal{L}_0 - \frac{1}{2\xi} \left(\frac{\partial A_\mu}{\partial x_\mu} \right)^2 + J_\mu A_\mu \right] = \frac{1}{2} \int d^4x \int d^4y \, A_\mu(x) K_{\mu\nu}(x,y) A_\nu(y) +$$

$$\frac{1}{2} \int d^4x \int d^4y \, J_\mu(x) D_{\mu\nu}(x-y) J_\nu(y)$$

In this way, the dependence on the external currents is singled out and the generating functional simply reads

$$\tilde{W}_0(J) = \exp \left\{ \frac{i}{2\hbar c} \int d^4x \int d^4y \, J_\mu(x) D_{\mu\nu}(x-y) J_\nu(y) \right\}$$

as anticipated.

The photon propagator in H-L units follows as

$$\left(\frac{\hbar c}{i} \right)^2 \left[\frac{\delta^2 \tilde{W}_0(J)}{\delta J_\mu(x) \delta J_\nu(y)} \right]_{J=0} = \frac{\hbar c}{i} D_{\mu\nu}(x-y)$$

---

[2]See Prob. H.3. Note also that $\int d^4x \, D_{\mu\rho}(x-z) K_{\mu\nu}(x,y) = -\delta^{(4)}(y-z) \delta_{\rho\nu}$.

where, from Eq. (H.26),

$$D_{\mu\nu}(x - y) = \int \frac{d^4k}{(2\pi)^4} \tilde{D}_{\mu\nu}(k) \, e^{ik \cdot (x-y)}$$

$$\tilde{D}_{\mu\nu}(k) = \frac{1}{k^2 - i\eta} \left[\delta_{\mu\nu} - (1 - \xi) \frac{k_\mu k_\nu}{k^2} \right]$$

The value $\xi = 1$ leads to the *Feynman gauge*, and the photon propagator $\tilde{D}_{\mu\nu}(k)$ of chapter 8, derived there starting in the Coulomb gauge. Other values of ξ lead to a photon propagator in other gauges, for example, $\xi = 0$ produces the *Landau gauge*.

Problem H.2 Re-express the result in Prob. H.1 in S-I units.

Solution to Problem H.2

The Lagrangian density in S-I units, with an external source $J_\mu(x)$, follows from Eqs. (8.3)

$$\mathcal{L}_{\text{EM}} = -\frac{\varepsilon_0}{4} F_{\mu\nu} F_{\mu\nu} + J_\mu A_\mu$$

$$= \varepsilon_0 c^2 \left[-\frac{1}{4} \left(\frac{\partial A_\nu}{\partial x_\mu} - \frac{\partial A_\mu}{\partial x_\nu} \right)^2 + \frac{1}{\varepsilon_0 c^2} J_\mu A_\mu \right]$$

Take $\varepsilon_0 c^2$ out, so the factor in front of the effective action is $i\varepsilon_0 c/\hbar$. The action is then analyzed in exactly the same manner as in Prob. H.1

$$\int d^4x \left[\mathcal{L}_0 - \frac{1}{2\xi} \left(\frac{\partial A_\mu}{\partial x_\mu} \right)^2 + J_\mu A_\mu \right] = \frac{1}{2} \int d^4x \int d^4y \, A_\mu(x) K_{\mu\nu}(x, y) A_\nu(y) +$$

$$\frac{1}{2} \int d^4x \int d^4y \, J_\mu(x) D_{\mu\nu}(x - y) J_\nu(y)$$

We need only make the simple substitution $J_\mu \to J_\mu/\varepsilon_0 c^2$. Hence the generating functional in S-I units reads

$$\tilde{W}_0(J) = \exp\left\{ \frac{i}{2\hbar c^3 \varepsilon_0} \int d^4x \int d^4y \, J_\mu(x) D_{\mu\nu}(x - y) \, J_\nu(y) \right\}$$

The photon propagator in S-I units is then obtained by taking variational derivatives of this generating functional with respect to $J_\mu(x)$

$$\left(\frac{\hbar c}{i} \right)^2 \left[\frac{\delta^2 \tilde{W}_0(J)}{\delta J_\mu(x) \delta J_\nu(y)} \right]_{J=0} = \frac{\hbar}{i c \varepsilon_0} D_{\mu\nu}(x - y)$$

This result agrees with that found in the solution to Prob. F.4(a).

Problem H.3 Show that the photon propagator in Eq. (H.26) satisfies the defining Eqs. (H.24)–(H.25).

Solution to Problem H.3

Consider the kernel in Eq. (H.24)

$$K_{\mu\nu}(x, y) = \left[\Box_x \delta_{\mu\nu} - \left(1 - \frac{1}{\xi} \right) \frac{\partial}{\partial x_\mu} \frac{\partial}{\partial x_\nu} \right] \delta^{(4)}(x - y)$$

and the photon propagator in Eqs. (H.26)

$$D_{\mu\nu}(x, y) = \int \frac{d^4 k}{(2\pi)^4} \tilde{D}_{\mu\nu}(k) e^{ik\cdot(x-y)}$$

where the Fourier transform is given by

$$\tilde{D}_{\mu\nu}(k) = \frac{1}{k^2 - i\eta} \left[\delta_{\mu\nu} - (1 - \xi) \frac{k_\mu k_\nu}{k^2} \right]$$

Equation (H.25) defining the inverse of the propagator reads

$$\int d^4 z \, K_{\mu\rho}(x, z) D_{\rho\nu}(z - y) = -\delta^{(4)}(x - y) \delta_{\mu\nu}$$

Let us work out the l.h.s. of Eq. (H.25) with the given $K_{\mu\rho}(x, z)$

$$\text{l.h.s.} = \int d^4 z \, K_{\mu\rho}(x, z) D_{\rho\nu}(z - y)$$

$$= \int d^4 z \int \frac{d^4 k}{(2\pi)^4} \left[\Box_x \delta_{\mu\rho} - \left(1 - \frac{1}{\xi} \right) \frac{\partial}{\partial x_\mu} \frac{\partial}{\partial x_\rho} \right] \times$$
$$\delta^{(4)}(x - z) \tilde{D}_{\rho\nu}(k) e^{ik\cdot(z-y)}$$

$$= \int \frac{d^4 k}{(2\pi)^4} \left[\Box_x \delta_{\mu\rho} - \left(1 - \frac{1}{\xi} \right) \frac{\partial}{\partial x_\mu} \frac{\partial}{\partial x_\rho} \right] \tilde{D}_{\rho\nu}(k) e^{ik\cdot(x-y)}$$

Now observe that

$$\Box_x e^{ik\cdot(x-y)} = -k^2 e^{ik\cdot(x-y)}$$
$$\frac{\partial}{\partial x_\mu} \frac{\partial}{\partial x_\rho} e^{ik\cdot(x-y)} = -k_\mu k_\rho e^{ik\cdot(x-y)}$$

Therefore

$$
\begin{aligned}
\text{l.h.s.} &= -\int \frac{d^4k}{(2\pi)^4} \left[k^2 \delta_{\mu\rho} - \left(1 - \frac{1}{\xi}\right) k_\mu k_\rho \right] \tilde{D}_{\rho\nu}(k) e^{ik\cdot(x-y)} \\
&= -\int \frac{d^4k}{(2\pi)^4} \frac{1}{k^2 - i\eta} e^{ik\cdot(x-y)} \times \\
&\quad \left[k^2 \delta_{\mu\rho} - \left(1 - \frac{1}{\xi}\right) k_\mu k_\rho \right] \left[\delta_{\rho\nu} - (1-\xi)\frac{k_\rho k_\nu}{k^2} \right] \\
&= -\int \frac{d^4k}{(2\pi)^4} \frac{1}{k^2 - i\eta} e^{ik\cdot(x-y)} \times \\
&\quad \left[k^2 \delta_{\mu\nu} - \left(1 - \frac{1}{\xi}\right) k_\mu k_\nu - (1-\xi)k_\mu k_\nu - \frac{(1-\xi)^2}{\xi} k_\mu k_\nu \right] \\
&= -\int \frac{d^4k}{(2\pi)^4} e^{ik\cdot(x-y)} \delta_{\mu\nu} = -\delta^{(4)}(x-y)\, \delta_{\mu\nu}
\end{aligned}
$$

This proves Eq. (H.25).

Problem H.4 Verify Eq. (H.21).

Solution to Problem H.4

Equation (H.21) for the Faddeev-Popov determinant expressed in terms of ghost fields reads

$$
\int\int \mathcal{D}(\bar{g})\mathcal{D}(g) \exp\left\{ \frac{i}{\hbar c} \int d^4x \, [\bar{g}\,\Box_x\, g] \right\} = \left(\frac{\epsilon^8}{i\hbar c} \right)^n \det[\underline{M}_f]
$$

where $M_f(x,y)$ is the function in Eq. (H.12)

$$
M_f(x,y) = \Box_x \delta^{(4)}(x-y) \equiv K(x,y)
$$

We work on the l.h.s. of the first relation

$$
\begin{aligned}
\text{l.h.s.} &= \int\int \mathcal{D}(\bar{g})\mathcal{D}(g) \exp\left\{ \frac{i}{\hbar c} \int d^4x \, [\bar{g}\Box_x g] \right\} \\
&= \int\int \mathcal{D}(\bar{g})\mathcal{D}(g) \exp\left\{ -\frac{1}{i\hbar c} \int d^4x \int d^4y \, [\bar{g}(x)\Box_x\delta(x-y)g(y)] \right\}
\end{aligned}
$$

Upon discretization of space-time, this expression reduces to

$$
\text{l.h.s.} = \int\int \mathcal{D}(\bar{g})\mathcal{D}(g) \exp\left\{ -\frac{\epsilon^8}{i\hbar c} \sum_i \sum_j \bar{g}_i [M_f]_{ij} g_j \right\}
$$

We now remember Eq. (10.121) to evaluate this expression

$$\int d\bar{c}_n \ldots \int d\bar{c}_1 \int dc_1 \ldots \int dc_n \; \exp\left\{ -\sum_{i=1}^{n}\sum_{j=1}^{n} \bar{c}_i N_{ij} c_j \right\} = \det \underline{N}$$

where c_i and $\bar{c}_i$ $(i = 1, \ldots, n)$ are Grassmann variables.

With the use of Eq. (10.121), we then have

$$\text{l.h.s.} = \det\left[\frac{\epsilon^8}{i\hbar c}\underline{M}_f\right] = \left[\frac{\epsilon^8}{i\hbar c}\right]^n \det \underline{M}_f$$

which proves Eq. (H.21).

Appendix I

Metric Conversion

There are no problems included with this appendix.

Bibliography

Abers, E. S., and Lee, B. W., (1973). *Phys. Rep.* **9**, 1

Amore, P., and Walecka, J. D., (2013). *Introduction to Modern Physics: Solutions to Problems*, World Scientific Publishing Company, Singapore

Amore, P., and Walecka, J. D., (2014). *Topics in Modern Physics: Solutions to Problems*, World Scientific Publishing Company, Singapore

Banks, T., (2008). *Modern Quantum Field Theory: A Concise Introduction*, Cambridge U. Press, New York, NY

Bardeen, J., Cooper, L. N., and Schrieffer, J. R., (1957). *Phys. Rev.* **106**, 162; *Phys. Rev.* **108**, 1175

Bethe, H. A., and Goldstone, J., (1957). *Proc. Roy. Soc. (London)* **A238**, 551

Binosi, D., and Theussi, L., (2004). *Comput. Phys. Comm.* **161**, 76

Binosi, D., Collins, J., Kaufhold, C., and Theussi, L., (2009). *Comput. Phys. Comm.* **180**, 1709

Bjorken, J. D., and Drell, S. D., (1964). *Relativistic Quantum Mechanics*, McGraw-Hill, New York, NY

Bjorken, J. D., and Drell, S. D., (1965). *Relativistic Quantum Fields*, McGraw-Hill, New York, NY

Blatt, J. M., and Weisskopf, V. F., (1952). *Theoretical Nuclear Physics*, John Wiley and Sons, New York, NY

Bloch, F., and Nordsieck, A., (1937). *Phys. Rev.* **73**, 54

Bogoliubov, N. N., (1947). *J. Phys. (USSR)* **11**, 23

Bogoliubov, N. N., (1958). *Sov. Phys. JETP* **7**, 41

Bohr, A., Mottelson, B. R., and Pines, D, (1958). *Phys. Rev.* **110**, 936

Brueckner, K. A., and Sawada, K., (1957). *Phys. Rev.* **106**, 1117

Chen, J.-Q., Wang, P.-N., Lü, Z.-M., and Wu, X.-B., (1987) *Tables of Clebsch-Gordan Coefficients of SU(n) groups*, World Scientific Publishing Company, Singapore

Cheng, T.-P., and Li, L.-F., (1984). *Gauge Theory of Elementary Particle Physics*, Clarendon Press, Oxford, UK

Cooper, L. N., (1956). *Phys. Rev.* **104**, 1189

Cvitanovik, P., and Kinoshita, T., (1974). *Phys. Rev.* **D10**, 4007

Dirac, P. A. M., (1947). *The Principles of Quantum Mechanics, 3rd ed.*, Oxford University Press, New York, NY

Donoghue, J. F., Golowich, E., and Holstein, B., (2014). *Dynamics of the Standard Model, 2nd ed.*, Cambridge University Press, New York, NY

Dyson, F. J., (1949). *Phys. Rev.* **75**, 486; *Phys. Rev.* **75**, 1736

Dyson, F. J., (1952). *Phys. Rev.* **85**, 631

Edmonds, A. R., (1974). *Angular Momentum in Quantum Mechanics*, 3rd printing, Princeton University Press, Princeton, NJ

Faddeev, L. D., and Popov, V. N., (1967). *Phys. Lett.* **25B**, 29

Fetter, A. L., and Walecka, J. D., (2003). *Theoretical Mechanics of Particles and Continua*, McGraw-Hill, New York, NY (1980); reissued by Dover Publications, Mineola, NY

Fetter, A. L., and Walecka, J. D., (2003a). *Quantum Theory of Many-Particle Systems*, McGraw-Hill, New York, NY (1971); reissued by Dover Publications, Mineola, NY

Feynman, R. P., (1949). *Phys. Rev.* **76**, 749; *Phys. Rev.* **76**, 769

Feynman, R. P., (1963). *Acta Phys. Polon.* **24**, 697

Feynman, R. P., and Hibbs, A. R., (1965). *Quantum Mechanics and Path Integrals*, McGraw-Hill, New York, NY

Fradkin, E., (2013). *Field Theories of Condensed Matter Physics*, Cambridge University Press, Cambridge UK

Gabrielse, G., (2009). *Measurements of the Electon Magnetic Moment*, to appear in *Lepton Dipole Moments: The Search for Physics Beyond the Standard Model*, eds. B. L. Roberts and W. J. Marciano, World Scientific Publishing Company, Singapore

Gell-Mann, M., and Goldberger, M. L., (1953). *Phys. Rev.* **91**, 398

Gell-Mann, M., and Levy, M., (1960). *Nuovo Cimento* **16**, 705

Gell-Mann, M., and Low, F. E., (1954). *Phys. Rev.* **95**, 1300

Gell-Mann, M., and Ne'eman, Y., (1963). *The Eightfold Way*, W. A. Benjamin, Reading, MA

Georgi, H., (1999). *Lie Algebras in Particle Physics: from Isospin to Unified Theories, 2nd ed.*, Westview Press, Boulder, CO

Giunti, C., and Kim, C. W., (2007). *Fundamentals of Neutrino Physics and Astrophysics*, Oxford University Press, Oxford, UK

Goldberger, M. L., and Watson, K. M., (2004). *Collision Theory*, Dover Publications, Mineola, NY

Gottfried, K., (1966). *Quantum Mechanics, Vol. I*, W. A. Benjamin, New York, NY

Hamermesh, M., (1989). *Group Theory and Its Applications to Physical Problems*, Dover Publications, Mineola, NY

Itzykson, C., and Zuber, J.-B., (1980). *Quantum Field Theory*, McGraw-Hill, New York, NY

Jacob, M., and Wick, G, C., (1959) *Ann. Phys.* **7**, 404

Lee, T. D., and Yang, C. N, (1957). *Phys. Rev.* **105**, 1119

Leibbrant, G., (1975). *Rev. Mod. Phys.* **47**, 849

Lippmann, B., and Schwinger, J., (1950). *Phys. Rev.* **79**, 469

MacDonald, J. K. L., (1933). *Phys. Rev.* **43**, 830

Merzbacher, M., (1998). *Quantum Mechanics, 3rd ed.*, John Wiley and Sons, New York, NY

Morse, P. M., and Feshbach, H., (1953). *Methods of Theoretical Physics, Vols. I-II*, McGraw-Hill, New York, NY

National Nuclear Data Center, (2009). *Nuclear Data*, http://www.nndc.bnl.gov/

Particle Data Group, (2009). *Particle Data Tables 2009*, http://pdg.lbl.gov/2009/tables/contents_tables.html

Pauli, W., and Weisskopf, V. F., (1934). *Helv. Phys. Acta* **7**, 709

Roos, M., (2015). *Introduction to Cosmology, 4th ed.*, John Wiley and Sons, New York, NY

Rotenberg, M., Bivens, R., Metropolis, N., and Wooten, J. K. Jr., (1959) *The 3-j and 6-j Symbols*, The Technology Press, M. I. T., Cambridge, MA

Schiff, L. I., (1968). *Quantum Mechanics, 3rd ed.*, McGraw-Hill, New York, NY

Schwartz, M. D., (2013). *Quantum Field Theory and the Standard Model*, Cambridge University Press, Cambridge UK

Schweber, S., (1961). *Relativistic Quantum Field Theory*, Row-Peterson, Chicago, IL

Schwinger, J., (1957). *Annals of Physics* **2**, 407

Schwinger, J., (1958). *Selected Papers on Quantum Electrodynamics*, ed. J. Schwinger, Dover Publications, Mineola, NY

Serot, B. D., and Walecka, J. D., (1986). *The Relativistic Nuclear Many-Body Problem, Adv. Nucl. Phys.* **16**, Plenum Press, New York, NY

t'Hooft, G., and Veltman, M., (1972) *Nucl. Phys.* **B44**, 189; *Nucl. Phys.* **B62**, 444

Tilley, D. R., Weller, H. R., and Hale, G. M., (1992). *Nucl. Phys.* **A541**, 1

Titchmarsh, E. C., (1976). *Theory of Functions, 2nd ed.*, Oxford, New York, NY

Valatin, J. G., (1958) *Nuovo Cimento* **7**, 843

Van Dyck, R. S., Schwinberg, P. B., and Dehmelt, H. G., (1977). *Phys. Rev. Lett.* **38**, 310

Vermaseren, J. A. N., (1994). *Comput. Phys. Comm.* **83**, 45

Walecka, J. D., (2001). *Electron Scattering for Nuclear and Nucleon Structure*, Cambridge University Press, Cambridge, UK

Walecka, J. D., (2004). *Theoretical Nuclear and Subnuclear Physics, 2nd ed.*, World Scientific Publishing Company, Singapore

Walecka, J. D., (2007). *Introduction to General Relativity*, World Scientific Publishing Company, Singapore

Walecka, J. D., (2008). *Introduction to Modern Physics: Theoretical Foundations*, World Scientific Publishing Company, Singapore

Walecka, J. D., (2010). *Advanced Modern Physics: Theoretical Foundations*, World Scientific Publishing Company, Singapore

Walecka, J. D., (2013). *Topics in Modern Physics: Theoretical Foundations*, World Scientific Publishing Company, Singapore

Ward, J. C., (1951). *Proc. Phys. Soc. London* **A64**, 54

Weinberg, S., (1960). *Phys. Rev.* **118**, 838

Weinberg, S., (2005). *The Quantum Theory of Fields, 3 Vols.*, Cambridge University Press, Cambridge UK

Wentzel, G., (1949). *Quantum Theory of Fields*, Interscience, New York, NY

Wick, G. C., (1950). *Phys. Rev.* **80**, 268

Wikipedia, (2009). http://en.wikipedia.org/wiki/(topic)

Wigner, E. P., (1937). *Phys. Rev.* **51**, 106

Wigner, E. P., (1939). *On Unitary Representations of the Inhomogeneous Lorentz Group*, *Annals Math.* **40**, 149

Wu, T.T., (1959). *Phys. Rev.* **115**, 1390

Yang, C. N., and Mills, R. L., (1954). *Phys. Rev.* **96**, 191

Index

Printed in the United States
by Bookmasters

Printed in the United States
By Bookmasters